LESSONS LEARNED FROM THE Fukushima Nuclear Accident FOR IMPROVING SAFETY OF U.S. Nuclear Plants

Committee on Lessons Learned from the Fukushima Nuclear Accident for Improving Safety and Security of U.S. Nuclear Plants

Nuclear and Radiation Studies Board

Division on Earth and Life Studies

NATIONAL RESEARCH COUNCIL
OF THE NATIONAL ACADEMIES

THE NATIONAL ACADEMIES PRESS
Washington, D.C.
www.nap.edu

THE NATIONAL ACADEMIES PRESS 500 Fifth Street, NW Washington, DC 20001

NOTICE: The project that is the subject of this report was approved by the Governing Board of the National Research Council, whose members are drawn from the councils of the National Academy of Sciences, the National Academy of Engineering, and the Institute of Medicine. The members of the committee responsible for the report were chosen for their special competences and with regard for appropriate balance.

This study was supported by Contract/Grant No. NRC-HQ-12-G-03-0002 between the National Academy of Sciences and the U.S. Nuclear Regulatory Commission. Any opinions, findings, conclusions, or recommendations expressed in this publication are those of the author(s) and do not necessarily reflect the views of the organizations or agencies that provided support for the project.

International Standard Book Number-13: 978-0-309-27253-7
International Standard Book Number-10: 0-309-27253-X

Library of Congress Catalog Card Number 2014954121

Cover images: (Front) Satellite image of the Fukushima Daiichi Nuclear Power Station taken on March 14, 2011, while the accident was still in progress. The Unit 1 reactor building (lower-left corner of image) was damaged by a hydrogen explosion at 15:36 (3:36 P.M.) Japan Standard Time on March 12. The Unit 3 reactor building (building with steam plume near center of image) was damaged by a hydrogen explosion at 11:01 the day the image was recorded. The Unit 4 reactor building, which is still intact in this image, would be damaged by a hydrogen explosion at 06:14 the next day (March 15). (Back) Satellite image of the harbor-front of the Fukushima Daiichi Nuclear Power Station showing extensive damage from the tsunami. Also shown is a seismogram of the Great East Japan Earthquake recorded at the Oshika K-Net Station (MYG011) in Miyagi Prefecture. Satellite image courtesy of DigitalGlobe©2014.

Additional copies of this report are available for sale from the National Academies Press, 500 Fifth Street, NW, Keck 360, Washington, DC 20001; (800) 624-6242 or (202) 334-3313; http://www.nap.edu/.

Printed in the United States of America

THE NATIONAL ACADEMIES

Advisers to the Nation on Science, Engineering, and Medicine

The **National Academy of Sciences** is a private, nonprofit, self-perpetuating society of distinguished scholars engaged in scientific and engineering research, dedicated to the furtherance of science and technology and to their use for the general welfare. Upon the authority of the charter granted to it by the Congress in 1863, the Academy has a mandate that requires it to advise the federal government on scientific and technical matters. Dr. Ralph J. Cicerone is president of the National Academy of Sciences.

The **National Academy of Engineering** was established in 1964, under the charter of the National Academy of Sciences, as a parallel organization of outstanding engineers. It is autonomous in its administration and in the selection of its members, sharing with the National Academy of Sciences the responsibility for advising the federal government. The National Academy of Engineering also sponsors engineering programs aimed at meeting national needs, encourages education and research, and recognizes the superior achievements of engineers. Dr. C. D. Mote, Jr., is president of the National Academy of Engineering.

The **Institute of Medicine** was established in 1970 by the National Academy of Sciences to secure the services of eminent members of appropriate professions in the examination of policy matters pertaining to the health of the public. The Institute acts under the responsibility given to the National Academy of Sciences by its congressional charter to be an adviser to the federal government and, upon its own initiative, to identify issues of medical care, research, and education. Dr. Victor J. Dzau is president of the Institute of Medicine.

The **National Research Council** was organized by the National Academy of Sciences in 1916 to associate the broad community of science and technology with the Academy's purposes of furthering knowledge and advising the federal government. Functioning in accordance with general policies determined by the Academy, the Council has become the principal operating agency of both the National Academy of Sciences and the National Academy of Engineering in providing services to the government, the public, and the scientific and engineering communities. The Council is administered jointly by both Academies and the Institute of Medicine. Dr. Ralph J. Cicerone and Dr. C. D. Mote, Jr., are chair and vice chair, respectively, of the National Research Council.

www.national-academies.org

COMMITTEE ON LESSONS LEARNED FROM THE FUKUSHIMA NUCLEAR ACCIDENT FOR IMPROVING SAFETY AND SECURITY OF U.S. NUCLEAR PLANTS

NORMAN P. NEUREITER, *Chair*, American Association for the Advancement of Science, Washington, D.C.
B. JOHN GARRICK, *Vice Chair*, Independent Consultant, Laguna Beach, California
ROBERT A. BARI, Brookhaven National Laboratory, Upton, New York
PERCY M. BEARD, JR., Institute of Nuclear Power Operations (retired), Atlanta, Georgia
JAN BEYEA, Consulting in the Public Interest, Lambertville, New Jersey
M. QUINN BREWSTER, University of Illinois at Urbana-Champaign
MICHAEL L. CORRADINI, University of Wisconsin–Madison
RICHARD S. DENNING, The Ohio State University, Columbus[1]
VIJAY K. DHIR, University of California, Los Angeles
MICHAEL W. GOLAY, Massachusetts Institute of Technology, Cambridge
BARBARA L. HAMRICK, University of California, Irvine Medical Center
SHELLEY A. HEARNE, Johns Hopkins Bloomberg School of Public Health, Baltimore, Maryland, and Big Cities Health Coalition, National Association of County and City Health Officials, Washington, D.C.
PAUL A. LOCKE, Johns Hopkins Bloomberg School of Public Health, Baltimore, Maryland
JAMES E. MATHESON, SmartOrg, Menlo Park, California
NAJMEDIN MESHKATI, University of Southern California, Los Angeles[2]
THOMAS G. MOSER, Osprey Global Solutions, Wilmington, North Carolina
ARTHUR T. MOTTA, Pennsylvania State University, University Park
JOHN A. ORCUTT, Scripps Institution of Oceanography, University of California at San Diego, La Jolla
EMILIE M. ROTH, Roth Cognitive Engineering, Menlo Park, California
JOSEPH E. SHEPHERD, California Institute of Technology, Pasadena
ELIZABETH Q. TEN EYCK, ETE Consulting, Inc., Great Falls, Virginia
MARK H. THIEMENS, University of California, San Diego[3]
FRANK N. VON HIPPEL, Princeton University, Princeton, New Jersey

---

[1]Separated from committee on November 21, 2012.
[2]Resigned from committee on June 12, 2013.
[3]Resigned from committee on July 26, 2012.

LORING A. WYLLIE, JR., Degenkolb Engineers, San Francisco, California

Technical Advisor

NAJMEDIN MESHKATI, University of Southern California, Los Angeles[4]

Staff

KEVIN D. CROWLEY, Study Director
OURANIA KOSTI, Senior Program Officer
DANIEL POMEROY, Postdoctoral Fellow (December 10, 2012, to September 1, 2013)
DARLENE GROS, Senior Program Assistant

[4]Effective July 10, 2013.

NUCLEAR AND RADIATION STUDIES BOARD

[1]Deceased on July 28, 2014.

Acknowledgments

The successful completion of this study would not have been possible without the cooperation and assistance of a very large number of organizations and individuals. The committee would especially like to thank the following organizations and individuals for providing logistical support, advice, and assistance during this study:

U.S. Nuclear Regulatory Commission, which sponsored the study and provided information on the agency's post-Fukushima activities and actions. The committee would especially like to thank former Executive Director of Operations Bill Borchardt and staff in the Japan Lessons-Learned Division, Office of Nuclear Reactor Regulation: David Skeen, Rob Taylor, Bill Reckley, and Kevin Witt. Mr. Witt served ably as the agency's liaison to the committee and was effective in responding to the committee's many requests for information.

U.S. Embassy in Tokyo, which assisted the committee with logistical arrangements for its meeting in Japan: former Ambassador John Roos and embassy staff Jeffrey Miller and Jen Ladlein.

Government of Japan for its full cooperation with the National Academy of Sciences (NAS) on this study and for providing technical and logistical assistance:

- Japanese Embassy in the United States: Ambassador Kenichiro Sasae, former Ambassador Ichiro Fujisaki, embassy staff members Takashi Toyota and Akira Tsugita, and former staff member Seiichi Shimasaki.

Embassy staff responded effectively to the committee's many requests for information and logistical support.

- Atomic Energy Commission of Japan: former Chairman Shunsuke Kondo.
- Japan Atomic Energy Agency: former President Atsuzuki Suzuki.
- Japan Science and Technology Agency: Center for Research and Development Strategy (CRDS) Director General Hiroyuki Yoshikawa; CRDS Deputy Director General Tateo Arimoto; Principal Fellow Kenkichi Hirose; and Washington, D.C. representative Takashi Ohama.
- Nuclear Regulation Authority: Commissioner Toyoshi Fuketa and staff member Masahiro Aoki.

National Graduate Institute for Policy Studies (Tokyo) for cosponsoring and supporting the committee's meeting at the institute's facilities in Tokyo: President Takashi Shiraishi and faculty members Tateo Arimoto, Kiyoshi Kurokawa, and Atsushi Sunami.

Science Council of Japan for assisting with the identification of technical experts for the committee's meeting in Tokyo: President Takashi Onishi and Vice President Fumiko Kasuga.

Tokyo Electric Power Company (TEPCO) for its full cooperation and support for this study, particularly for providing technical information about the company's response to the earthquake and tsunami at the Fukushima Daiichi and Fukushima Daini nuclear plants; for arranging and hosting tours of those plants; and for participating in the committee's meeting in Tokyo: Chief Nuclear Officer and Executive Vice President Zengo Aizawa, General Manager of International Relations and Strategy Group Akira Kawano; Fukushima Daiichi Superintendent Takeshi Takahashi; Fukushima Daini Superintendent Naohiro Masuda; TEPCO's Washington, D.C., representative Kenji Tateiwa; and the many other TEPCO staff listed in Appendix B. Mr. Tateiwa served effectively as a liaison between TEPCO and NAS staff.

Tohoku Electric Power Company for arranging and hosting the tour of the Onagawa nuclear plant and participating in the committee's Tokyo meeting: Managing Director Shigeru Inoue; Onagawa Superintendent Shun Tsubata; Deputy Superintendent Isao Kato; staff members Masayoshi Abe and Satoshi Otokita; Washington, D.C., representative Hiroya Shimanuki; and the other Tohoku staff listed in Appendix B.

Nuclear Energy Institute for providing information about the U.S. nuclear industry's response to the Fukushima Daiichi accident to improve safety of nuclear plants in the United States: President and Chief Executive Officer Marvin Fertel; Executive Director (Fukushima Response) Jim Scarola; Vice President of Nuclear Operations Joe Pollock; and Senior Technical Advisor Steven Kraft. Mr. Kraft provided lead support for the committee's information requests and visits to the Oyster Creek Generating Station and Edwin I. Hatch Nuclear Plant.

Institute of Nuclear Power Operations for providing information about the U.S. nuclear industry's response to the Fukushima Daiichi nuclear accident: President and CEO Robert Willard and Vice President David Hembree.

Southern Company for hosting a tour of the Edwin I. Hatch Nuclear Plant: Site Vice President Dennis Madison, staff member J. Derwood Tootle, and the many other plant staff who supported the plant tour.

Exelon Corporation for hosting a tour of the Oyster Creek Generating Station: Site Vice President Garey Stathes, Plant Manager Russell Peake, and the many other plant staff who supported the plant tour.

Speakers and invited participants at the committee's information-gathering meetings. These individuals are too numerous to be listed here. Their names are provided in Appendix B.

Other individuals and organizations who provided information and support for this study: Scott Deitchman, Centers for Disease Control and Prevention; Nancy Hedinger, Paula Gotsch, and Kerry Margaret Butch, League of Women Voters of New Jersey; Jaewon Hong, U.S. Food and Drug Administration; Kiyohiko Mabuchi, National Cancer Institute; Alan Morris, Anbex; Damian Peko, U.S. Department of Energy; Andrew Sowder, Electric Power Research Institute; Satoru Tanaka, Atomic Energy Society of Japan; and the World Health Organization, especially María del Rosario Pérez.

Najmedin Meshkati, University of Southern California, who served as a technical advisor to the committee for issues related to nuclear safety culture.

And last, but certainly not least, **National Academy of Sciences' staff** for organizing and facilitating this study. Study Director Kevin Crowley and Senior Program Officer Ourania Kosti organized the committee meetings and assisted the committee with report writing and review. They also provided valuable technical and policy insights and perspectives. Postdoctoral Fellow Daniel Pomeroy assisted with research and report writing. Program Assistant Darlene Gros managed the logistics of the meetings, report review, and publication. Senior Program Officer Jennifer Heimberg and Program Assistant Erin Wingo assisted with production of the final report.

Reviewer Acknowledgments

This report has been reviewed in draft form by individuals chosen for their diverse perspectives and technical expertise in accordance with procedures approved by the Report Review Committee of the National Research Council. The purpose of this independent review is to provide candid and critical comments that will assist the National Research Council in making this published report as sound as possible and will ensure that this report meets institutional standards for objectivity, evidence, and responsiveness to the study charge. The review comments and draft manuscript remain confidential to protect the integrity of the deliberative process. We wish to thank the following individuals for their participation in the review of this report:

Brian Atwater, U.S. Geological Survey and University of Washington
Lake Barrett, L. Barrett Consulting
Dennis Bley, Buttonwood Consulting, Inc.
Daniel Blumenthal, U.S. Department of Energy
Robert Budnitz, Lawrence Berkeley National Laboratory
Cynthia Costello, New York State Department of Health
Jeffrey Gabor, ERIN Engineering and Research, Inc.
Randall Gauntt, Sandia National Laboratories
Robert Geller, University of Tokyo
David Gertman, Idaho National Laboratory
Sherrell Greene, Advanced Technology Insights, LLC
Dave Klimek, Nexus Technical Services Corporation
Joseph Klinger, Illinois Emergency Management Agency

Shunsuke Kondo, Japan Atomic Energy Commission (retired)
Kiyoshi Kurokawa, National Graduate Institute for Policy Studies
David Lochbaum, Union of Concerned Scientists
Thomas McKenna, International Atomic Energy Agency
Richard Meserve, Carnegie Institution for Science
Randall Mumaw, Boeing Commercial Airplanes
Per Peterson, University of California, Berkeley
William Radasky, Metatech Corporation
Scott Sagan, Stanford University
Craig Sawyer, General Electric Company (retired)
Fred Tollison, Institute of Nuclear Power Operations (retired)

Although the reviewers listed above provided many constructive comments and suggestions, they were not asked to endorse the conclusions and recommendations of this report, nor did they see the final draft of the report before its release. The review of this report was overseen by Marvin Adams, Texas A&M University, and Chris Whipple, ENVIRON. Appointed by the National Research Council, Drs. Adams and Whipple were responsible for making certain that an independent examination of this report was carried out in accordance with institutional procedures and that all review comments were considered carefully. Responsibility for the final content of this report rests entirely with the authoring committee and the institution.

Contents

Synopsis

The U.S. Congress asked the National Academy of Sciences (NAS) to examine the causes of the March 11, 2011, accident at the Fukushima Daiichi nuclear plant and identify lessons learned for the United States. Brief descriptions of key selected findings and recommendations are provided below.

Causes of the Fukushima Daiichi Accident: The Fukushima Daiichi nuclear accident was initiated by the March 11, 2011, Great East Japan Earthquake and tsunami. Personnel at the plant responded to the accident with courage and resilience; their actions likely reduced its severity and the magnitude of offsite radioactive material releases. However, several factors relating to the management, design, and operation of the plant prevented plant personnel from achieving greater success and contributed to the overall severity of the accident.

Lessons Learned from the Fukushima Daiichi Accident for the United States: NAS recommends that several actions be taken to improve the resilience of U.S. nuclear plants and enhance U.S. emergency response. These actions are summarized below.

- Nuclear plant licensees and their regulators must actively seek out and act on new information about hazards that have the potential to affect nuclear plant safety.

- The U.S. nuclear industry and its regulator (the U.S. Nuclear Regulatory Commission) should improve specific nuclear plant systems, resources, and training to enable effective responses to severe accidents.
- The U.S. nuclear industry and the U.S. Nuclear Regulatory Commission should strengthen their capabilities for assessing risks from events that could challenge the design of nuclear plant structures and components and lead to a loss of critical safety functions. The U.S. Nuclear Regulatory Commission should support industry's efforts to strengthen its capabilities by providing guidance on approaches and by overseeing rigorous peer review.
- The U.S. Nuclear Regulatory Commission should further incorporate modern risk concepts into its nuclear safety regulations using these strengthened capabilities.
- The U.S. nuclear industry and U.S. emergency response organizations should examine and, as needed, revise their emergency response plans, including the balance among protective actions, to enable effective responses to severe nuclear accidents.
- The U.S. Nuclear Regulatory Commission and the U.S. nuclear power industry must maintain and continuously monitor a strong nuclear safety culture in their safety-related activities and should examine opportunities to increase the transparency of and communication about their efforts to assess and improve nuclear safety.

Summary

The March 11, 2011, Great East Japan Earthquake and tsunami sparked a humanitarian disaster in northeastern Japan and initiated a severe nuclear accident at the Fukushima Daiichi nuclear plant. Three of the six reactors at the plant sustained severe core damage and released hydrogen and radioactive materials. Explosion of the released hydrogen damaged three reactor buildings and impeded onsite emergency response efforts.

At the time of the Fukushima Daiichi accident, the Blue Ribbon Commission on America's Nuclear Future was completing an assessment of options for managing spent nuclear fuel and high-level radioactive waste in the United States (BRC, 2012). The Commission recommended that the National Academy of Sciences (NAS) conduct an assessment of lessons learned from the Fukushima Daiichi accident. This recommendation was taken up by the U.S. Congress, which subsequently directed the U.S. Nuclear Regulatory Commission to contract with NAS for this study.

The statement of task for this study is shown in Sidebar S.1. Study charges 1, 3, and 4 are addressed in this report; study charge 2 (on spent fuel safety and security) will be addressed in a future report.

A committee of 21 experts was appointed by NAS to carry out this study (see Appendix A). The committee held 39 meetings during the course of this study to gather information and develop this report (see Appendix B for a list of the committee's information-gathering meetings). One of these meetings was held in Tokyo, Japan, to enable in-depth discussions about the accident with Japanese technical experts from industry, academia, and government. The committee also visited the Fukushima Daini, Fukushima Daiichi, and Onagawa nuclear plants (see Chapter 3) to learn about their

SIDEBAR S.1
Statement of Task for This National Research Council Study

The National Research Council will provide an assessment of lessons learned from the Fukushima nuclear accident for improving the safety and security of nuclear plants in the United States. This assessment will address the following issues:

1. Causes of the Fukushima nuclear accident, particularly with respect to the performance of safety systems and operator response following the earthquake and tsunami.
2. Reevaluation of the conclusions from previous National Academy of Sciences studies on safety and security of spent nuclear fuel and high-level radioactive waste storage, particularly with respect to the safety and security of current storage arrangements and alternative arrangements in which the amount of commercial spent fuel stored in pools is reduced.[a]
3. Lessons that can be learned from the accident to improve commercial nuclear plant safety and security systems and operations.
4. Lessons that can be learned from the accident to improve commercial nuclear plant safety and security regulations, including processes for identifying and applying design-basis events for accidents and terrorist attacks to existing nuclear plants.

The study may examine policy options related to these issues but should not make policy recommendations that involve nontechnical value judgments.

[a] This task will be addressed in a subsequent report. It is not addressed in this report.

designs, operations, and responses to the earthquake and tsunami. Subgroups of the committee visited two nuclear plants in the United States that are similar in design to the Fukushima Daiichi plant to learn about their designs and operations.

S.1 CAUSES OF THE FUKUSHIMA DAIICHI ACCIDENT (Study Charge 1)

NAS's examination of the Fukushima Daiichi accident is provided in Chapters 3 and 4 of this report. Chapter 3 describes the March 11, 2011, Great East Japan Earthquake and tsunami and their impacts on Japanese nuclear plants. Chapter 4 describes the accident at the Fukushima Daiichi plant, including the accident time line, key actions taken by plant personnel, and challenges faced in taking those actions. One finding emerged from this examination:

FINDING 4.1[1]: The accident at the Fukushima Daiichi nuclear plant was initiated by the March 11, 2011, Great East Japan Earthquake and tsunami. The earthquake knocked out offsite AC power to the plant and the tsunami inundated portions of the plant site. Flooding of critical plant equipment resulted in the extended loss of onsite AC and DC power with the consequent loss of reactor monitoring, control, and cooling functions in multiple units. Three reactors sustained severe core damage (Units 1, 2, and 3); three reactor buildings were damaged by hydrogen explosions (Units 1, 3, and 4); and offsite releases of radioactive materials contaminated land in Fukushima and several neighboring prefectures. The accident prompted widespread evacuations of local populations and distress of the Japanese citizenry, large economic losses, and the eventual shutdown of all nuclear power plants in Japan.

Personnel at the Fukushima Daiichi plant responded with courage and resilience during the accident in the face of harsh circumstances; their actions likely reduced the severity of the accident and the magnitude of offsite radioactive material releases. Several factors prevented plant personnel from achieving greater success—in particular, averting reactor core damage—and contributed to the overall severity of the accident:

1. Failure of the plant owner (Tokyo Electric Power Company) and the principal regulator (Nuclear and Industrial Safety Agency) to protect critical safety equipment at the plant from flooding in spite of mounting evidence that the plant's current design basis for tsunamis was inadequate.
2. The loss of nearly all onsite AC and DC power at the plant—with the consequent loss of real-time information for monitoring critical thermodynamic parameters in reactors, containments, and spent fuel pools and for sensing and actuating critical valves and equipment—greatly narrowed options for responding to the accident.
3. As a result of (1) and (2), the Unit 1, 2 and 3 reactors were effectively isolated from their ultimate heat sink (the Pacific Ocean) for a period of time far in excess of the heat capacity of the suppression pools or the coping time of the plant to station blackout.
4. Multiunit interactions complicated the accident response. Unit operators competed for physical resources and the attention and services of staff in the onsite emergency response center.
5. Operators and onsite emergency response center staff lacked adequate procedures and training for accidents involving extended loss

[1] The first digit denotes the chapter in which the finding (or recommendation) appears; the second digit denotes the serial order of the finding (or recommendation) in the chapter.

of all onsite AC and DC power, particularly procedures and training for managing water levels and pressures in reactors and their containments and hydrogen generated during reactor core degradation.

6. Failures to transmit information and instructions in an accurate and timely manner hindered responses to the accident. These failures resulted partly from the loss of communications systems and the challenging operating environments throughout the plant.

7. The lack of clarity of roles and responsibilities within the onsite emergency response center and between the onsite and headquarters emergency response centers may have contributed to response delays.

8. Staffing levels at the plant were inadequate for managing the accident because of its scope (affecting several reactor units) and long duration.

S.2 LESSONS LEARNED FOR THE UNITED STATES (Study Charges 3 and 4)

Findings and recommendations on lessons learned from the Fukushima Daiichi accident are provided in Chapters 3-7. They are organized into five sections in this summary.

1. Seek out and act on new information about hazards.
2. Improve nuclear plant systems, resources, and training to enable effective ad hoc responses to severe accidents.
3. Strengthen capabilities for assessing risks from beyond-design-basis events.
4. Further incorporate modern risk concepts into nuclear safety regulations.
5. Examine offsite emergency response capabilities and make necessary improvements.
6. Improve the nuclear safety culture.

S.2.1 Seek Out and Act on New Information About Hazards

FINDING 3.1: The overarching lesson learned from the Fukushima Daiichi accident is that nuclear plant licensees and their regulators must actively seek out and act on new information about hazards that have the potential to affect the safety of nuclear plants. Specifically,

1. Licensees and their regulators must continually seek out new scientific information about nuclear plant hazards and methodologies for estimating their magnitudes, frequencies, and potential impacts.

2. Nuclear plant risk assessments must incorporate new information and methodologies as they become available.

3. Plant operators and regulators must take *timely* actions to implement countermeasures when such new information results in substantial changes to risk profiles at nuclear plants.

S.2.2 Improve Nuclear Plant Systems, Resources, and Training

Many national governments and international bodies initiated reviews of nuclear plant safety following the Fukushima Daiichi accident (see Table 1.1 in Chapter 1). Two major initiatives are now under way in the United States—one by the U.S. Nuclear Regulatory Commission and the other by the U.S. nuclear industry—and are resulting in changes to U.S. nuclear plant systems, operations, and regulations.

FINDING 5.1: Nuclear plant operators and regulators in the United States and other countries have identified and are taking useful actions to upgrade nuclear plant systems, operating procedures, and operator training in response to the Fukushima Daiichi accident. In the United States, these actions include the nuclear industry's FLEX (diverse and flexible coping strategies) initiative as well as regulatory changes proposed by the U.S. Nuclear Regulatory Commission's Near-Term Task Force. Implementation of these actions is still under way; consequently, it is too soon to evaluate their comprehensiveness, effectiveness, or status in the regulatory framework.

RECOMMENDATION 5.1A: As the nuclear industry and its regulator implement the actions referenced in Finding 5.1, they should give specific attention to improving plant systems in order to enable effective responses to beyond-design-basis events, including, when necessary, developing and implementing ad hoc responses[2] to deal with unanticipated complexities. Attention to availability, reliability, redundancy, and diversity of plant systems and equipment is specifically needed for

- **DC power for instrumentation and safety system control;**
- **Tools for estimating real-time plant status during loss of power;**
- **Decay-heat removal and reactor depressurization and containment venting systems and protocols;**
- **Instrumentation for monitoring critical thermodynamic parameters in reactors, containments, and spent fuel pools;**

[2] The term "ad hoc" refers to responses that are not planned and trained on in advance but rather are developed on the spot.

- Hydrogen monitoring (including monitoring in reactor buildings) and mitigation;
- Instrumentation for both onsite and offsite radiation and security monitoring; and
- Communications and real-time information systems to support communication and coordination between control rooms and technical support centers, between control rooms and the field, and between onsite and offsite support facilities.

The quality and completeness of the changes that result from this recommendation should be adequately peer reviewed.

RECOMMENDATION 5.1B: As the nuclear industry and its regulator implement the actions referenced in Finding 5.1, they should give specific attention to improving resource availability and operator training to enable effective responses to beyond-design-basis events, including, when necessary, developing and implementing ad hoc responses to deal with unanticipated complexities. Attention to the following is specifically needed:

1. Staffing levels for emergencies involving multiple reactors at a site, that last for extended durations, and/or that involve stranded-plant conditions.[3]
2. Strengthening and better integrating emergency procedures, extensive damage mitigation guidelines, and severe accident management guidelines, in particular for
   - Coping with the complete loss of AC and DC power for extended periods,
   - Depressurizing reactor pressure vessels and venting containments when DC power and installed plant air supplies (i.e., compressed air and gas) are unavailable,
   - Injecting low-pressure water when plant power is unavailable,
   - Transitioning between reactor pressure vessel depressurization and low-pressure water injection while maintaining sufficient water levels to protect the core from damage,
   - Preventing and mitigating the effects of large hydrogen explosions on cooling systems and containments, and
   - Maintaining cold shutdown in reactors that are undergoing maintenance outages when critical safety systems have been disabled.
3. Training of operators and plant emergency response organizations, in particular,

[3] That is, when the plant is cut off from outside supply of materials and personnel.

- **Specific training on the use of ad hoc responses for bringing reactors to safe shutdown during extreme beyond-design-basis events, and**
- **More general training to reinforce understanding of nuclear plant system design and operation and enhance operators' capabilities for managing emergency situations.**

The quality and completeness of the changes that result from this recommendation should be adequately peer reviewed.

S.2.3 Strengthen Capabilities for Assessing Risks from Beyond-Design-Basis Events

A "design-basis event" is a postulated event that a nuclear plant system, including its structures and components, must be designed and constructed to withstand without a loss of functions necessary to protect public health and safety. An event that is "beyond design basis" has characteristics that could challenge the design of plant structures and components and lead to a loss of critical safety functions. The Great East Japan Earthquake and tsunami were beyond-design-basis events.

FINDING 5.2: Beyond-design-basis events—particularly low-frequency, high-magnitude[4] (i.e., extreme) events—can produce severe accidents at nuclear plants that damage reactor cores and stored spent fuel. Such accidents can result in the generation and combustion of hydrogen within the plant and release of radioactive material to the offsite environment. There is a need to better understand the safety risks[5] that arise from such events and take appropriate countermeasures to reduce them.

RECOMMENDATION 5.2A: The U.S. nuclear industry and the U.S. Nuclear Regulatory Commission should strengthen their capabilities for identifying, evaluating, and managing the risks from beyond-design-basis events. Particular attention is needed to improve the identification of such events; better account for plant system interactions and the performance of plant operators and other critical personnel in responding to such events; and better estimate the broad range of offsite health,

[4] The term "extreme event" refers to high-magnitude environmental events, such as large earthquakes or floods, that occur very infrequently, for example, on the order of once every few centuries to millennia. The Great East Japan Earthquake and tsunami are examples of extreme events.

[5] Risk is defined and discussed in Appendix I.

environmental, economic, and social consequences that can result from such events.

RECOMMENDATION 5.2B: The U.S. Nuclear Regulatory Commission should support industry's efforts to strengthen its capabilities by providing guidance on approaches and by overseeing independent review by technical peers (i.e., peer review).

RECOMMENDATION 5.2C: As the U.S. nuclear industry and the U.S. Nuclear Regulatory Commission carry out the actions in Recommendation 5.2A, they should pay particular attention to the risks from beyond-design-basis events that have the potential to affect large geographic regions and multiple nuclear plants. These include earthquakes, tsunamis and other geographically extensive floods, and geomagnetic disturbances.

S.2.4 Further Incorporate Modern Risk Concepts into Nuclear Safety Regulations

A design-basis accident is a stylized accident—for example, a loss-of-coolant accident or transient overpower accident—that is required (by regulation) to be considered in a reactor system's design. The Fukushima Daiichi accident was a beyond-design-basis accident. Other major nuclear accidents (Three Mile Island in 1979 and Chernobyl in 1986) are also considered to be beyond-design-basis accidents.

FINDING 5.3: Four decades of analysis and operating experience have demonstrated that nuclear plant core-damage risks are dominated by beyond-design-basis accidents. Such accidents can arise, for example, from multiple human and equipment failures, violations of operational protocols, and extreme external events. Current approaches for regulating nuclear plant safety, which traditionally have been based on deterministic concepts such as the design-basis accident, are clearly inadequate for preventing core-melt accidents and mitigating their consequences. Modern risk assessment principles are beginning to be applied in nuclear reactor licensing and regulation. The more complete application of these principles in licensing and regulation could help to further reduce core-melt risks and their consequences and enhance the overall safety of all nuclear plants, especially currently operating plants.

RECOMMENDATION 5.3: The U.S. Nuclear Regulatory Commission should further incorporate modern risk concepts into its nuclear reactor safety regulations. This effort should utilize the strengthened

capabilities for identifying and evaluating risks that are described in Recommendation 5.2A.

The committee uses the term "modern risk concepts" to mean risk that is defined in terms of the risk triplet (What can go wrong? How likely is that to happen? What are the consequences if it does happen?) and subject to the limitations for quantitative analyses discussed in Section 5.2 in Chapter 5. Implementing this recommendation fully would likely require changes to some current U.S. Nuclear Regulatory Commission regulatory procedures, for example, those used for backfit analyses.

S.2.5 Examine Offsite Emergency Response Capabilities and Make Necessary Improvements

Emergency response to the Fukushima Daiichi accident was greatly inhibited by the widespread and severe destruction caused by the March 11, 2011, earthquake and tsunami. Japan is known to be well prepared for natural hazards; however, the earthquake and tsunami caused devastation on a scale beyond what was expected and prepared for. Twenty prefectures on three of Japan's major islands (Hokkaido, Honshu, and Shikoku) were affected by the earthquake and tsunami.

FINDING 6.1: The Fukushima Daiichi accident revealed vulnerabilities in Japan's offsite emergency management. The competing demands of the earthquake and tsunami diminished the available response capacity for the accident. Implementation of existing nuclear emergency plans was overwhelmed by the extreme natural events that affected large regions, producing widespread disruption of communications, electrical power, and other critical infrastructure over an extended period of time. Additionally:

- **Emergency management plans in Japan at the time of the Fukushima Daiichi accident were inadequate to deal with the magnitude of the accident, requiring emergency responders to improvise.**
- **Decision-making processes by government and industry officials were challenged by the lack of reliable, real-time information on the status of the plant, offsite releases, accident progression, and projected doses to nearby populations.**
- **Coordination among the central and local governments was hampered by limited and poor communications.**
- **Protective actions were improvised and uncoordinated, particularly when evacuating vulnerable populations (e.g., the elderly and sick) and providing potassium iodide.**

- Different and revised radiation standards and changes in decontamination criteria and policies added to the public's confusion and distrust of the Japanese government.
- Cleanup of contaminated areas and possible resettlement of populations are ongoing efforts 3 years after the accident with uncertain completion time lines and outcomes.
- Failure to prepare and implement an effective strategy for communication during the emergency contributed to the erosion of trust among the public for Japan's government, regulatory agencies, and the nuclear industry.

FINDING 6.2: The committee did not have the time or resources to perform an in-depth examination of U.S. preparedness for severe nuclear accidents. Nevertheless, the accident raises the question of whether a severe nuclear accident such as occurred at the Fukushima Daiichi plant would challenge U.S. emergency response capabilities because of its severity, duration, and association with a regional-scale natural disaster. The natural disaster damaged critical infrastructure and diverted emergency response resources.

RECOMMENDATION 6.2A: The nuclear industry and organizations with emergency management responsibilities in the United States should assess their preparedness for severe nuclear accidents associated with offsite regional-scale disasters. Emergency response plans, including plans for communicating with affected populations, should be revised or supplemented as necessary to ensure that there are scalable and effective strategies, well-trained personnel, and adequate resources for responding to long-duration accident and/or disaster scenarios involving

- Widespread loss of offsite electrical power and severe damage to other critical offsite infrastructure, for example, communications, transportation, and emergency response infrastructure;
- Lack of real-time information about conditions at nuclear plants, particularly with respect to releases of radioactive material from reactors and/or spent fuel pools; and
- Dispersion of radioactive materials beyond the 10-mile emergency planning zones for nuclear plants that could result in doses exceeding one or more of the protective action guidelines.

RECOMMENDATION 6.2B: The nuclear industry and organizations with emergency management responsibilities in the United States should assess the balance of protective actions (e.g., sheltering in place,

evacuation, relocation, and distribution of potassium iodide) for off-site populations affected by severe nuclear accidents and revise the guidelines as appropriate. Particular attention should be given to the following issues:

- **Protective actions for special populations (children, ill, elderly) and their caregivers;**
- **Long-term impacts of sheltering in place, evacuation and/or relocation, including social, psychological and economic impacts; and**
- **Decision making for resettlement of evacuated populations in areas contaminated by radioactive material releases from nuclear plant accidents.**

S.2.6 Improve the Nuclear Safety Culture

The term "safety culture" is generally understood to encompass a set of attitudes and practices that emphasize safety over competing goals such as production or costs. There is universal acceptance by the nuclear community that safety culture practices need to be adopted by regulatory bodies and other organizations that set nuclear power policies; by senior management of organizations operating nuclear power plants; and by individuals who work in those plants.

FINDING 7.1: While the Government of Japan acknowledged the need for a strong nuclear safety culture prior to the Fukushima Daiichi accident, TEPCO and its nuclear regulators were deficient in establishing, implementing, and maintaining such a culture. Examinations of the Japanese nuclear regulatory system following the Fukushima Daiichi accident concluded that regulatory agencies were not independent and were subject to regulatory capture.[6]

FINDING 7.2: The establishment, implementation, maintenance, and communication of a nuclear safety culture in the United States are priorities for the U.S. nuclear power industry and the U.S. Nuclear Regulatory Commission. The U.S. nuclear industry, acting through the Institute of Nuclear Power Operations, has voluntarily established nuclear safety culture programs and mechanisms for evaluating their implementation at nuclear plants. The U.S. Nuclear Regulatory Commission has published a policy statement on nuclear safety culture,

---

[6] The term "regulatory capture" refers to the processes by which regulated entities manipulate regulators to put their interests ahead of public interests.

but that statement does not contain implementation steps or specific requirements for industry adoption.

RECOMMENDATION 7.2A: The U.S. Nuclear Regulatory Commission and the U.S. nuclear power industry must maintain and continuously monitor a strong nuclear safety culture in all of their safety-related activities. Additionally, the leadership of the U.S. Nuclear Regulatory Commission must maintain the independence of the regulator. The agency must ensure that outside influences do not compromise its nuclear safety culture and/or hinder its discussions with and disclosures to the public about safety-related matters.

RECOMMENDATION 7.2B: The U.S. nuclear industry and the U.S. Nuclear Regulatory Commission should examine opportunities to increase the transparency of and communication about their efforts to assess and improve their nuclear safety cultures.

All committee members agree with the safety culture findings and recommendations (i.e., Findings 7.1-7.2 and Recommendations 7.2A, B), but members have a range of views about the current status of the nuclear safety culture in the United States. A selection of views is provided in Section 7.4 in Chapter 7.

1

Introduction

The March 11, 2011, Great East Japan Earthquake and tsunami created a humanitarian and material disaster in northeastern Japan. These natural events caused extensive damage to coastal communities in Iwate, Miyagi, and Fukushima Prefectures (Figure 1.1) and were responsible for about 15,900 deaths and 2,600 missing persons (National Police Agency of Japan , 2014); untold human suffering, especially of injured and displaced persons; and physical infrastructure losses exceeding $200 billion[1] (~¥17 trillion).

The earthquake and tsunami were also responsible for initiating a severe nuclear accident at the Fukushima Daiichi Nuclear Power Station[2] (Figure 1.2) located in east-central Fukushima Prefecture about 180 km southwest of the earthquake hypocenter (see Figure 1.1). The accident was rated as a Level 7 (major accident) event on the International Nuclear and Radiological Event Scale of the International Atomic Energy Agency, on par with the 1986 Chernobyl accident. However, releases of radioactive material to the atmosphere (mainly noble gases, iodine-131, and cesium-134/ cesium-137) from the Fukushima Daiichi accident were less than 15 percent of the Chernobyl releases.[3]

---

[1] Cabinet Office of Japan: http://www.meti.go.jp/english/earthquake/nuclear/japan-challenges/pdf/japan-challenges_c.pdf. Accessed on June 3, 2014.

[2] When the formal names of nuclear plants are used in this report they are capitalized, for example the Fukushima Daiichi Nuclear Power Station. Lower case is used for informal names, for example, the Fukushima Daiichi plant.

[3] Based on TEPCO release estimates. See http://www.tepco.co.jp/en/press/corp-com/release/betu12_e/images/120524e0201.pdf. Accessed on June 3, 2014.

FIGURE 1.1 Map of the Tohoku region of Japan (northern Honshu) showing the epicenter of the Great East Japan Earthquake (yellow star) and location of the Fukushima Daiichi Nuclear Power Station.

FIGURE 1.2 Oblique aerial photo of the Fukushima Daiichi Nuclear Power Station prior to the March 11, 2011, Great East Japan Earthquake and tsunami. Locations of reactor units are indicated by their numbers (i.e., Units 1-6). Locations of the turbine buildings, seawall, and common spent fuel storage (pool and dry) are also indicated. SOURCE: Courtesy of TEPCO.

Three of the six reactors at the Fukushima Daiichi plant sustained severe core damage during the accident and released hydrogen and radioactive materials. Explosion of the released hydrogen in three reactor buildings (Figure 1.3) caused severe structural damage and impeded onsite emergency response efforts. Offsite transport of the released radioactive materials by winds contaminated parts of Fukushima Prefecture and smaller regions of neighboring prefectures (Chiba, Gunma, Ibaraki, Miyagi, and Tochigi Prefectures) (Figure 1.4). About 78,000 residents were evacuated from a 20-km-radius exclusion zone established around the station and 62,000 from a 20- to 30-km-radius from the plant (UNSCEAR, 2013b). A large portion of this exclusion zone will likely remain off-limits to full-time reoccupation for the foreseeable future.

1.1 BACKGROUND ON THE STUDY CHARGE

At the time of the Fukushima Daiichi accident, the Blue Ribbon Commission on America's Nuclear Future was completing an assessment for the U.S. Secretary of Energy of options for managing spent nuclear fuel and high-level radioactive waste in the United States. In the weeks following the accident, concerns were raised about the condition of the spent fuel pools

FIGURE 1.3 March 16, 2011, photo of the Fukushima Daiichi Nuclear Power Station showing damaged Units 1 (foreground), 3, and 4 reactor buildings. SOURCE: Courtesy of TEPCO (http://photo.tepco.co.jp/library/110316/110316_1f_chijou_2.jpg).

in the damaged reactor buildings at the Fukushima Daiichi plant and the potential for large-scale releases of radioactive materials from the stored spent fuel.

In view of these concerns, the Blue Ribbon Commission recommended that the National Academy of Sciences (NAS) conduct an assessment of lessons learned from the Fukushima Daiichi accident (BRC, 2012, pp. xii-xiii):

> [T]he [Blue Ribbon] Commission recommends that the National Academy of Sciences (NAS) conduct a thorough assessment of lessons learned from Fukushima and their implications for conclusions reached in earlier NAS studies [NRC, 2004b[4]] on the safety and security of current storage arrangements for spent nuclear fuel and high-level waste in the United States. This effort would complement investigations already under way by the [U.S. Nuclear Regulatory Commission] and other organizations.

[4] An abbreviated public version of this report was issued (NRC, 2006b).

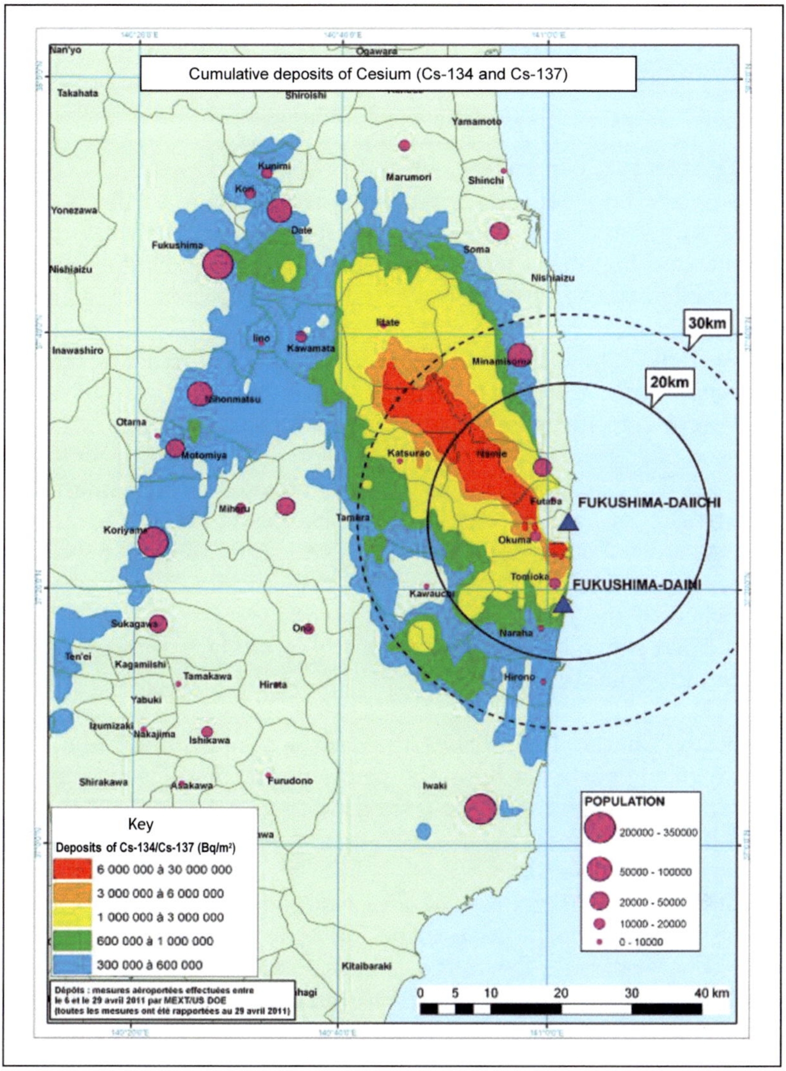

FIGURE 1.4 Map showing cumulative ground deposition of cesium-134 and cesium-137 (becquerels per square meter) in northeastern Japan. The figure was produced by the Institut de Radioprotection et de Sûreté Nucléaire based on airborne surveys carried out in April 2011 and published by the Japanese Ministry of Education, Culture, Sports, Science and Technology. The concentric circles demarcate the 20- and 30-km-radius zones around the Fukushima Daiichi Nuclear Power Station. SOURCE: IRSN (2011, Figure 7 [http://hps.org/documents/irsn_fukushima_report.pdf]).

This recommendation was taken up by the U.S. Congress, which subsequently directed[5] the U.S. Nuclear Regulatory Commission (USNRC) to contract with NAS for a study focused on five issues:

1. Causes of the crisis at Fukushima;
2. Lessons that can be learned;
3. Lessons' implications for conclusions reached in earlier NAS studies on the safety and security of current storage arrangements for spent nuclear fuel and high-level waste in the United States, including an assessment of whether the amount of spent fuel currently stored in reactor pools should be reduced;
4. Lessons' implications for commercial nuclear reactor safety and security regulations; and
5. Potential to improve design-basis threats assessment.

Congress directed that this study "be conducted in coordination with the Department of Energy and, if possible, the Japanese Government" and that the study "build upon the 2004 NAS study of storage issues and complement the other efforts to learn from Fukushima that have already been launched by the [US]NRC and industry."

The formal statement of task for this NAS study is shown in Sidebar 1.1. It contains four study charges:

- Study Charge 1 addresses the causes of the Fukushima Daiichi accident, focusing particularly on the performance of safety systems at the Fukushima Daiichi plant and the responses of its operators following the earthquake and tsunami. This study charge maps directly to the first issue in the congressional mandate.
- Study Charge 2[6] focuses on a reevaluation of conclusions from the 2004 NAS report on spent nuclear fuel safety and security (NRC, 2004b, 2006b [public version]). This charge maps to the third issue of the congressional mandate. Moreover, it calls for an evaluation of current storage arrangements for spent fuel in the context of the 2004 NAS study, rather than a de novo assessment of current storage arrangements and whether they should be changed.
- The remaining two study charges map to the second, fourth, and fifth issues of the congressional mandate. They focus on lessons learned from the Fukushima Daiichi accident for improving safety and security of plant systems and operations (Charge 3) and regulations (Charge 4).

[5] This directive was contained in the conference report from the Consolidated Appropriations Act of 2012 (Public Law 112-74).

[6] For reasons explained later in this chapter, this task is not addressed in this report.

SIDEBAR 1.1
Statement of Task for This National Research Council Study

The National Research Council will provide an assessment of lessons learned from the Fukushima nuclear accident for improving the safety and security of nuclear plants in the United States. This assessment will address the following issues:

1. Causes of the Fukushima nuclear accident, particularly with respect to the performance of safety systems and operator response following the earthquake and tsunami.
2. Reevaluation of the conclusions from previous National Academy of Sciences studies on safety and security of spent nuclear fuel and high-level radioactive waste storage, particularly with respect to the safety and security of current storage arrangements and alternative arrangements in which the amount of commercial spent fuel stored in pools is reduced.[a]
3. Lessons that can be learned from the accident to improve commercial nuclear plant safety and security systems and operations.
4. Lessons that can be learned from the accident to improve commercial nuclear plant safety and security regulations, including processes for identifying and applying design-basis events for accidents and terrorist attacks to existing nuclear plants.

The study may examine policy options related to these issues but should not make policy recommendations that involve nontechnical value judgments.

[a] This task will be addressed in a subsequent report. It is not addressed in this report.

Study Charge 4 also calls for an assessment of approaches used to identify and apply design-basis events[7] (see Sidebar 1.2) for accidents and terrorist attacks to existing nuclear plants.

- An additional sentence was added to the end of the statement of task by NAS to preclude policy recommendations that involve nontechnical value judgments. Such nontechnical factors, for example, cost and public acceptability, can be as important as technical factors in the policy-making process. Policy recommendations are well beyond the technical scope of this study.

[7] A "design-basis event" is a postulated event that a nuclear plant system, including its structures and components, must be designed and constructed to withstand without a loss of functions necessary to protect public health and safety.

SIDEBAR 1.2
Nuclear Plant Accident Terminology

Several terms are used throughout this report to describe accidents at nuclear power plants (referred to in this report as "nuclear plants") and the events that initiate them. These terms have specific meanings when applied to nuclear plant safety as described in this sidebar.

Nuclear plant accidents are classified according to their implications for safety and the specific type of events that initiate them, known as an "accident sequence." Nuclear plants are designed with extensive safety features and operators are trained to handle a wide range of normal and abnormal conditions, including accidents caused by equipment failure, loss of power, and loss of reactor core cooling capability.

There is extensive guidance from the U.S. Nuclear Regulatory Commission (USNRC) in the form of General Design Criteria (Title 10, Part 50, Appendix A, of the *Code of Federal Regulations*) to cover a specified set of failures or abnormal events, referred to collectively as "design-basis accidents." A plant design must include specific engineering safety features such as emergency core cooling systems (see Chapter 2) so that the plant operators can recover the plant to a safe state following such accidents. The safety systems for design-basis accidents are designed to limit the damage to the fuel in the reactor core and minimize the release of radioactive material from the plant's containment to levels that do not affect the health and safety of the general public.

Accidents that are not anticipated by the General Design Criteria specifications are known as "beyond-design-basis accidents." Such accidents can be initiated

Because the final statement of task for this study differs in wording from the congressional mandate, NAS shared it with appropriate congressional staff prior to initiation of the study to confirm its acceptability.

Given the charge by Congress to focus on lessons learned from the Fukushima accident for U.S. nuclear plants, some explicit choices were made to narrow the study focus before the committee was assembled. In particular, an explicit decision was made not to focus the study on the geological and geophysical processes that produced the earthquake and tsunami. While these are no doubt important to Japan, they have limited relevance to nuclear plant safety in the United States.

1.2 STUDY PROCESS

The study was carried out using established NAS procedures. The committee appointments were designed to provide diverse expertise and by a range of events originating inside the plant, referred to as "internal events," or outside the plant, referred to as "external events." Examples of internal events include equipment failures such as stuck valves (e.g., a stuck-open valve was the initiator of the 1979 Three Mile Island accident), pipe breaks, and human error (e.g., the 1986 Chernobyl accident was initiated by operator actions that had unforeseen consequences). Examples of external events include terrorist attacks as well as natural events such as large earthquakes and tsunamis (e.g., as discussed in Chapter 3 of this report, an earthquake and tsunami initiated the Fukushima Daiichi accident). Beyond-design-basis accidents can challenge the engineering safety systems at nuclear plants and require improvised operator actions and resources beyond the standard design features of the plant to recover a safe operational state.

If a beyond-design-basis accident results in excessive loss of reactor cooling and heatup of the reactor core, significant core damage can occur, resulting in a "severe accident." The USNRC defines a severe accident as a "type of accident that may challenge safety systems at a level much higher than expected." According to the International Atomic Energy Agency, a severe accident involves significant degradation of the reactor core (IAEA, 2007).

Severe accidents are associated with the release of fission products from the reactor fuel and the production of hydrogen from metal–water reactions in the reactor core. In the most extreme cases the fuel in the reactor core can melt, flow to the bottom of the steel vessel that holds the reactor core, and melt through the vessel onto the concrete floor of the plant's containment. This can result in elevated temperatures, pressures, radiation levels, and combustible gas concentrations, such as hydrogen and carbon monoxide, inside containment.

experience in technical disciplines relevant to the study task; these include geophysics, health physics, human factors, law and regulation, materials sciences, mechanical and structural engineering, nuclear engineering, nuclear power plant operations, nuclear safety and security, public health, and risk analysis. The committee chair is an NAS member with demonstrated leadership capabilities and strong knowledge of Japan and its culture; however, he has no experience with the nuclear power industry. The vice chair, a member of the National Academy of Engineering, has devoted his career to the development and application of risk assessment to improve nuclear plant safety. Biographical sketches of the committee members are provided in Appendix A.

Committee members hold a range of views on the desirability of nuclear energy, as well as other energy-generating technologies, as energy sources for the United States. These views are not relevant to this study because it does not address energy policy issues.

The committee held 39 in-person and conference-call meetings during the course of this study to gather information and develop this report. Information about the committee's information-gathering meetings is provided in Appendix B. One of these meetings was held in Tokyo, Japan, to enable in-depth discussions about the Fukushima Daiichi accident with Japanese technical experts from industry, academia, and government. The committee also visited the Fukushima Daini, Fukushima Daiichi, and Onagawa plants (see Figure 3.1 in Chapter 3) to learn about their designs, operations, and responses to the earthquake and tsunami. Additionally, subgroups of the committee visited two nuclear plants in the United States that are similar in design to the Fukushima Daiichi plant to learn about their designs and operations: Oyster Creek Generating Station in Forked River, New Jersey, and the Edwin I. Hatch Nuclear Plant in Baxley, Georgia.

1.3 STRATEGY TO ADDRESS THE STUDY CHARGE

The initial strategy for this NAS study was to address all four charges of the study task (Sidebar 1.1) in a single report. However, NAS encountered unanticipated administrative delays in obtaining national security clearances for the committee. (These clearances are needed to address Charge 2 of the statement of task on spent fuel safety and security.) Additionally, once the necessary security clearances were obtained, the committee had to cancel two of its meetings (in April and November 2013) owing to the federal budget sequester and federal government shutdown. These meetings were to have been devoted to gathering information to address Charge 2.

NAS determined that it was not possible to complete the entire study on the original schedule because of these delays. However, because work on the other three study charges was proceeding on schedule, NAS decided to issue the results of that work in the present report and to negotiate a new schedule and budget with the study sponsor (USNRC) for addressing Charge 2 of the study task (Sidebar 1.1) and issuing the results in a separate report. Consequently, with one exception in Chapter 5 (see Section 5.1.1.6), the security portion of the study task is not addressed in this report.

This NAS study is one of many investigations or assessments initiated in the wake of the Fukushima Daiichi accident. Some key written products from these activities are listed in Table 1.1. They include, for example, four accident investigations in Japan: Two by the Japanese government (one each by the executive and legislative branches), one by a private organization, and one by Tokyo Electric Power Company, owner and operator of the Fukushima Daiichi plant. Nuclear plant operators and regulators in several countries have also conducted assessments to determine if operational or regulatory changes are needed to cope with extreme natural events (e.g., earthquakes and floods) that could occur at nuclear plants. The Interna-

TABLE 1.1 Selected Key Reports from Fukushima Daiichi Accident-Related Investigations and Assessments (as of June 2014)

Japanese Government and Related

- Report of Japanese Government to the IAEA Ministerial Conference on Nuclear Safety (June 2011) (Government of Japan, 2011b)
- Additional Report of the Japanese Government to the IAEA—The Accident at TEPCO's Fukushima Nuclear Power Stations (Second Report) (September 2011) (Government of Japan, 2011a)
- Investigation Committee on the Accident at Fukushima Nuclear Power Stations of Tokyo Electric Power Company (Established by Japanese Cabinet): Interim Report (December 2011) (Investigation Committee, 2011)
- Nuclear and Industrial Safety Agency, Technical Knowledge of the Accident at Fukushima Dai-ichi Nuclear Power Station of Tokyo Electric Power Co., Inc. (Provisional Translation) (March 2012) (NISA, 2012)
- Investigation Committee on the Accident at Fukushima Nuclear Power Stations of Tokyo Electric Power Company (Established by Japanese Cabinet), Final Report (July 2012) (Investigation Committee, 2012)
- National Diet of Japan Fukushima Nuclear Accident Independent Investigation Commission (NAIIC), The Official Report of the Fukushima Nuclear Accident Independent Investigation Commission (July 2012) (NAIIC, 2012)

Japanese Industry

- Japan Nuclear Technology Institute, Examination of Accident at Tokyo Electric Power Co., Inc.'s Fukushima Daiichi Nuclear Power Station and Proposal of Countermeasures (October 2011) (JANTI, 2011)
- TEPCO, Fukushima Nuclear Accident Analysis Report: Interim Report (December 2011) (TEPCO, 2011a)
- TEPCO, Fukushima Nuclear Accident Investigation Report: Interim Report, Supplementary Volume (December 2011) (TEPCO, 2011b)
- Tokyo Electric Power Company (TEPCO), The Nuclear Safety and Quality Assurance Meeting's Accident Investigation Examination Committee's Opinion of the Tokyo Electric Power Company's "Fukushima Nuclear Accident Investigation Report" (Midterm Report) (November 2011) (TEPCO, 2011d)
- TEPCO, Mid-and-Long-Term Roadmap Towards the Decommissioning of Fukushima Daiichi Nuclear Power Units 1-4 (December 2011) (TEPCO, 2011c) [Note: this document was updated in 2012 and 2013.]
- TEPCO, Estimation of the Released Amount of Radioactive Materials into the Atmosphere as a Result of the Accident in the Fukushima Daiichi Nuclear Power Station (May 2012) (TEPCO, 2012a)
- TEPCO, Fukushima Nuclear Accident Analysis Report (June 2012) (TEPCO, 2012b)
- TEPCO, Evaluation of the Situation of Cores and Containment Vessels of Fukushima Daiichi Nuclear Power Station Units-1 to 3 and Examination into Unsolved Issues in the Accident Progression: Progress Report No. 1 (December 2013) (TEPCO, 2013)

continued

TABLE 1.1 Continued

Other Japanese Organizations

- Atomic Energy Society of Japan (AESJ). Estimated Current Status of Fukushima-Daiichi Nuclear Power Plant Units 1-3 (April 18, 2011) (AESJ, 2011a)
- AESJ. Lessons Learned from the Accident at the Fukushima Daiichi Nuclear Power Plant (May 9, 2011) (AESJ, 2011b)
- Science Council of Japan, Report to the Foreign Academies from Science Council of Japan on the Fukushima Daiichi Nuclear Power Plant Accident (May 2011) (SCJ, 2011)
- Rebuild Japan Initiative Foundation, Independent Investigation Commission on the Fukushima Nuclear Accident (February 2012) (RJIF, 2014)

International Organizations

- International Atomic Energy Agency, IAEA International Fact Finding Expert Mission of the Fukushima Dai-Ichi NPP Accident Following the Great East Japan Earthquake and Tsunami (June 2011) (IAEA, 2011)
- International Atomic Energy Agency, International Nuclear Safety Advisory Group Annual Assessment Letters for 2011, 2012, and 2013 (INSAG, 2011, 2012, 2013)
- Nuclear Energy Agency, The Fukushima Daiichi Nuclear Power Plant Accident: OECD/NEA Nuclear Safety Response and Lessons Learnt (September 2013) (NEA, 2013)

U.S. Government and Related

- U.S. Nuclear Regulatory Commission (The Near-Term Task Force Review of Insights from the Fukushima Dai-ichi Accident), Recommendations for Enhancing Reactor Safety in the 21st Century (July 2011) (USNRC NTTF, 2011)
- Blue Ribbon Commission on America's Nuclear Future, Report to the Secretary of Energy (January 2012) (BRC, 2012)
- Sandia National Laboratories, Fukushima Daiichi Accident Study (Status as of April 2012) (July 2012) (Gauntt et al., 2012b)
- Government Accountability Office, Nuclear Safety: Countries' Regulatory Bodies Have Made Changes in Response to the Fukushima Daiichi Accident (March 2014) (USGAO, 2014)

U.S. Industry and Related

- Institute of Nuclear Power Operations (INPO), Special Report on the Nuclear Accident at the Fukushima Daiichi Nuclear Power Station (November 2011) (INPO, 2011)
- Electric Power Research Institute (EPRI), Fukushima Daiichi Accident–Technical Causal Factor Analysis (March 2012) (EPRI, 2012a)
- EPRI, Summary of the EPRI Early Event Analysis of the Fukushima Daiichi Spent Fuel Pools Following the March 11, 2011 Earthquake and Tsunami in Japan (May 2012) (EPRI, 2012c)
- American Nuclear Society, Fukushima Daiichi: ANS Committee Report (June 2012) (ANS, 2012)

TABLE 1.1 Continued

- INPO, Lessons Learned from the Nuclear Accident at the Fukushima Daiichi Nuclear Power Station (August 2012) (INPO, 2012)
- EPRI, Fukushima Technical Evaluation Phase 1—MAAP5 Analysis (April 2013) (EPRI, 2013)
- Lochbaum et al., Fukushima: The Story of a Nuclear Disaster (February 2014) (Lochbaum et al., 2014)

Other Governments and Related

- Government of the United Kingdom, Japanese Earthquake and Tsunami: Implications for the UK Nuclear Industry Final Report (September 2011) (ONR, 2011)
- Canadian Nuclear Safety Commission, CNSC Fukushima Task Force Report (October 2011) (CNSC, 2011)
- Swiss Federal Nuclear Safety Inspectorate, Lessons Fukushima 11032011: Lessons Learned Checkpoints Based on the Nuclear Accidents at Fukushima (October 2011) (ENSI, 2011)

tional Atomic Energy Agency (IAEA) and Nuclear Energy Agency have organized meetings of member countries to share information and best practices from these assessments. The IAEA plans to issue a report in 2015 on the causes of the Fukushima Daiichi accident and lessons learned.

In the United States, the nuclear industry launched a fast-track effort to understand the Fukushima Daiichi accident and identify and implement appropriate countermeasures at U.S. nuclear plants. This effort is being led by the Institute of Nuclear Power Operations and Nuclear Energy Institute with technical support from plant operators and the Electric Power Research Institute.

The U.S. government, primarily through the USNRC with technical support from the Department of Energy and its national laboratories, launched a parallel effort to reevaluate nuclear plant safety regulations in light of the accident. Initially, the USNRC established a task force of senior agency staff to review current USNRC processes and regulations and make recommendations to improve them. The USNRC subsequently created the Japan Lessons-Learned Project Directorate,[8] overseen by a steering committee of senior agency officials, to implement the task force's recommendations. Several of the task force's recommendations were implemented by the USNRC while this NAS study was under way and work to implement others was proceeding.

The committee did not have the time to perform an in-depth evaluation of these industry, government, and international efforts. However, the committee used the written products from these activities to inform its

[8] This directorate became the Japan Lessons-Learned Division effective June 1, 2014.

own work. The committee has provided a cross-walk between its findings and recommendations and key findings and recommendations from other investigations and assessments for the benefit of readers (see Appendix E).

The peer-reviewed literature also served as an important source of information for this study. This literature was particularly important for understanding, for example, the earthquake and tsunami; Japanese laws, regulations, and nuclear safety culture; human factors for responses to emergencies; and emergency preparedness and response.

The committee relied almost exclusively on English-language information sources for this study. Fortunately, English translations of key Japanese government and industry reports were readily available to the committee for this purpose (e.g., see Table 1.1). However, the committee did not have access to the full range of Japanese-language papers, reports, and analyses of the Fukushima Daiichi accident.

Most of the industry, government, and international activities described previously were undertaken under demanding schedules, typically a few months to a year, and were intended to implement safety improvements to existing nuclear plants on an accelerated schedule. This NAS study is being carried out on a longer (2-year) schedule and has a different scope: *It is intended to be a broad-scope and high-level review of lessons learned from the Fukushima Daiichi accident to improve safety and security of U.S. nuclear plants, taking into account where possible the results of these other investigations and assessments* (e.g., Table 1.1). This NAS study is intended to complement the efforts by industry and regulators to learn from the Fukushima Daiichi accident, as Congress directed when it issued the study mandate.

A great deal of additional information about the Fukushima Daiichi accident—for example, the status of currently inaccessible plant components, the location and characteristics of the damaged reactor cores, and pathways for hydrogen and radioactive material migration—will likely be uncovered as the reactors are dismantled and studied over the next four decades.[9] As understanding of accident progression and phenomenology improve, new lessons will likely be learned and some existing lessons, including those in this report, may require revision.

The NAS was asked to carry out a technical assessment of lessons learned from the Fukushima Daiichi accident. NAS was not asked to:

- *Assign blame for the accident.* The reports from the Japanese accident investigations, which are referenced in Table 1.1, address this issue.

[9] TEPCO's roadmap for decommissioning the Fukushima Daiichi plant can be found at http://www.tepco.co.jp/en/nu/fukushima-np/roadmap/images/t120730_03-e.pdf. Accessed on June 3, 2014.

- *Recommend changes to nuclear plant operations or regulations in Japan or other foreign countries.* The mandate from Congress directed NAS to focus on U.S. nuclear plants. However, the committee hopes that the results of this NAS study will be useful to other countries.
- *Recommend specific changes to U.S. laws or regulations, for example, to shut down or impose additional operating requirements on reactors in the United States.* Such changes are the responsibility of the U.S. government, require the participation of affected stakeholders, and involve consideration of nontechnical factors that are beyond the scope of this study.
- *Recommend specific changes to the designs or operations of U.S. nuclear plants.* Such changes are the responsibility of the nuclear industry and its regulator, acting in response to their own assessments and with input from interested organizations and individuals, and require plant design-specific information that is unavailable to the committee.
- *Assess whether U.S. nuclear plants are safe.* The primary focus of this study is on how nuclear plant safety and security can be improved based on lessons learned from the Fukushima Daiichi accident. This focus should not be construed to suggest that nuclear plants are currently unsafe. Nuclear plant operators and regulators strive to make continuous improvements to nuclear plant safety (see Chapter 7).

The committee focused its information-gathering efforts on boiling water reactor (BWR) plants having designs similar to the Fukushima Daiichi plant. Some of the findings and recommendations in this report apply specifically to those plants. However, many of the findings and recommendations apply to both BWR and pressurized water reactor plants. Unless otherwise noted in individual findings and recommendations, they are intended to apply to both plant types.

1.4 REPORT ORGANIZATION

This report is organized into seven chapters:

- Chapter 1 (this chapter) describes the study task and process.
- Chapter 2 provides information on nuclear plant design and operations in Japan and the United States. It is intended to provide background information to support the more detailed discussions of the Fukushima Daiichi accident that appear in Chapters 3 and 4.
- Chapter 3 describes the Great East Japan Earthquake and tsunami and their impacts on nuclear plants in Japan.
- Chapter 4 describes the accident at the Fukushima Daiichi plant.

- Chapter 5 presents the committee's lessons learned from the Fukushima Daiichi accident for nuclear plant operations and regulations in the United States.
- Chapter 6 describes the offsite emergency response associated with the Fukushima Daiichi accident and lessons learned from that response for the United States.
- Chapter 7 describes the nuclear safety culture in Japan and lessons learned for the United States.

The appendixes provide additional information to support the discussions in the report chapters.

2

Background on Japanese and U.S. Nuclear Plants

This chapter is intended to provide non-expert readers with basic information about nuclear power plant[1] design, operation, and regulation in Japan and the United States. This information will be useful for understanding the technical discussions in subsequent report chapters. Expert readers may wish to skip ahead to Chapter 3.

This chapter is organized into five sections.

- Section 2.1 provides an overview of nuclear plant design and operation.
- Section 2.2 describes the design of boiling water reactors (BWRs) and their safety systems. (The reactors at the Fukushima Daiichi nuclear plant were BWRs.)
- Sections 2.3 and 2.4 describe nuclear plants and regulation of nuclear power in Japan and the United States, respectively.
- Section 2.5 describes some key differences in BWR designs in Japan and the United States.

[1] The terms "nuclear power plant" and "nuclear plant" are used interchangeably in this report.

2.1 NUCLEAR PLANT DESIGN AND OPERATION

Nuclear plants are used in the United States and many other countries primarily to meet baseload[2] demands for electricity. These plants are especially well suited for this purpose because they can be operated for long periods without maintenance outages and can produce electricity at constant rates.

Nuclear plants generate electricity using the Rankine thermal cycle: the plant's nuclear reactors produce heat that is used to convert water to steam. The steam drives a turbine that spins a generator to produce electricity. After passing through the turbine, the steam is cooled, condensed, and recirculated. This "steam engine" cycle is also used to produce electricity in other types of thermal power plants, particularly coal- and gas-fired plants.

The primary fuel for nuclear plants is slightly enriched uranium,[3] usually in the form of 1-cm-long cylindrical uranium dioxide pellets. These pellets are encased in metal tubes, referred to as nuclear *fuel cladding*, each ~10 mm in diameter and about 4 m in length, made of various zirconium alloys (Zircaloy, ZIRLO, M5) containing 98 percent or more zirconium. The cladding provides structural support for the fuel pellets, serves as a barrier to the release of radioactive material from the fuel, and provides an efficient geometry for cooling. The ensemble of pellets and cladding is referred to as a nuclear *fuel rod*. Fuel rods are grouped into bundles, or *fuel assemblies* (Figure 2.1 [top]), each containing between about 64 and 300 rods.

The entire set of fuel assemblies, along with control rods and associated structural supports (Figure 2.1 [bottom]), constitutes the *reactor core*. The control rods contain materials that are highly neutron absorbing, such as hafnium, boron, or silver. The control rods can be used to shut down the reactor when fully inserted.

The reactor core is enclosed in a robust steel pressure vessel, the *reactor pressure vessel* (RPV) (Figure 2.2). The RPV contains numerous penetrations for steam and water lines, instrumentation, and controls. The robust RPV design allows the reactor to operate at high temperature and pressure to increase its thermal efficiency.[4] It also provides a major barrier to the release of radioactive material from the reactor during an accident. Water

---

[2] That is, the continuous demand for electricity from customers in regions served by the plant. Countries such as France have such a high percentage of nuclear power that the output of some of their plants is varied according to demand.

[3] Enriched uranium contains uranium-235 in higher-than-natural abundances. Natural uranium contains about 0.7 percent uranium-235. Uranium used in most reactors contains 3-5 percent uranium-235.

[4] BWRs operate at pressures and temperatures of about 7 MPa and 285°C. Pressurized water reactors (PWRs) operate at pressures and temperatures of about 15 MPa and 315°C. See Footnote 15 for a definition of MPa.

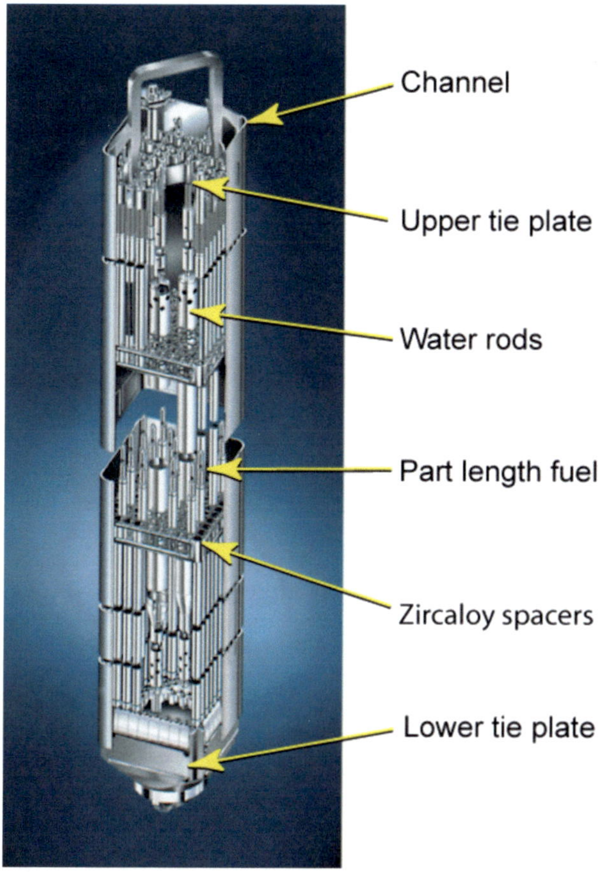

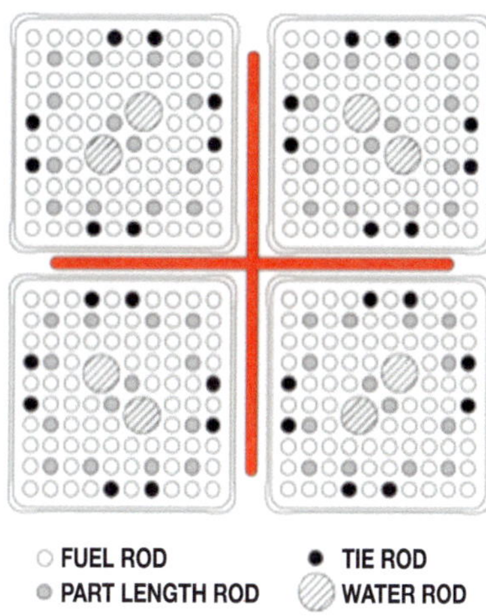

FIGURE 2.1 (Top) Side view of a BWR fuel assembly. Some fuel rods have been removed to reveal construction details. (Bottom) Cross-section showing four BWR fuel assemblies and a control rod. The control rod consists of four blades in the shape of a cross. It can be seen in cross section (red cross) in the figure. The control rods are moved into and out of the reactor core to control its power. SOURCE: ANS (2012, Fig. 4).

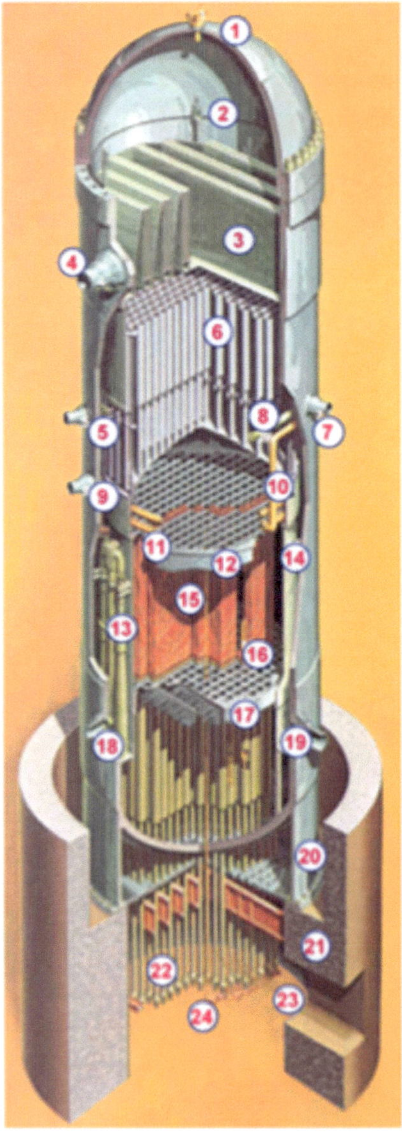

1 Vent and Head Spray
2 Steam Dryer Lifting Lug
3 Steam Dryer Assembly
4 Steam Outlet
5 Core Spray Inlet
6 Steam Separator Assembly
7 Feedwater Inlet
8 Feedwater Sparger
9 Low Pressure Coolant Injection Inlet
10 Core Spray Line
11 Core Spray Sparger
12 Top Guide
13 Jet Pump Assembly
14 Core Shroud
15 Fuel Assemblies
16 Control Blade
17 Core Plate
18 Jet Pump/Recirculation Water Inlet
19 Recirculation Water Outlet
20 Vessel Support Skirt
21 Shield Wall
22 Control Rod Drives
23 Control Rod Drive Hydraulic Lines
24 In-Core Flux Monitor

FIGURE 2.2 Schematic of a BWR/5 or BWR/6 reactor pressure vessel. SOURCE: ANS (2012, Fig. 5).

circulation through the RPV is used to control reactor pressure and temperature and generate steam for electricity production. The movement of water and/or steam out of the RPV is pressure-driven and is controlled by opening and closing valves.

The RPV is located within the *containment* of the building that houses the reactor (Figure 2.3). The containment is constructed of reinforced concrete a meter in thickness or a carbon steel shell a few centimeters thick and contains heavy metal bulkheads to allow access for maintenance work. Like the RPV, the containment also has numerous penetrations for steam and water lines, instrumentation, and controls. The containment serves as a barrier to the release of radioactive material to the environment during a severe accident (Sidebar 2.1).

Reactor power is regulated by manipulating the positions of the control rods in the reactor core.[5] The reactor can be "started" by partially withdrawing the control rods from the core. This allows a sustained nuclear fission chain reaction to be initiated in the uranium fuel, which generates large quantities of heat. This heat is removed by the constant circulation of cooling water through the core. As noted previously, the reactor can be shut down by fully inserting the control rods into the core. A reactor is said to be *scrammed* when all of the reactor's control rods are fully inserted and the fission process is halted after an off-normal condition is detected. Shutdown may occur automatically or can be initiated by reactor operators.

The operation of a reactor produces a wide range of radioactive isotopes:

- Fission of the uranium fuel results in the production of dozens of highly radioactive fission products, for example, cesium-137, iodine-131, and strontium-90. Some of these fission products, notably cesium and iodine, are volatile.
- Absorption of neutrons by materials in the reactor core produces transuranic elements such as plutonium-239 as well as neutron activation products such as cobalt-60.

These isotopes continue to decay and generate heat (referred to as *decay heat*) even after the reactor is shut down. Decay-heat generation immediately following reactor shutdown can be up to about 6 percent of the reactor's licensed power. Heat generation decreases rapidly as short-lived isotopes (primarily fission products) decay (see Figure 2.4). Cooling is crucial in the first few days after the reactor is shut down and continues

---

[5] Other means are used to regulate reactor power as well. The power in BWRs can be regulated by varying water flow through the core. The power in PWRs can be regulated by varying the concentration of boron, a neutron absorber, in reactor cooling water.

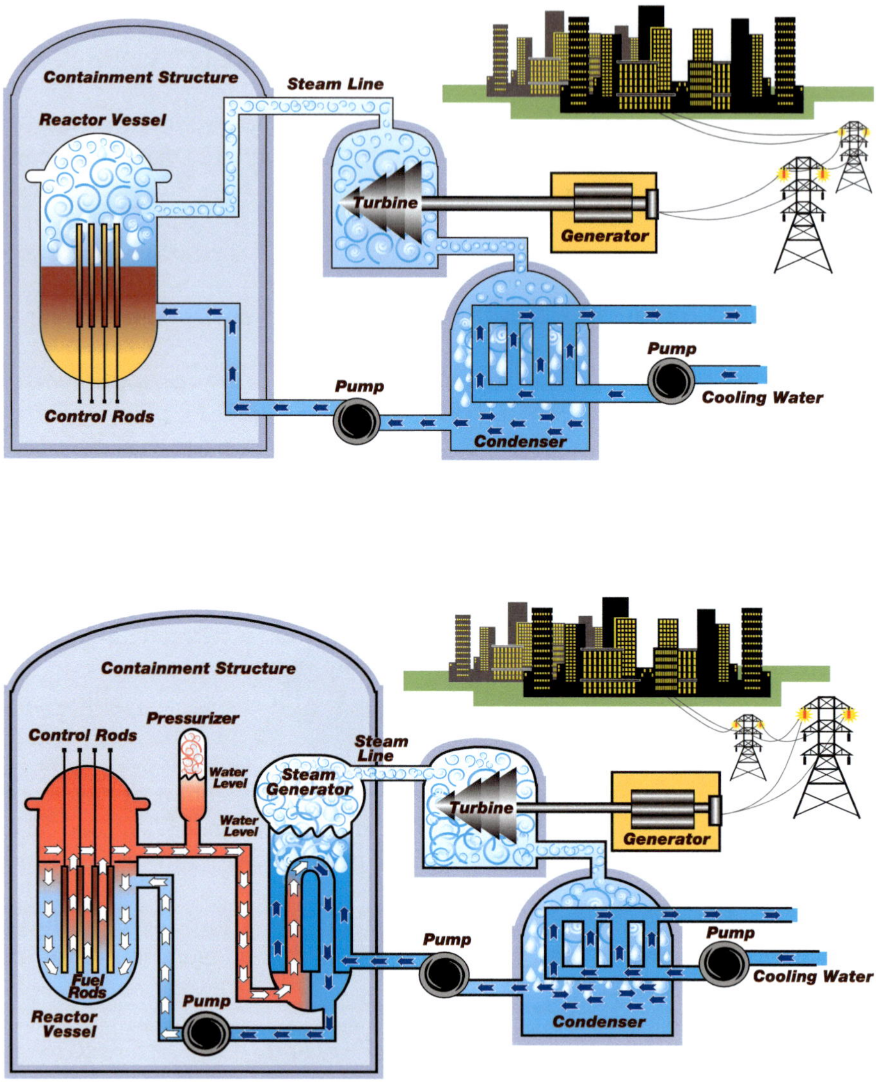

FIGURE 2.3 Simplified schematics for nuclear power reactors. (Top) Boiling water reactor. (Bottom) Pressurized water reactor. SOURCE: ANS (2012, Fig. 2 [Top] and Fig. 3 [Bottom]).

to be required for years (but at lower levels) to remove the heat generated from the decay of long-lived fission products. Reactor cooling systems are designed to remove this heat so as to prevent excessive temperature rise.

Reactor cooling is provided by several safety systems. Some safety systems operate during normal conditions to maintain RPV pressures and temperatures and water levels within a set range. Other safety systems are part of the *emergency core cooling system* (ECCS). These systems are used to cool the core during off-normal conditions. The effectiveness of these systems depends on their ability to remove decay heat through a combination of heating and boiling of water in the reactor while maintaining the water level in the RPV above the top of the reactor core.

In the United States, nuclear fuel in a reactor must be replaced every 4 to 6 years depending on the reactor's design and operation. U.S. reactors are typically shut down every 18-24 months[6] for replacing a portion of the reactor fuel. The *used* (or *spent*) *fuel* is transferred from the reactor to a spent fuel pool. The pool has its own cooling system (the pool water is circulated through a heat exchanger) to remove decay heat from the fuel.

Nuclear plants can contain one or more reactors and their support systems, including water and electrical supplies, mechanical systems, and spent fuel pools. All nuclear plant sites in Japan and most in the United States have multiple reactors (see Sections 2.3 and 2.4 in this chapter).

More than 400 nuclear power reactors[7] are currently operating throughout the world and 70 more are currently under construction. The large majority of nuclear power plants in the world and all plants in the United States and Japan are *light-water reactors*; these reactors are cooled and moderated by regular water.[8] Two types of light-water reactors have been deployed worldwide for electricity production, including in Japan and the United States: BWRs and pressurized water reactors (PWRs). The design of these reactors is illustrated in Figure 2.3.

The primary difference between BWRs and PWRs is the mechanism for generating steam to produce electricity. BWRs produce steam directly in the core; that steam is separated, dried, and used to drive turbines and the electrical generators connected to them. PWRs produce high-temperature water that is circulated through a heat exchanger (referred to as the *steam generator*) to produce steam in a secondary water circulation loop. The steam in this secondary loop drives the turbines and their associated electrical generators. In both plant designs, the steam is condensed to water after

[6] Japanese and some European reactors are shut down every 12 months for refueling.

[7] Prior to the Fukushima nuclear accident, there were 442 operating power reactors in 30 countries (World Energy Council, 2012).

[8] Moderation refers to the slowing down of fission neutrons to thermal energies to increase their nuclear fission cross-section. The CANDU (Canadian Deuterium Uranium) reactor uses heavy water as moderator and accounts for about 10 percent of reactors worldwide.

SIDEBAR 2.1
Radioactive Material Releases in Severe Accidents

Nuclear reactors generate radioactive by-products, primarily *fission products* (e.g., iodine and cesium) and transuranic isotopes (e.g., plutonium), that build up inside the fuel and fuel rods during the course of operations. This radioactive material is contained completely within the fuel rods under normal operating conditions. In a severe reactor accident, however, the zirconium fuel rods can oxidize and rupture, the uranium fuel can overheat and melt, and radionuclides can be released into the reactor pressure vessel. If the reactor's containment fails, radionuclides can be released into the reactor building and possibly into the environment. This is exactly the scenario that occurred in Units 1, 2, and 3 at the Fukushima Daiichi plant following the March 11, 2011, earthquake and tsunami.

Fission product releases, which normally constitute the largest fraction of radioactivity released during a severe accident, are usually in the form of gases (xenon and krypton), which are released when fuel rods rupture, and aerosols (iodine[a] and cesium), which are formed by condensation after vaporizing from hot fuel. Other fission products (strontium) and associated radioactive materials (uranium, plutonium) have very high vaporization temperatures and are largely retained in the reactor fuel, even when molten. Release of iodine-131 in any form—aerosol, molecular, or organic compound—is of particular concern because of its high activity (it has an 8-day half-life) and its ability to concentrate in the human thyroid gland if ingested. Children are particularly at risk of developing thyroid cancer as a result of exposure to radioactive iodine.

Once released to the atmosphere, cesium and iodine are transported by prevailing winds and can travel for considerable distances before wet or deposi-

passing through the turbines and the condensed water is recirculated. The water used to condense the steam is taken from a nearby ocean, river, or other water supply.

The reactors at the Fukushima Daiichi plant are BWRs. Consequently, the discussion in the remainder of this chapter focuses primarily on the design and operation of this reactor type.

2.2 BOILING WATER REACTORS

BWRs were initially developed by General Electric Company during the 1950s and have evolved through "generations," with each generation representing iterative evolutions in the design of steam systems, water recirculation systems, safety systems, and containments. In the United States, BWR containments are designated Mark I, Mark II, and Mark III (the oldest to most recent designs). These designs are illustrated schematically in tion brings them to the ground or the surface of a water body such at a river, lake, or ocean. Prevailing winds at the time of the Fukushima Daiichi accident appear to have transported most of the radioactivity released from the damaged reactors out to the Pacific Ocean (Kawamura et al., 2011; Morino et al., 2011). However, sufficient quantities were also transported and deposited inland to contaminate large land areas (see Figure 1.4 in Chapter 1) to levels requiring long-term human use restrictions. The cesium isotopes Cs-134 (2-year half-life) and, more importantly, Cs-137 (30-year half-life) are the most important sources of long-term contamination. These isotopes have also contaminated the Fukushima Daiichi reactor buildings and will impede efforts to remove damaged fuel from the Unit 1, 2, and 3 reactors.

The most important function of the reactor containment during a severe accident is to prevent the release of iodine and cesium aerosols. If containment has to be deliberately vented to prevent excess pressure (see Sidebar 2.2), vented gases can be filtered through sand or water, if such filters are available, to reduce the quantity of aerosols that are released into the environment.

Venting BWR containments through the suppression chamber is preferred because the vent gases can be passed through the suppression pool to scrub out aerosols. The reactor building, which serves as a secondary containment, can also be used to reduce aerosol releases when the containment is bypassed or develops a leak as happened at the Fukushima Daiichi plant (see Chapter 4). However, hydrogen explosions in the Fukushima Daiichi reactor buildings reduced their effectiveness in filtering out radioactive aerosols.

[a] Iodine can also exist in other forms as well, for example, as elemental or organic gases (e.g., Gavrilin et al., 2004).

Figure 2.5. Additionally, there are six reactor generations denoted BWR/1,[9] BWR/2, BWR/3, BWR/4, BWR/5, and BWR/6. The design of reactor cooling systems (see Section 2.2.3) has also evolved in these BWR reactor generations.

Newer BWR designs, the advanced boiling water reactor (ABWR) and economic simplified boiling water reactor (ESBWR), have been developed by General Electric. The ABWR design has been approved by the U.S. Nuclear Regulatory Commission (USNRC), and several ABWRs have been constructed in Japan (see Section 2.3). The ESBWR design has been submitted to the USNRC for approval and its review is nearing completion. The discussion in the remainder of this section focuses on first-generation BWRs (BWR/1-BWR/6).

Unit 1 at the Fukushima Daiichi plant is a BWR/3 with a Mark I containment: Units 2-5 are BWR/4 with Mark I containments; and Unit 6 is a

[9] No first-generation (BWR/1) reactors are operating today.

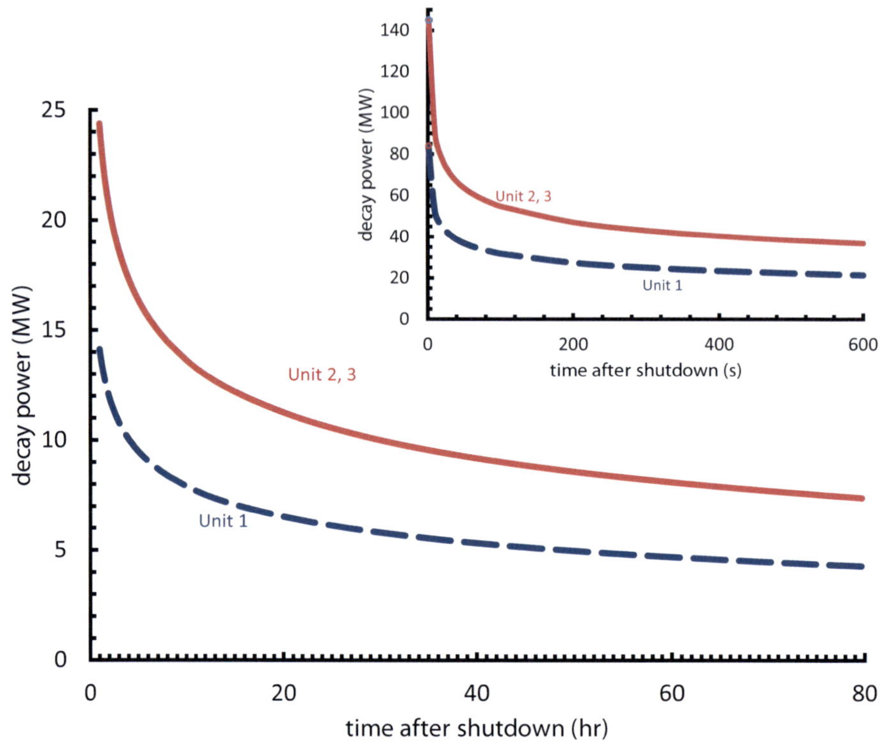

FIGURE 2.4 Estimated thermal power output of reactor cores at the Fukushima Daiichi plant following shutdown. SOURCE: Based on methodology used in Gauntt et al. (2012b) and Phillips et al. (2012).

BWR/5 with a Mark II containment (Table 2.1). A number of U.S. nuclear plants have reactors and containments that are similar to those in Units 1-4 at the Fukushima Daiichi plant (Table 2.2).

The discussion of containment systems below focuses on the Mark I containment because of its relevance to the Fukushima Daiichi accident.

2.2.1 Containment System

The Mark I containment comprises the structure, referred to as the *drywell*, that houses the RPV. The drywell is connected to a water-filled chamber, referred to as the *suppression chamber*.[10] The water pool in this

[10] Also sometimes referred to as the *wetwell* and *torus*.

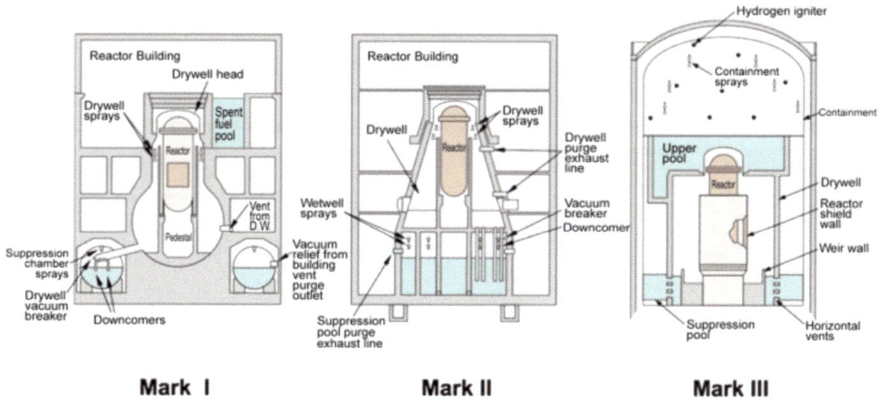

FIGURE 2.5 BWR containment designs. The location of the spent fuel pool is not shown in the Mark II containment. SOURCE: ANS (2012, Fig. 16).

chamber is referred to as the *suppression pool* and is designed to condense steam that is released from the RPV if it becomes overpressured. The pool is also used to remove (i.e., *scrub*) fission products in the vented gases when the reactor fuel is damaged. The RPV can be depressurized by opening safety relief valves (SRVs).

The suppression chamber can be cooled using various systems to maintain it within design pressures and temperatures. If cooling is lost, the suppression chamber can be vented to the atmosphere to reduce pressures and temperatures. The suppression pool water can be used to filter out radioactive material before venting (Sidebar 2.2).

The spent fuel pool resides outside of containment but inside the reactor building (Figure 2.5). It is located near the top of the drywell to allow fuel unloading to be performed under water. This requires that the spent fuel pool be elevated above ground level.

2.2.2 Pressure Control System

During normal BWR operations, steam produced in the reactor exits the RPV and flows to the main turbine. If the reactor is shut down, the main steam isolation valves (MSIVs) are closed, isolating the RPV from the power conversion system. Depending on the nature of the shutdown and associated operating procedures, the RPV can be depressurized by opening the SRVs; this allows steam to flow from the RPV into the suppression pool where it is condensed. (This cooling pathway is shown in the Sidebar 2.2 figure.) Depressurization is required before operators can activate the

TABLE 2.1 BWR Reactor Designs

	Reactor Type and Fukushima Daiichi Unit		
Reactor Function	BWR/3, Unit 1	BWR/4, Units 2, 3, 4, 5	BWR/5, Unit 6
Reactor isolation pressure control	Isolation condenser and SRVs	All use SRVs; some have steam condensing mode of RHR	All use SRVs; some have steam condensing mode of RHR
Reactor isolation inventory control	Isolation condenser	RCIC	RCIC
ECCS high-pressure pumping	HPCI	HPCI	HPCS
ECCS high-pressure pump type	Turbine-driven HPCI	Turbine-driven	Motor-driven
ECCS low-pressure flooding delivery point	Recirculation pump discharge pipe	Recirculation pump discharge pipe or inside shroud (core region)	Inside core shroud, core region
Containment type	Mark I	Mark I	Mark II

NOTES: The reactor designs included in this table are most pertinent to the Fukushima Daiichi and Daini plants. The table does not include advanced BWR designs such as the Advanced Boiling Water Reactor (ABWR) or Economic Simplified Boiling Water Reactor (ESBWR). The table shows general design features and may not be applicable to every BWR reactor. ECCS = emergency core cooling system, HPCI = high-pressure coolant injection, HPCS = high-pressure core spray, RFP = reactor feedwater pump, RCIC = reactor core isolation cooling, and SRV = safety relief valve.
SOURCE: Information taken from Boiling Water Reactor GE BWR/4 Technology Advanced Manual (available at http://pbadupws.nrc.gov/docs/ML0230/ML023010606.pdf).

low-pressure cooling systems to cool the reactor (see Section 2.2.3.1 in this chapter for a discussion of these cooling systems).

The SRVs will also automatically actuate through a purely mechanical function when pressures exceed preset values. This is a passive safety feature designed to protect the RPV from excessive pressures if operators are unable to actuate the SRVs.

The containment can be vented from the suppression chamber (see Sidebar 2.2) or drywell. This venting capability was enhanced (i.e., hardened) for BWR Mark I systems in the United States following the 1979 Three Mile Island nuclear accident[11]; hardened vents were also installed in

[11] The reactors at the Three Mile Island plant are PWRs.

TABLE 2.2 Mark I-BWRs in the United States That Are Similar to Units at the Fukushima Daiichi Plant

Plant Name	Location (State)
BWR/2, Mark I with IC	
Nine Mile Point 1	New York
Oyster Creek	New Jersey
BWR/3, Mark I with IC (similar to Fukushima Daiichi Unit 1)	
Dresden 2	Illinois
Dresden 3	Illinois
BWR/3, Mark I with RCIC	
Monticello	Minnesota
Pilgrim 1	Massachusetts
Quad Cities 1	Illinois
Quad Cities 2	Illinois
BWR/4, Mark I (similar to Fukushima Daiichi Units 2-4)	
Browns Ferry 1	Alabama
Browns Ferry 2	Alabama
Browns Ferry 3	Alabama
Brunswick 1	North Carolina
Brunswick 2	North Carolina
Cooper	Nebraska
Duane Arnold	Iowa
Fermi 2	Michigan
FitzPatrick	New York
Hatch 1	Georgia
Hatch 2	Georgia
Hope Creek 1	New Jersey
Peach Bottom 2	Pennsylvania
Peach Bottom 3	Pennsylvania
Vermont Yankee	Vermont

BWR Mark I reactors in Japan (see Section 2.5.2). This provided operators with a means to control containment pressures if they became elevated due to accident conditions. The enhancement was to typically install piping instead of sheet metal ducting as the pressure relief pathway. However, this enhancement was not made at all BWR plants.

Containment venting requires manual operator action using emergency operating procedures. In U.S. nuclear plants the venting path is established through piping from above the suppression chamber that passes through the reactor building and exhausts into the atmosphere.

SIDEBAR 2.2
Venting

Venting involves the controlled release of gases from the containment of a nuclear plant to the environment in emergency situations, for example, after the failure of a reactor's emergency core cooling system. In such situations, steam buildup in the containment can raise temperatures and pressures above design levels which, if unvented, could result in containment leakage or failure and uncontrolled releases of hydrogen and radioactive material into the reactor building and from there into the environment. Venting reduces containment pressures and temperatures; this reduces the potential for uncontrolled containment leakage and enables the reactor pressure vessel to be depressurized so that alternative means can be deployed to cool the core, for example, the injection of low-pressure cooling water from diesel-driven fire pumps or from fire trucks.

To vent a BWR reactor, operators must open motor-operated and air-operated valves (Figure S2.1). Motor-operated valves are typically opened (or "lined up") using either AC or DC power; they can also be opened manually if operators can physically access them. Air-operated valves can be opened using compressed air and DC power. Once valve lineup is complete and containment pressure is high enough, a rupture disk in the vent line (if present) will activate and containment gases will be vented through the plant's exhaust stack. These vented materials can contain radioactive materials (radioactive gases and fine particulate materials) and hydrogen from fuel cladding–steam reactions (see Sidebar 4.1 in Chapter 4).

In BWRs, gases can be vented through the suppression pool prior to release to "scrub out" some of their radioactive constituents. (In PWRs, gases can be vented into containment and scrubbed using water sprays.) Scrubbing is not 100 percent effective in removing radioactive constituents from the vented gases, however. Consequently, the venting of a reactor with damaged fuel would likely result in the release of some radioactive materials into the environment (e.g., noble gases). Decisions on venting and appropriate protective actions for the public need to balance the benefits of maintaining the integrity of containment to prevent large-scale radioactive releases with the consequences of immediate but smaller releases.

At present, no U.S. reactors have filtered vents, but the USNRC has initiated a rulemaking process to determine whether such vents should be required for BWR Mark I and Mark II reactors. Eighty non-BWR nuclear plants in Western Europe have filtered vents. Eighteen non-BWR reactors in Canada have filtered vents or have committed to installing them. Only 13 BWRs in the world have filtered vents (Borchardt, 2012a, Enclosure 3). Only a few nuclear plants in Japan have filtered vents but all have now committed to install them.

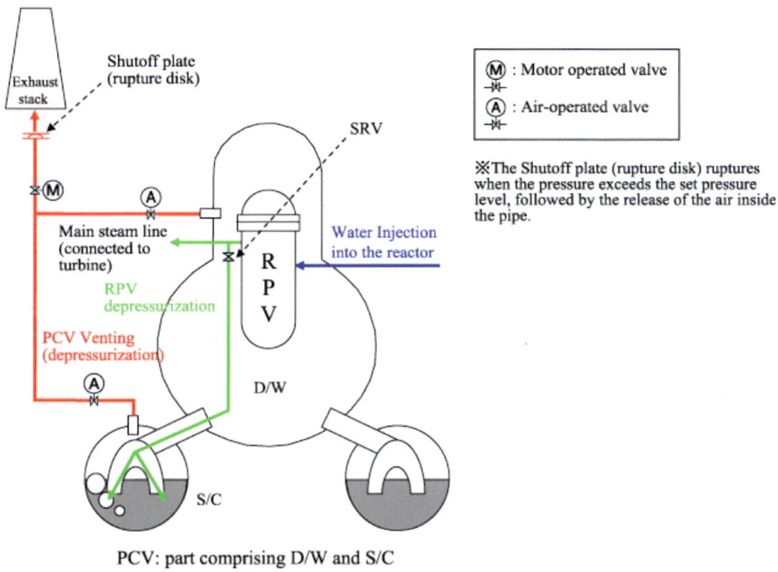

FIGURE S2.1 Simplified illustration of the containment venting system for Units 1-3 at the Fukushima Daiichi plant. The SRVs are opened to depressurize the RPV into the suppression pool in the suppression chamber (green line). The suppression chamber is vented to the exhaust stack by opening two valves and activating the rupture disk, if present (red line). NOTE: D/W = drywell, PCV = primary containment vessel, RPV = reactor pressure vessel, S/C = suppression chamber, SRV = safety relief valve. SOURCE: Courtesy of TEPCO (Available at http://www.tepco.co.jp/en/nu/fukushima-np/review/review1_2-e.html. Accessed on June 3, 2014.)

2.2.3 Core Cooling System

BWRs have various engineered safety features to cool their cores depending on their generation. Three systems played key roles in the Fukushima nuclear accident (see Chapter 4):

- Isolation condenser (IC) system: Used in BWR/2s and BWR/3s, including Unit 1 at the Fukushima Daiichi plant.
- Reactor core isolation cooling (RCIC) system: Used in BWR/4s, including Units 2-4 at the Fukushima Daiichi plant, BWR/5s, BWR/6s, and the Advanced Boiling Water Reactor.
- High-pressure coolant injection (HPCI) system: Used in BWR/3s and BWR/4s, including Units 2-4 at the Fukushima Daiichi plant.

These systems are designed to remove decay heat from the reactor in the absence of AC power. They require DC power for control purposes but in some situations can operate for extended periods without any power. These systems are described in subsequent sections of this chapter. More complete descriptions can be found in technical information documents such as the Reactor Concepts Training Manual (USNRC, 2012b).

AC power is required to operate other safety systems. These include the core spray, residual heat removal (RHR), and containment spray systems. Containment and suppression pool spray systems also can be powered by the diesel-driven fire protection system or emergency water sources. These systems played little or no role in the Fukushima nuclear accident and so they are described only briefly in the next section.

2.2.3.1 Low-Pressure Core Cooling Systems

Low-pressure core cooling systems comprise two separate and independent systems: the core spray system and the low-pressure coolant injection (LPCI) system of the RHR system (Figure 2.6). These systems require AC power to operate pumps, controls, and valves.

The core spray system pumps water from the suppression pool into the RPV (to remove decay heat) using two separate and independent pumping loops. The core spray system sprays water from above the core onto the tops of the fuel assemblies. Water is supplied by AC-powered, high-volumetric flow pumps. The core spray system and the LPCI mode of the RHR system operate only when the RPV is at low pressure.

The RHR system is a multipurpose system that uses AC-powered, high-volumetric flow pumps in different configurations to supply plant needs. The RHR system is normally aligned in the LPCI configuration to supply water makeup to the RPV for core cooling under loss-of-coolant conditions.[12] During LPCI operation, RHR pumps take water from the suppression pool and discharge it into the RPV directly or after flowing through a heat exchanger that transfers heat to the ultimate heat sink.

2.2.3.2 Isolation Condenser System (Fukushima Daiichi Unit 1 Reactor)

The IC system (Figure 2.7) is used to remove decay heat and conserve reactor water inventory when the RPV becomes isolated from the power conversion system (i.e., the turbine and condenser; see Figure 2.3) at or near operating pressures. It has two trains of equipment (labeled "Train A" and "Train B" in Figure 2.7), each consisting of a large heat exchanger and associated piping. The secondary (shell) side of the heat exchanger, basically a large tank, contains enough water to remove decay heat from the RPV for several hours. The shell-side water can be replenished using the makeup-water or fire protection systems or fire trucks.

The system can operate without electrical power or operator intervention as long as the system valves are open and there is water in the shell side of the heat exchanger. The system operates by gravity flow: Steam enters the heat exchanger via a steam line from the RPV, and condensate is returned to the RPV through a recirculation pump line.

As shown in Figure 2.7, there are four valves for each IC train. The two valves outside of containment are operated by DC power from batteries; the two valves inside containment are operated by AC power. If DC power is lost, a separate DC-powered interlocking logic circuit causes all four valves in each train to close, effectively shutting down the IC system. Once closed, the valves inside containment cannot be reopened unless AC power is available.[13] This system logic affected the operation of the valves for the IC in Unit 1 of the Fukushima Daiichi plant during the accident (see Chapter 4).

2.2.3.3 Reactor Core Isolation Cooling System (Fukushima Daiichi Unit 2 and 3 Reactors)

The RCIC system (Figure 2.8) is designed to make up water inventory losses from the RPV caused by water boil-off when the RPV is isolated from the turbine and condenser. It is designed to operate independently of auxiliary AC power, service air, or external cooling-water systems and can provide adequate makeup water to the RPV in the following circumstances:

---

[12] For example, when water levels in the RPV drop below acceptable levels because of a pipe break or other off-normal condition.

[13] There are several possible sources of AC power: offsite AC power, onsite emergency diesel generators, and onsite DC sources via inverters.

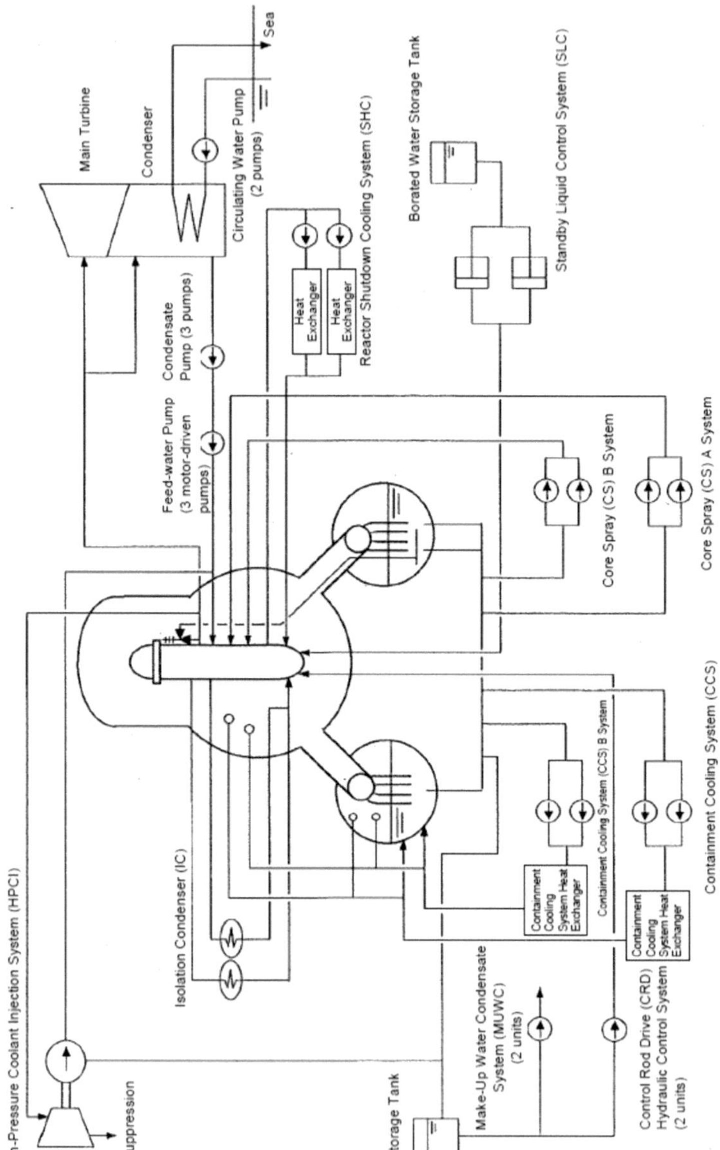

Main Turbine
Condenser
Sea
Circulating Water Pump (2 pumps)
Condensate Pump (3 pumps)
Feed-water Pump (3 motor-driven pumps)
Heat Exchanger
Heat Exchanger
Reactor Shutdown Cooling System (SHC)
Borated Water Storage Tank
Standby Liquid Control System (SLC)
Core Spray (CS) B System
Core Spray (CS) A System
Containment Cooling System (CCS) B System
Containment Cooling System (CCS) A System
Containment Cooling System Heat Exchanger
Containment Cooling System Heat Exchanger
High-Pressure Coolant Injection System (HPCI)
To Suppression Pool
Isolation Condenser (IC)
Condensate Storage Tank
Make-Up Water Condensate System (MUWC) (2 units)
Control Rod Drive (CRD) Hydraulic Control System (2 units)

A

B

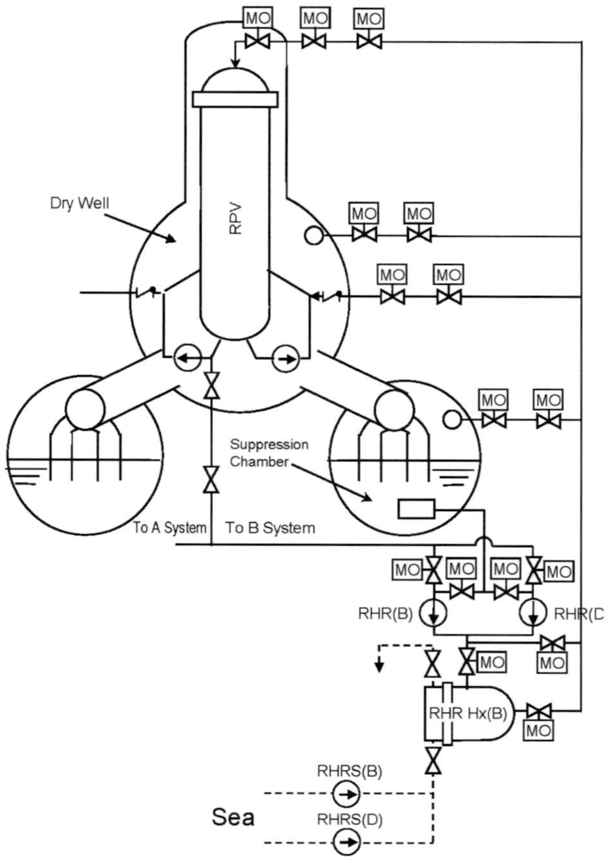

FIGURE 2.6 Schematic of the (A) core spray system for Unit 1 and (B) residual heat removal (RHR) system for Units 2 and 3 at the Fukushima Daiichi plant. Motor-operated (MO) valves are indicated by connected triangles. SOURCE: Government of Japan (2011b, Figs. IV-2-1 and IV-2-9).

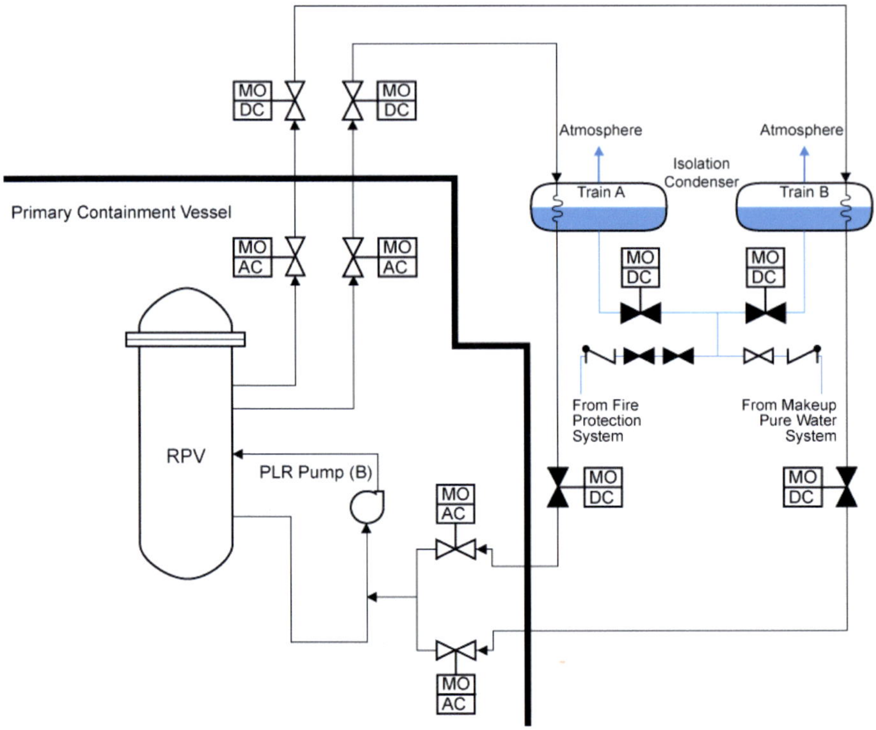

FIGURE 2.7 Schematic of the isolation condenser (IC) system for Unit 1 at the Fukushima Daiichi plant. Motor-operated (MO) valves are indicated by connected triangles. Black indicates valve closed during normal operations; white indicates valve open during normal operations. Power sources to operate the valves (AC or DC power) are indicated. SOURCE: ANS (2012, Appendix F, Fig. 1).

- RPV is isolated from the power conversion system (turbine and condenser) and is being maintained at operational pressures and temperatures.
- Reactor is shut down and at high pressure[14] with loss of normal feedwater.
- AC power is lost.

The RCIC system consists of a steam-driven turbine pump and associated piping, valves, and instrumentation necessary to implement several

[14] That is, before the reactor is depressurized to a level where the low-pressure cooling systems can be operated. However, it is also possible to lower the reactor pressure to below the shutoff head of the RHR/LPCI or LPCS so that one of these sources of water can be injected to the RPV while still having enough RPV pressure to provide steam for RCIC operation (150- to 200-psi range).

flow paths. The system is driven by steam produced by decay heat in the RPV. Steam exits through isolation valves and is routed through the turbine pump to provide the motive force for pumping makeup water into the reactor. The steam is exhausted to the suppression pool after exiting the turbine. Makeup water can be supplied from either the condensate storage tank (CST) or the suppression pool, with the preferred source being the CST. Makeup water enters the RPV through the feedwater injection line (see Figure 2.8).

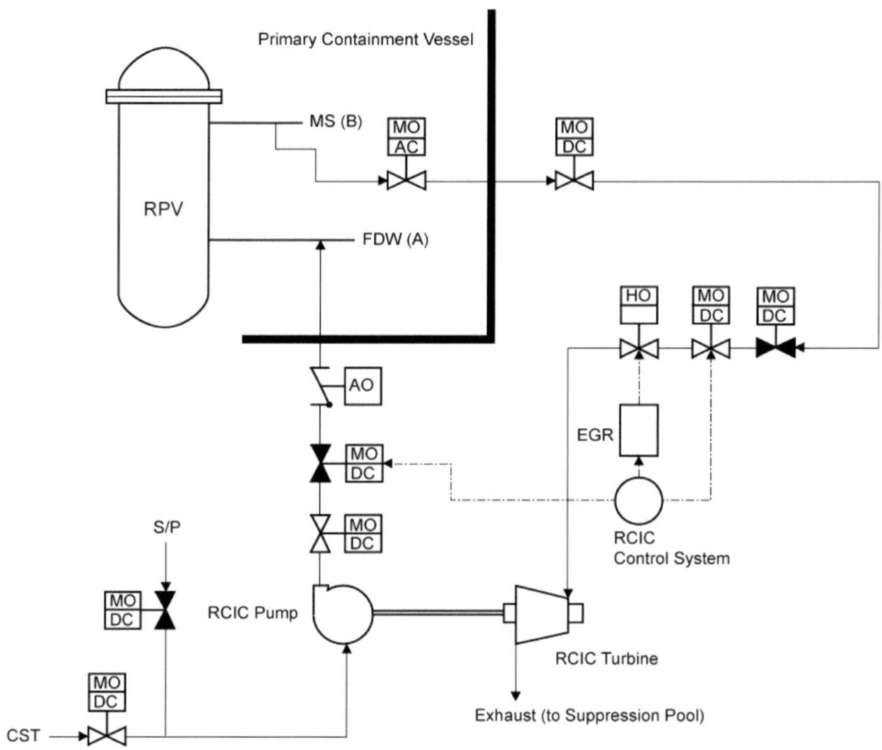

FIGURE 2.8 Schematic of the reactor core isolation cooling (RCIC) system for Units 2 and 3 at the Fukushima Daiichi plant. Valves are indicated by connected triangles. Black indicates valve closed; white indicates valve open. Power sources (AC or DC power) for motor-operated (MO) valves are indicated. Hydraulically operated (HO) valves are controlled either automatically or manually via a DC-powered control system. The electronic governor regulator (EGR) controls the HO valve and throttles steam flow to the reactor core isolation cooling (RCIC) turbine. Water is supplied from the suppression pool (S/P) or condensate storage tank (CST). SOURCE: ANS (2012, Appendix F, Fig. 2).

As shown in Figure 2.8, the valve outside of containment is operated by DC power supplied by batteries; but the valve inside containment is operated by AC power. If DC power is lost, a separate DC-powered interlocking logic circuit causes both the DC and AC valves to close, effectively shutting down the RCIC system. This logic circuitry was specifically intended as an isolation function to prevent leakage from the containment if a break occurs in the RCIC piping. Once closed, the valve inside containment cannot be reopened unless AC power is available (see Footnote 13).

The RCIC system is designed to operate over a wide range of RPV pressures—from full operating pressures (~7 MPa[15]) to ~1 MPa. The suppression pool acts as the heat sink for steam generated by reactor decay heat. Decay heat can be removed from the suppression pool using the heat exchangers in the RHR system when AC power is available.

The continued operation of the RCIC system following a loss of DC and AC power depends on the timing of the power losses in the AC and DC circuits that control the valves and the "failsafe" control logic—similar to the IC system operation that was described in Section 2.2.3.2. In the case of an extended loss of AC power, such as occurred at the Fukushima Daiichi plant following the tsunami, the RCIC system may stop operation for the following reasons:

- DC power for the failsafe logic control has failed, causing the system's valves to close (if motive power for the valves is still available).
- Suppression pool temperature is too high, possibly leading to failure of the turbine and pump bearings.
- Containment pressure is too high, causing the RCIC system turbine to shut down.[16]

2.2.3.4 High-Pressure Coolant Injection System (All Fukushima Daiichi Reactors)

The HPCI system (Figure 2.9) is similar to the RCIC system in function except that it has about seven times the flow capacity (680-1,270 m^3/h). It is designed to operate when the RPV remains at high pressure; such conditions might occur when a small pipe break causes water levels in the RPV to drop, but the diameter of the broken pipe is not large enough to depressurize the RPV. The HPCI can also act as a backup to the RCIC system. The same types of actuation signals initiate and terminate both the HPCI

---

[15] Megapascals (10^6 pascals). Pascal is the SI-derived unit for pressure and is equal to 1 N/m^2; 1 MPa ≈ 145 pounds per square inch (psi).

[16] BWR emergency operating procedures are being rewritten to override the high containment backpressure trip for the RCIC turbine.

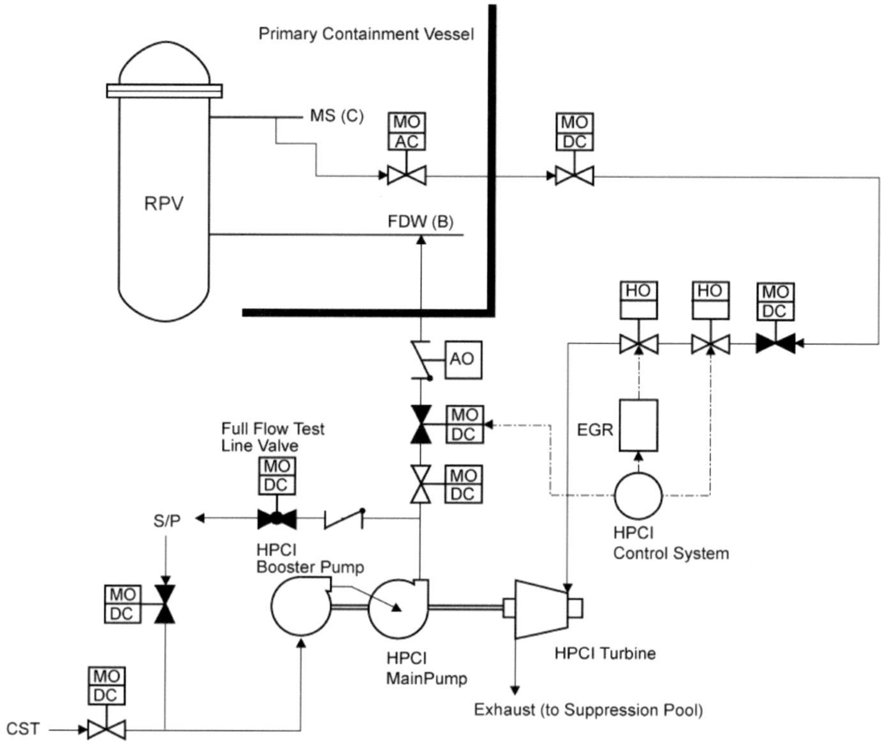

FIGURE 2.9 Schematic of the high-pressure core injection (HPCI) system for Units 1-3 at the Fukushima Daiichi plant. Valves are indicated by connected triangles. Black indicates valve closed; white indicates valve open. Power sources (AC or DC power) for motor-operated (MO) valves are indicated. Hydraulically operated (HO) valves are controlled either automatically or manually via a DC-powered control system. The electronic governor regulator (EGR) controls the HO valve and throttles steam flow to the high-pressure coolant injection (HPCI) turbine. Water is supplied from the suppression pool (S/P) or condensate storage tank (CST). SOURCE: ANS (2012, Appendix F, Fig. 3).

and RCIC, and DC power is needed to operate the HPCI pump and some HPCI system valves.

2.2.4 Emergency Power Systems

Nuclear plants are designed with multiple power sources to run pumps, valves, and controls to remove the decay heat from the reactor core. AC power is normally provided from offsite sources and is brought into the plant through multiple independent power lines. If offsite power is lost, AC power can be generated by onsite emergency diesel generators. These generators are designed to start up automatically in the first minute following a loss of offsite power. Each reactor at a nuclear plant has at least two diesel generators for redundancy. There is enough fuel onsite to last for several days if power and operable pumps are available to move it from large onsite storage tanks to smaller tanks that supply the diesel generators.

Large batteries (or banks of batteries) are situated onsite to provide emergency DC power for a select set of valves, instruments, lighting, and communications; these batteries are designed to supply power for about 8 hours under typical load conditions. As noted previously, DC power is used to operate critical valves and monitoring instrumentation for the IC and RCIC systems. Consequently, it is essential to protect the batteries and circuits used to carry DC power through the plant so that these will continue to function even when AC power is lost.

2.3 NUCLEAR PLANTS IN JAPAN

Prior to the Fukushima Daiichi accident, Japan had 54 operating nuclear power reactors at 16 sites (see Figure 2.10 and Table 2.3). These reactors provided about 30 percent of Japan's electricity supply. In early 2011, Japan was the world's third-largest producer of electricity from nuclear power, after the United States and France. Tokyo Electric Power Company, the owner/operator of the Fukushima Daiichi plant, owns 17 nuclear reactors at three sites: Fukushima Daiichi (6 reactors), Fukushima Daini (4 reactors), and Kashiwazaki Kariwa (7 reactors) (see Figure 2.10). Collectively, these reactors supplied about a third of Japan's nuclear power–generated electricity before the accident.

The nuclear plant fleet in Japan consists of 24 PWRs and 26 BWRs. All but four of these plants are Generation II designs.[17] Four ABWRs, at Hamaoka, Kashiwazaki-Kariwa, and Shika (Figure 2.10), are Generation III designs.

Figure 2.11 shows the operating electrical generating capacity of nuclear plants in Japan in 2011 and 2012. There was a decrease in capacity following the Fukushima Daiichi accident in March 2011 as reactors were

[17] Reactor generation terminology was developed by the U.S. Department of Energy. Generation II reactors were constructed beginning in the 1960s. They have mechanically and electrically operated safety systems that can be started automatically or by operator control. Most of the world's current reactor fleet consists of Generation II reactors. Generation III reactors were constructed beginning in the 1990s. They incorporate more passive safety systems and have other design improvements. See Goldberg and Rosner (2011) for additional information.

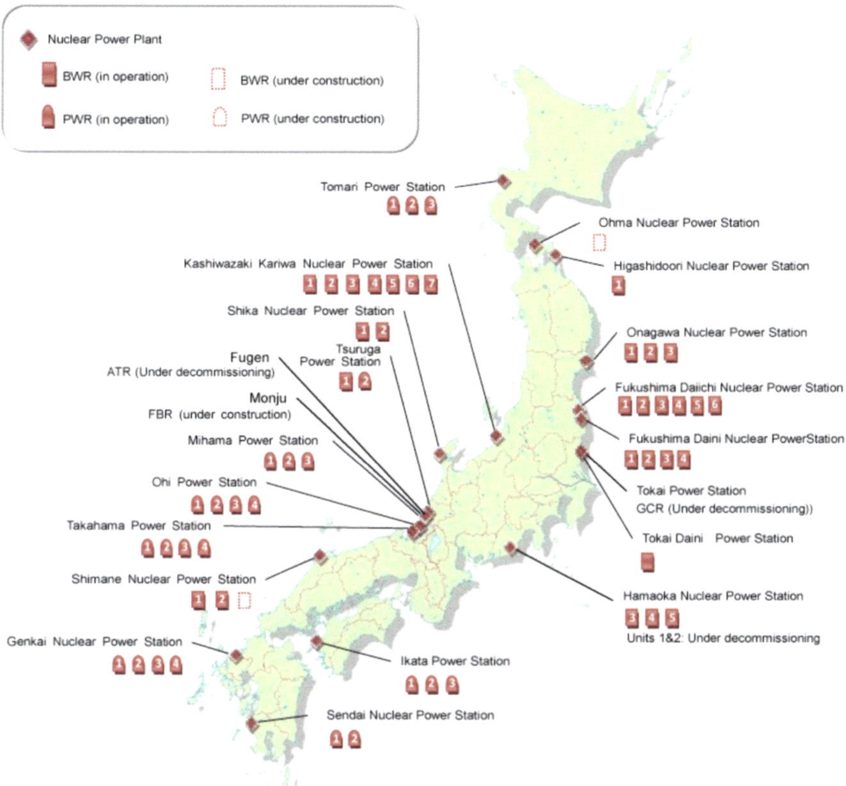

FIGURE 2.10 Locations of nuclear power plants in Japan. SOURCE: Nuclear Energy Agency (available at http://www.oecd-nea.org/press/2011/NEWS-02.html. Accessed on June 3, 2014).

taken offline for scheduled maintenance and were not allowed to restart. All Japanese reactors were shut down by April 2012. Two of the reactors at the Kansai Electric Power Co.'s Ohi plant in western Japan (Figure 2.10) were allowed to restart in July 2012 because of concerns about power shortages in the Kyoto region. These reactors were subsequently shut down for scheduled maintenance in September 2013 and were not allowed to restart.

All nuclear reactors in Japan must undergo a safety review by the new nuclear plant regulator (Nuclear Regulation Authority; see next section) before they can be restarted. These reviews are currently under way, and no completion date has been announced.

TABLE 2.3 Operating Nuclear Plants in Japan Prior to the Fukushima Daiichi Accident

Plant Name	Unit	Reactor and Containment Type	Thermal Capacity (MWt)	Commercial Operation	Licensee
Fukushima Daiichi	1	BWR-Mark I	1,380	1971	TEPCO
	2	BWR-Mark I	2,381	1974	
	3	BWR-Mark I	2,381	1976	
	4	BWR-Mark I	2,381	1978	
	5	BWR-Mark I	3,293	1978	
	6	BWR-Mark II		1979	
Fukushima Daiini	1	BWR-Mark II	3,293	1982	TEPCO
	2	BWR-Mark II (Improved)	3,293	1984	
	3	BWR-Mark II (Improved)	3,293	1985	
	4	BWR-Mark II (Improved)	3,293	1987	
Genkai	1	PWR	1,650	1975	Kyushu
	2	PWR	1,650	1981	
	3	PWR	3,423	1994	
	4	PWR	3,423	1997	
Hamaoka	3	BWR-Mark I (Improved)	3,293	1987	Chubu
	4	BWR-Mark I (Improved)	3,293	1993	
	5	ABWR	3,926	2005	
Higashidori	1	BWR-Mark I (Improved)	3,293	2005	Tohoku
Ikata	1	PWR	1,650	1977	Shikoku
	2	PWR	1,650	1982	
	3	PWR	2,660	1994	

Kashiwazaki-Kariwa	1	BWR-Mark II	3,293	1985	TEPCO
	2	BWR-Mark II (Improved)	3,293	1990	
	3	BWR-Mark II (Improved)	3,293	1993	
	4	BWR-Mark II (Improved)	3,293	1994	
	5	BWR-Mark II (Improved)	3,293	1990	
	6	ABWR	3,926	1996	
	7	ABWR	3,926	1997	
Mihama	1	PWR	1,031	1970	KEPCO
	2	PWR	1,456	1972	
	3	PWR	2,440	1976	
Ohi	1	PWR-ICECND	3,423	1979	KEPCO
	2	PWR-ICECND	3,423	1979	
	3	PWR	3,423	1991	
	4	PWR	3,423	1993	
Onagawa	1	BWR-Mark I	1,593	1984	Tohoku
	2	BWR-Mark I (Improved)	2,436	1995	
	3	BWR-Mark I (Improved)	2,436	2002	
Sendai	1	PWR	2,660	1984	Kyushu
	2	PWR	2,660	1985	
Shika	1	BWR-Mark I (Improved)	1,593	1993	Hokuriku
	2	ABWR	3,926	2006	
Shimane	1	BWR-Mark I	1,380	1974	Chugoku
	2	BWR-Mark I (Improved)	2,436	1989	
Takahama	1	PWR	2,440	1974	KEPCO
	2	PWR	2,440	1975	
	3	PWR	2,660	1985	
	4	PWR	2,660	1985	

continued

TABLE 2.3 Continued

Plant Name	Unit	Reactor and Containment Type	Thermal Capacity (MWt)	Commercial Operation	Licensee
Tokai	2	BWR-Mark II	3,293	1978	JAPCO
Tomari	1	PWR	1,650	1989	HEPCO
	2	PWR	1,650	1991	
	3	PWR	2,660	2009	
Tsuruga	1	BWR-Mark I	1,070	1970	JAPCO
	2	PWR	3,411	1987	

NOTES: ABWR = advanced boiling water reactor; BWR = boiling water reactor; PWR = pressurized water reactor.
SOURCES: Nuclear Regulation Authority, written communication; IAEA (2014b).

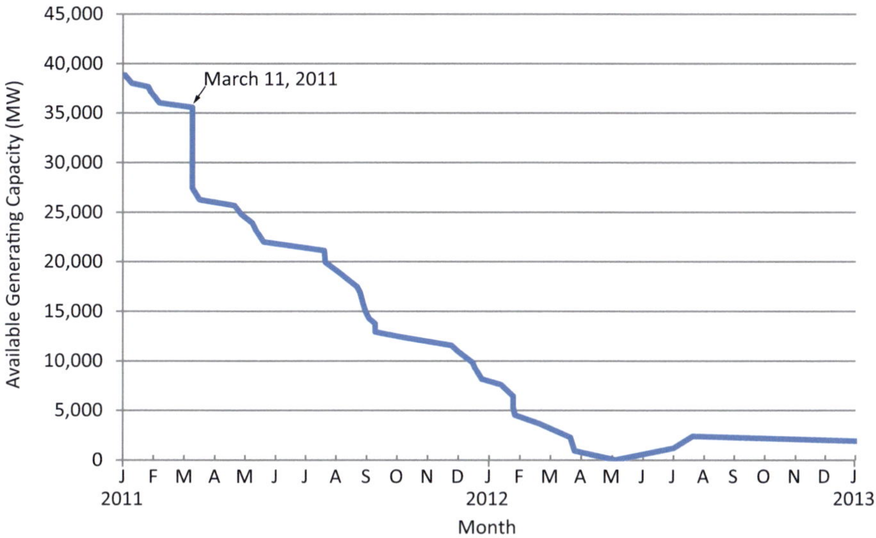

FIGURE 2.11 Electrical generating capacity from operating Japanese nuclear power plants prior to and following the Fukushima Daiichi accident. SOURCE: Electrical generating capacity from Nuclear Regulation Authority, written communication; reactor shutdown dates from NRA (2013a).

2.3.1 Regulation of Nuclear Plants in Japan

Prior to the Fukushima Daiichi accident, the Nuclear and Industrial Safety Agency (NISA) within the Ministry of Economy, Trade and Industry (METI) was responsible for nuclear plant regulation in Japan. NISA was overseen by the Nuclear Safety Commission (NSC), a senior government body responsible for formulating safety policy, and the Atomic Energy Commission (AEC), which was responsible for nuclear power and research policy. Both the NSC and AEC were part of the Cabinet Office[18]; however, they were advisory and neither had direct authority over nuclear plant regulation.

Following the Fukushima Daiichi accident, NISA's association with METI was seen to compromise its independence and pose a conflict of interest because METI also promotes nuclear energy. The Japanese government decided to eliminate NISA and establish a new organization in its place. This new organization, the Nuclear Regulation Authority (NRA), was

[18] The Cabinet Office is Japan's executive branch of government. It consists of the Japanese prime minister and other state ministers.

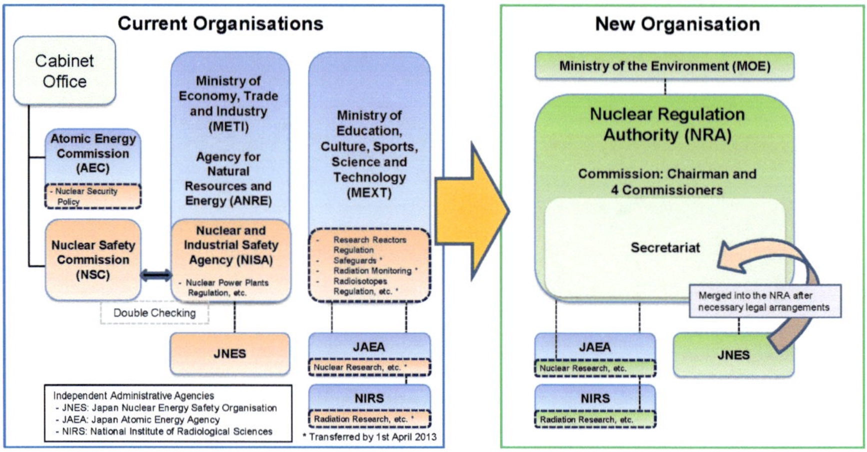

FIGURE 2.12 (Left) Japanese government nuclear organizations prior to the Fukushima Daiichi nuclear accident. (Right) Organization of new Japanese nuclear regulator (NRA) effective September 2012. JAEA = Japan Atomic Energy Agency. JNES = Japan Nuclear Energy Safety Organization. NIRS = National Institute of Radiological Sciences. SOURCE: Task Force for the Reform of Nuclear Safety Regulations and Organisations (n.d).

established as an extra-ministerial organization of the Ministry of Environment in September 2012. NRA combines the roles of NISA and NSC and also assumed the nuclear-related functions of the Ministry of Education and Science (see Figure 2.12).

The NRA is headed by a five-member commission composed of a chairman and four commissioners who are appointed by the Japanese prime minister and confirmed by the National Diet for 5-year terms. A secretary general directs the activities of the Secretariat of the NRA and carries out the policies and decisions of the commission. Most of the staff of NRA were transferred from METI and the Ministry of Education, Culture, Sports, and Science and Technology (MEXT); they will not be allowed to return to METI or MEXT in the future because they were hired by the NRA under a "no-return" rule.

2.4 NUCLEAR PLANTS IN THE UNITED STATES

There are 100 nuclear power reactors currently licensed to operate at 65 sites in 31 states (Figure 2.13, Table 2.4). Collectively, these reactors

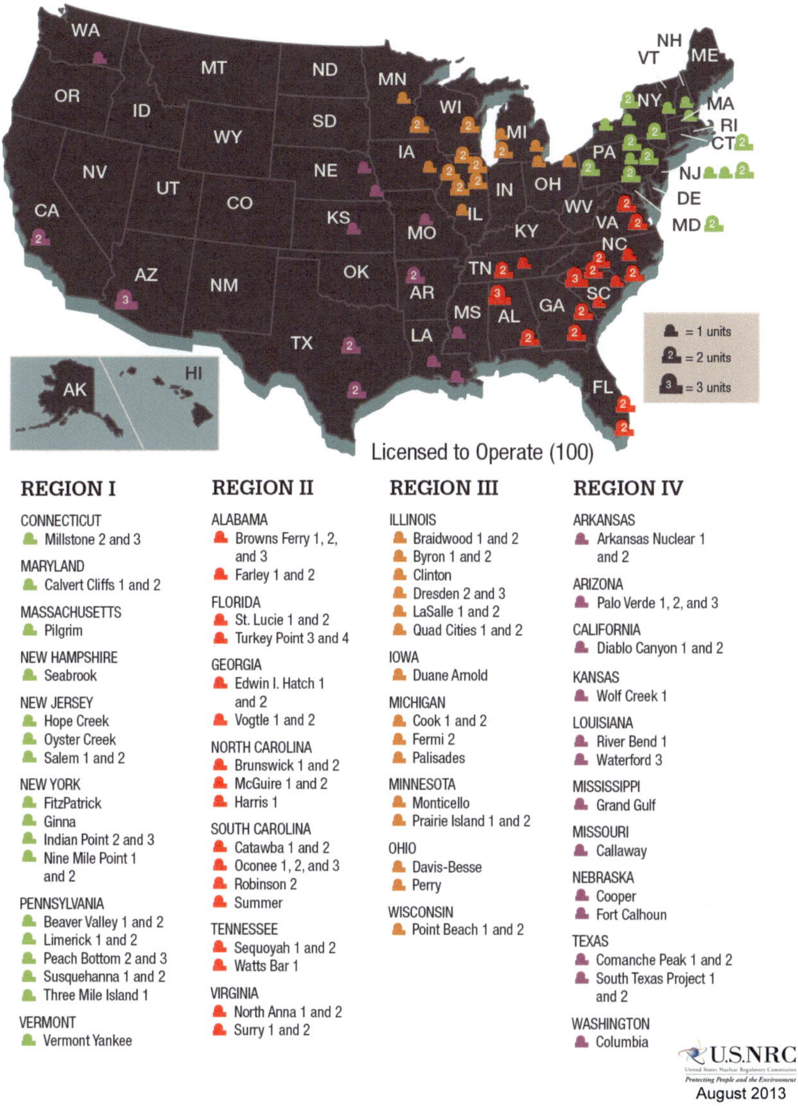

FIGURE 2.13 Locations and names of currently operating nuclear plants in the United States. SOURCE: USNRC (2013b, Fig. 16).

TABLE 2.4 Operating Nuclear Plants in the United States (June 2014)

Plant Name	Unit	Reactor and Containment Type	Thermal Capacity (MWt)	Commercial Operation	Licensee
Arkansas Nuclear	1	PWR-DRYAMB	2,568	1974	Entergy
	2	PWR-DRYAMB	3,026	1980	
Beaver Valley	1	PWR-DRYAMB	2,900	1976	FirstEnergy
	2	PWR-DRYAMB	2,900	1987	
Braidwood	1	PWR-DRYAMB	3,587	1988	Exelon
	2	PWR-DRYAMB	3,587	1988	
Browns Ferry	1	BWR-Mark I	3,458	1974	Tennessee Valley Authority
	2	BWR-Mark I	3,458	1975	
	3	BWR-Mark I	3,458	1977	
Brunswick	1	BWR-Mark I	2,923	1977	Carolina Power & Light
	2	BWR-Mark I	2,923	1975	
Byron	1	PWR-DRYAMB	3,587	1985	Exelon
	2	PWR-DRYAMB	3,587	1987	
Callaway		PWR-DRYAMB	3,565	1984	Union Electric
Calvert Cliffs	1	PWR-DRYAMB	2,737	1975	Calvert Cliffs
	2	PWR-DRYAMB	2,737	1977	
Catawba	1	PWR-ICECND	3,411	1985	Duke Energy Carolinas
	2	PWR-ICECND	3,411	1986	
Clinton	1	BWR-Mark III[a]	3,473	1987	Exelon
Columbia	2	BWR-Mark II	3,486	1984	Energy Northwest

Comanche Peak	1	PWR-DRYAMB	3,612	1990	Luminant
	2	PWR-DRYAMB	3,612	1993	
Cooper		BWR-Mark I	2,419	1974	Nebraska Public Power
Davis-Besse	1	PWR-DRYAMB	2,817	1978	FirstEnergy
Diablo Canyon	1	PWR-DRYAMB	3,411	1985	Pacific Gas & Electric
	2	PWR-DRYAMB	3,411	1986	
D. C. Cook	1	PWR-ICECND	3,304	1975	Indiana Michigan Power
	2	PWR-ICECND	3,468	1978	
Dresden	2	BWR-Mark I	2,957	1970	Exelon
	3	BWR-Mark I	2,957	1971	
Duane Arnold		BWR-Mark I	1,912	1975	FPL Energy Duane Arnold
E. I. Hatch	1	BWR-Mark I	2,804	1975	Southern
	2	BWR-Mark I	2,804	1979	
Fermi	2	BWR-Mark I	3,430	1988	DTE Electric
Fort Calhoun	1	PWR-DRYAMB	1,500	1973	Omaha Public Power
Grand Gulf	1	BWR-Mark III[a]	4,408	1985	Entergy
H. B. Robinson	2	PWR-DRYAMB	2,339	1971	Carolina Power & Light
Hope Creek	1	BWR-Mark I[b]	3,840	1986	PSEG Nuclear
Indian Point	2	PWR-DRYAMB	3,216	1974	Entergy
	3	PWR-DRYAMB	3,216	1976	
J. A. FitzPatrick		BWR-Mark I	2,536	1975	Entergy FitzPatrick
J. M. Farley	1	PWR-DRYAMB	2,775	1977	Southern
	2	PWR-DRYAMB	2,775	1981	

continued

TABLE 2.4 Continued

Plant Name	Unit	Reactor and Containment Type	Thermal Capacity (MWt)	Commercial Operation	Licensee
LaSalle	1	BWR-Mark II	3,546	1984	Exelon
	2	BWR-Mark II	3,546	1984	
Limerick	1	BWR-Mark II	3,515	1986	Exelon
	2	BWR-Mark II	3,515	1990	
McGuire	1	PWR-ICECND	3,411	1981	Duke Energy Carolinas
	2	PWR-ICECND	3,411	1984	
Millstone	2	PWR-DRYAMB	2,700	1975	Dominion
	3	PWR-DRYSUB	3,650	1986	
Monticello	1	BWR-Mark I	1,775	1971	NSP Minnesota
Nine Mile Point	1	BWR-Mark I	1,850	1969	Nine Mile Point Nuclear
	2	BWR-Mark II	3,988	1988	
North Anna	1	PWR-DRYSUB	2,940	1978	Virginia Electric & Power
	2	PWR-DRYSUB	2,940	1980	
Oconee	1	PWR-DRYAMB	2,568	1973	Duke Energy Carolinas
	2	PWR-DRYAMB	2,568	1974	
	3	PWR-DRYAMB	2,568	1974	
Oyster Creek	1	BWR-Mark I	1,930	1969	Exelon
Palisades		PWR-DRYAMB	2,565	1971	Entergy
Palo Verde	1	PWR-DRYAMB	3,990	1986	Arizona Public Service
	2	PWR-DRYAMB	3,990	1986	
	3	PWR-DRYAMB	3,990	1988	

Peach Bottom	2	BWR-Mark I	3,514	1974	Exelon
	3	BWR-Mark I	3,514	1974	
Perry	1	BWR-Mark III[a]	3,758	1987	FirstEnergy
Pilgrim	1	BWR-Mark I	2,028	1972	Entergy
Point Beach	1	PWR-DRYAMB	1,800	1970	NextEra Point Beach
	2	PWR-DRYAMB	1,800	1972	
Prairie Island	1	PWR-DRYAMB	1,677	1973	NSP Minnesota
	2	PWR-DRYAMB	1,677	1974	
Quad Cities	1	BWR-Mark I	2,957	1973	Exelon
	2	BWR-Mark I	2,957	1973	
River Bend	1	BWR-Mark III[a]	3,091	1986	Entergy
R. E. Ginna		PWR-DRYAMB	1,775	1970	R. E. Ginna Nuclear
Saint Lucie	1	PWR-DRYAMB	3,020	1976	Florida Power & Light
	2	PWR-DRYAMB	3,020	1983	
Salem	1	PWR-DRYAMB	3,459	1977	PSEG Nuclear
	2	PWR-DRYAMB	3,459	1981	
Seabrook	1	PWR-DRYAMB	3,648	1990	NextEra Seabrook
Sequoyah	1	PWR-ICECND	3,455	1981	Tennessee Valley Authority
	2	PWR-ICECND	3,455	1982	
Shearon Harris	1	PWR-DRYAMB	2,900	1987	Carolina Power & Light
South Texas	1	PWR-DRYAMB	3,853	1988	STP Nuclear
	2	PWR-DRYAMB	3,853	1989	
Surry	1	PWR-DRYSUB	2,857	1972	Virginia Electric & Power
	2	PWR-DRYSUB	2,857	1973	

continued

TABLE 2.4 Continued

Plant Name	Unit	Reactor and Containment Type	Thermal Capacity (MWt)	Commercial Operation	Licensee
Susquehanna	1	BWR-Mark II	3,952	1983	PPL Susquehanna
	2	BWR-Mark II	3,952	1985	
Three Mile Island	1	PWR-DRYAMB	2,568	1974	Exelon
Turkey Point	3	PWR-DRYAMB	2,644	1972	Florida Power & Light
	4	PWR-DRYAMB	2,644	1973	
Vermont Yankee	1	BWR-Mark I	1,912	1972	Entergy
V. C. Summer	1	PWR-DRYAMB	2,900	1984	South Carolina Electric & Gas
Vogtle	1	PWR-DRYAMB	3,626	1987	Southern Nuclear
	2	PWR-DRYAMB	3,626	1989	
Waterford	3	PWR-DRYAMB	3,716	1985	Entergy
Watts Bar	1	PWR-ICECND	3,459	1996	Tennessee Valley Authority
Wolf Creek	1	PWR-DRYAMB	3,565	1985	Wolf Creek Nuclear

[a] Mark III containments have a concrete secondary containment (also known as a Shield Building).
[b] Has concrete secondary containment unlike other BWRs of this type.

NOTES: Reactor types: BWR = boiling water reactor; PWR = pressurized water reactor. Containment types: DRYAMB = dry, ambient pressure; DRYSUB = dry, subatmospheric pressure; ICECND = wet, ice condenser; Mark I = wet, Mark I; Mark II = wet, Mark II; Mark III = wet, Mark III.

SOURCES: USNRC (2013b), IAEA (2014b).

provide about 20 percent of U.S. electricity supply. Thirty-five of these reactors are BWRs and 65 are PWRs, all of Generation II design. One Generation II reactor (Watts Bar Unit 2) and four Generation III reactors (Vogtle Units 3 and 4 and V.C. Summer Units 2 and 3) are under construction. These Generation III plants are PWR designs (AP1000).

2.4.1 Regulation of Nuclear Power in the United States

The U.S. Nuclear Regulatory Commission (USNRC) is responsible for nuclear reactor and materials safety in the United States and U.S. territories. The USNRC was established by the Energy Reorganization Act of 1974 to be an independent[19] agency in the executive branch of the U.S. government. Before the USNRC was established, nuclear safety regulation and nuclear power promotion were the responsibility of the Atomic Energy Commission (AEC). The Energy Reorganization Act dissociated AEC's responsibilities: USNRC assumed the AEC's regulatory responsibilities and the Energy Research and Development Administration assumed AEC's responsibilities for nuclear promotion. ERDA was later reorganized into the U.S. Department of Energy (USDOE).

The USNRC is overseen by five Commissioners, one of whom is designated as chairman, who are appointed by the president of the United States and confirmed by the U.S. Senate to serve 5-year terms. The Commission formulates policies and regulations for nuclear reactor safety, issues orders to licensees, and adjudicates legal matters brought before it. The USNRC is headquartered in Rockville, Maryland, and has four regional offices (in Pennsylvania, Georgia, Illinois, and Texas) to provide direct links to individual nuclear plants through resident inspectors.

The Atomic Energy Act of 1954 specifies that U.S. nuclear energy facilities can be licensed for an initial period of 40 years and that such licenses are renewable. USNRC regulations permit licenses to be renewed for periods not to exceed 20 years (10 CFR § 54.31[20]). Most of the currently operating nuclear plants in the United States have received or are seeking 20-year license renewals, which would extend their operating lives to 60 years. The USNRC and nuclear industry are examining the feasibility of an additional 20-year renewal to extend plant operating lives to 80 years.

---

[19] Independent agencies in the U.S. government are run by commissions or boards with oversight from the U.S. Congress. The members of the commissions and boards are appointed by the president; some appointments require U.S. Senate confirmation.

[20] This section of Title 10 of the Code of Federal Regulations (CFR) is available at http://www.nrc.gov/reading-rm/doc-collections/cfr/part054/part054-0031.html. Accessed on June 3, 2014.

2.5 COMPARISON OF JAPANESE AND U.S. BWR PLANTS

Twenty-one BWRs in the United States have the same reactor and containment designs as the Fukushima Daiichi units (see Table 2.2).[21] Six are of the same design as the Fukushima Unit 1 (BWR/3 Mark I) and 15 are the same design as Fukushima Units 2, 3, and 4 (BWR/4 Mark I).[22] Several safety enhancements have been made to Japanese and U.S. Mark I BWRs since they began operating; some of these enhancements are described in the following sections.

2.5.1 Fire Protection

After a 1975 fire in Unit 1 at the Browns Ferry Nuclear Plant, fire protection requirements in the United States were enhanced. Reactor safety shutdown systems were physically separated to provide redundancy and independence during any single fire event. However, as discussed in Chapter 7 (see Section 7.3.3), not all U.S. reactors have adopted these measures. These measures also have not been adopted in Japan.

2.5.2 Hardened Containment Vents

Installation of hardened containment vents in Mark 1 BWRs was recommended by the USNRC[23] following the 1979 Three Mile Island accident; the U.S. nuclear industry committed to voluntarily comply with this recommendation. All but one[24] Mark I BWR in the United States are currently equipped with hardened vents, but vent designs are plant specific. Hardened vents were also installed at all eight of the Mark I BWR plants in Japan. Following the Fukushima Daiichi accident, the USNRC issued a new order[25] to Mark I and Mark II BWR licensees to design and install "Reliable Hardened Containment Vents Capable of Operation Under Severe Accident Conditions" (see Chapter 5 and especially Appendix F).

---

[21] Although the reactors and containments have a standard design, the design, of the remainder of the plant, including reactor buildings, control rooms, and locations of safety systems, are not standardized.

[22] In addition, there are two BWR/2 Mark I plants in the United States: Nine Mile Point Unit 1 (New York) and Oyster Creek (New Jersey).

[23] Installation of a Hardened Wetwell Vent (Generic Letter 89-16). Available at http://i2.cdn.turner.com/cnn/2012/images/02/16/nrc.gl.89.16.pdf. Accessed on June 3, 2014.

[24] The exception is the James A. FitzPatrick plant, which is located in New York.

[25] Compliance with Order EA-13-109, Order Modifying Licenses with Regard to Reliable Hardened Containment Vents Capable of Operation Under Severe Accident Conditions (JLD-ISG-2013-02). Available at http://pbadupws.nrc.gov/docs/ML1330/ML13304B836.pdf. Accessed on June 3, 2014.

2.5.3 Containment Inerting

The USNRC also required the inerting[26] of containments in BWRs with Mark I and Mark II designs following the 1979 Three Mile Island accident. This practice was adopted worldwide. Following the Fukushima Daiichi accident, the USNRC examined the need for additional hydrogen control measures but decided not to take immediate action (see Chapter 5).

2.5.4 Other Containment Modifications

In the early 1980s, Mark I BWR containment systems were modified to improve their safety margins in loss-of-coolant accidents. Modifications included reinforcements to the suppression chamber and associated structures.[27] Japanese regulations and plant modifications closely followed U.S. practice, and so these changes may have been implemented in Japanese plants as well.

2.5.5 Control Room Improvements

The 1979 Three Mile Island accident prompted the nuclear industry to enhance control room process and design. Access to control rooms was limited, safety alarms were improved, and changes in control and display systems were made. Some of these changes were likely implemented in Japan.[28]

2.5.6 Station Blackout

In 1988, the USNRC issued a station blackout rule[29] that required nuclear plants to maintain highly reliable onsite AC power; ensure that plants can cope with station blackout (defined as the loss of both offsite AC power and onsite emergency AC power) for a predetermined period of time using battery backup power; develop procedures and training for restoring offsite AC power and onsite emergency AC power; and make other modifications to plants as needed.

---

[26] Nitrogen is used to displace air within the primary containment vessel when the reactor is operating. By reducing the concentration of oxygen to less than 4 percent, it is possible to prevent explosions or fires within the containment even if hydrogen is generated and released from the RPV into containment. BWR Mark III containments are not inerted. They have hydrogen control systems that are designed to burn hydrogen at low concentrations.

[27] The modifications are described in Borchardt (2012a, Enclosure 2).

[28] It is difficult to make a direct comparison between Japan and the United States; control rooms in Japan are different in terms of reliance on computer controls and number of staff.

[29] 10 CFR § 50.63, Loss of all alternating current power. Available at http://www.nrc.gov/reading-rm/doc-collections/cfr/part050/part050-0063.html. Accessed on June 3, 2014.

The Japanese Nuclear Safety Commission's Regulatory Guide for Reviewing Safety Design of Light Water Nuclear Power Reactor Facilities (August 1990) provides the following guidance on station blackout:

> Guideline 27. Design Considerations against Loss of Power
> The nuclear reactor facilities shall be so designed that safe shutdown and proper cooling of the reactor after shutting down can be ensured in case of a short-term total AC power loss.[30]

According to an unofficial translation of the Nuclear Safety Commission's Special Committee on Nuclear Safety Standards and Guides, "short term" has been routinely interpreted as meaning DC battery capacity to maintain residual heat removal for 30 minutes under station blackout conditions (NSCJ, 2012).

Nuclear plants throughout Japan have AC emergency power and coping capabilities similar to U.S. nuclear plants with multiple redundant AC power sources (e.g., 13 emergency diesel generators at the Fukushima Daiichi plant) and backup batteries (trains of battery-powered 125VDC and 250VDC power sources). The battery coping time (i.e., the length of time that station batteries can provide power under a specified load) at Japanese nuclear plants is comparable to the 4- to 8-hour coping time typical for U.S. nuclear plants.

The USNRC recently issued an additional order (USNRC, 2012d) that requires nuclear plants to cope with a station blackout for an indefinite length of time (see Chapter 5 and Appendix F).

2.5.7 Improved Mitigation Capabilities

The September 11, 2001, terrorist attacks led to an extensive review of accident scenarios beyond then-current plant design standards. The USNRC issued an Interim Compensatory Measures Order[31] in 2002 that directed nuclear plant licensees to develop mitigation strategies to cope with large fires and explosions from any cause, including aircraft impacts. These strategies are intended to use readily available resources to maintain or restore core cooling, containment, and spent fuel pool cooling. A final rule was issued in March 2009 (USNRC, 2009a).

Japanese utilities did not implement these measures because they were unaware of the details of the U.S. program. Japanese regulatory agencies that were aware of this program did not discuss it with utilities or impose

---

[30] The quoted material is taken from an unofficial translation of the guide.

[31] Interim Compensatory Measures Order EA-02-026 (this document is designated as Safeguards Information and is not available to the public).

similar requirements.[32] However, following the 2007 Niigata Chuetsu-Oki Earthquake and fire at the Kashiwazaki-Kariwa plant, additional water supplies, fire pumper trucks, and external connections to the reactor building fire protection system were required at all nuclear plants in Japan.

[32] Section B.5.b of the USNRC Order for Interim Safeguards and Security Compensatory Measures was designated by the USNRC as Safeguards Information, and so it was exempt from public release. Consequently, TEPCO would not have had direct access to this information. However, as discussed in Chapter 7, the USNRC shared some B.5.b information with Japanese government authorities. Moreover, the USNRC requirements were made publicly available in 2009 (USNRC, 2009a) and were incorporated into reactor designs that were being developed by Japanese vendors for sale in the United States.

3

Great East Japan Earthquake and Tsunami and Impacts on Japanese Nuclear Plants

The objective of this chapter is to provide an overview of the March 11, 2011, Great East Japan Earthquake and tsunami and their key impacts on Japanese nuclear plants located along the northeastern coast of Honshu, Japan's main island. The chapter is intended to provide background information to support the committee's detailed discussions of the Fukushima nuclear accident, which appear in subsequent chapters of this report.

This chapter is organized into five sections.

- Section 3.1 describes the Great East Japan Earthquake and tsunami.
- Section 3.2 describes some key impacts of the earthquake and tsunami on the five nuclear plants located on the northeastern coast of Japan.
- Section 3.3 identifies some key differences in impacts among these plants and reasons for these differences.
- Section 3.4 focuses on tsunami hazards at the Fukushima Daiichi and Daini nuclear plants, in particular, how the plant operator's (TEPCO's) understanding of these hazards evolved over time.
- Section 3.5 provides an initial discussion of lessons learned to support more detailed discussions in subsequent chapters.

3.1 GREAT EAST JAPAN EARTHQUAKE AND TSUNAMI

The Great East Japan Earthquake[1] occurred off the northeast coast of Honshu (see Figure 3.1) beginning at 14:46[2] on March 11, 2011. The earthquake was produced by rupture of a large fault at the Japan Trench. The dip of the fault, based on the analyses of the seismic source and previous multichannel seismic measurements of the depth of the fault (Chester et al., 2013; Nakamura et al., 2013), was approximately 10°-13° at the event hypocenter, increasing to about 20° near Japan[3] (Gusman et al., 2012). The epicenter was determined by the U.S. Geological Survey to be located at 38.297°N, 142.372°E, 129 km east of Sendai.

3.1.1 Moment Magnitude

Seismogram amplitudes, at least at low frequencies, are proportional to seismic moment.[4] For this earthquake, estimates of moments were 2.9 × 10^{22} newton-meters (N·m) (Lay and Kanamori, 2011), 3.8 × 10^{22} N·m (Fujii et al., 2011), 4.2 × 10^{22} N·m (Satake et al., 2013; Yue and Lay, 2013), 4.3 × 10^{22} N·m (Frankel, 2013b), 4.42 × 10^{22} N·m (Suzuki et al., 2011), and 5.3 × 10^{22} N·m (Nettles et al., 2011[5]; Gusman et al., 2012).

When the seismic moment is known from the analysis of seismic and geodetic data, the calculation of a moment magnitude is straightforward.[6] Estimates of the moment magnitude were largely 9.0 (e.g., Lay and Kanamori, 2011; Frankel, 2013b; Maeda et al., 2013; Satake et al., 2013; Sugawara et al., 2013; Yamazaki et al., 2013). The exception to this was Nettles et al. (2011), who reported the moment magnitude to be 9.1.[7] This was the fourth-largest earthquake in the past 60 or so years and the largest instrumentally recorded earthquake to ever strike Japan.

---

[1] This earthquake is also referred to as the Tohoku Earthquake by some investigators. See, for example, Noggerath et al. (2011).

[2] The 24-hour clock is used for time notation in this report.

[3] That is, the dip of the fault increases toward the west nearer to Japan (e.g., Nakahara, 2013).

[4] Seismic moment, M_o, is expressed as μAD where μ is the rigidity at the fault, A is the area of the fault that ruptured, and D is the fault displacement.

[5] Nettles et al. (2011) is a standard product of the Global Centroid Moment Tensor (GCMT) Project (http://www.globalcmt.org), which has an extensive catalog.

[6] Moment magnitude, M_w, is determined using the equation $M_w = 2/3 \log10(M_o) - 6.0$, where M_o is the seismic moment. The constants in the equation were chosen to be consistent with earlier scales at magnitudes less than 8.

[7] The latter analysis differed substantially from the others in that a Centroid Moment Tensor was calculated with filtered low-frequency seismograms 8.5 hours long from 100 seismic stations distributed globally.

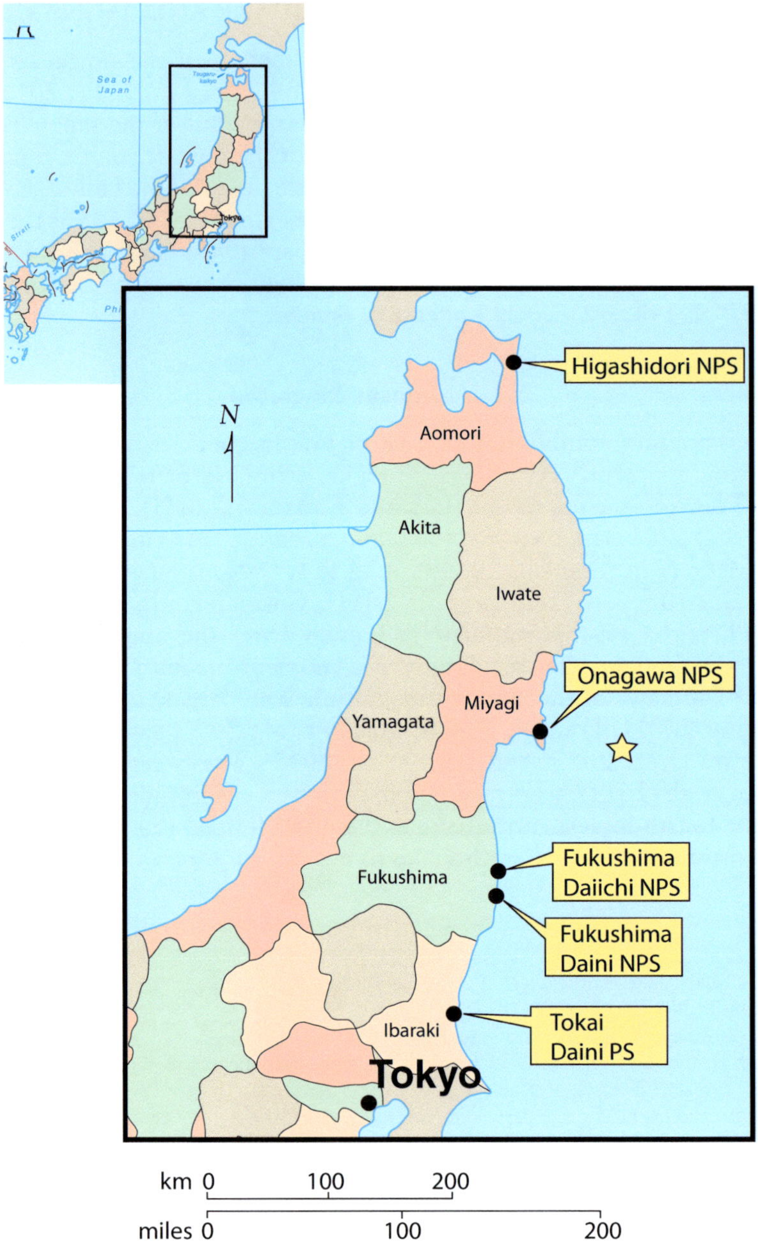

FIGURE 3.1 Map of the Tohoku region of Japan (shaded area in inset) showing the hypocenter of the Great East Japan Earthquake and the locations of the five nuclear plants discussed in this chapter.

3.1.2 Hypocenter Depth

While the epicenter was well determined, the hypocenter depth is poorly constrained from seismic data because it was a shallow event. Estimated depths of 10-25 km were common, assuming the hypocenter was at the boundary between the overlying Eurasian plate and the Pacific plate being thrust beneath (e.g., Chester et al., 2013; Ghofrani et al., 2013; Nakahara, 2013; Tajima et al., 2013).

3.1.3 Fault Dimensions and Displacements

Fault dimension estimates also varied but were generally constrained in a box approximately 600 km along strike[8] and 200 km across strike (e.g., Suzuki et al., 2011; Lay et al., 2013; Nakahara, 2013). The fault rupture itself extended from the trench to beneath Japan at the Honshu shoreline, a distance of at least 200 km. The fault, delineated by the trench axis, broke for approximately 500 km. The average fault slip over this surface was 10 m, but the major contribution to the seismic moment was more compact, perhaps 200 km × 200 km.

Kubo and Kakehi (2013) inverted seismological data with offshore seafloor static displacements to determine that 43-m displacements occurred on the shallow part of the fault seaward of the hypocenter, whereas displacements of 4 m extended for 300 km along the fault (Kozdon and Dunham, 2013). Yue and Lay (2013) inverted teleseismic P-wave, short-period Rayleigh waves, permanent GPS stations on Japan, and seafloor displacements to find that the peak slip at the trench was 60 m and slip at the hypocenter was 25 m (Lay et al., 2013). Slip greater than 20 m extended over areas as large as 200 × 200 km.

The earthquake produced strong (exceeding 1*g* and as large as 3*g*), long-duration (exceeding 6 minutes) ground shaking in some regions of Japan (Figure 3.2). Seismic waves were generated from several distinct ruptures on the fault (Figure 3.3). Suzuki et al. (2011) have observed that the fault between the hypocenter and trench broke upward for 60-100 s, emitting low frequencies, which were responsible for tsunami excitation. After 100 s, the southwestward propagation with displacements on the order of 5 m provided the bulk of strong motion in Japan.

Intense seismic waves from two or three parts of the fault zone that were inferred to be asperities (i.e., locked sections) are quite visible in the accelerogram (Figure 3.3). Nakahara (2013) found evidence for three sources, two of which led to shaking in the Miyagi region. Energy from a source at greater depth near the border of Fukushima and Ibaraki Prefec-

[8] Where strike refers to the trend of the trench or fault axis.

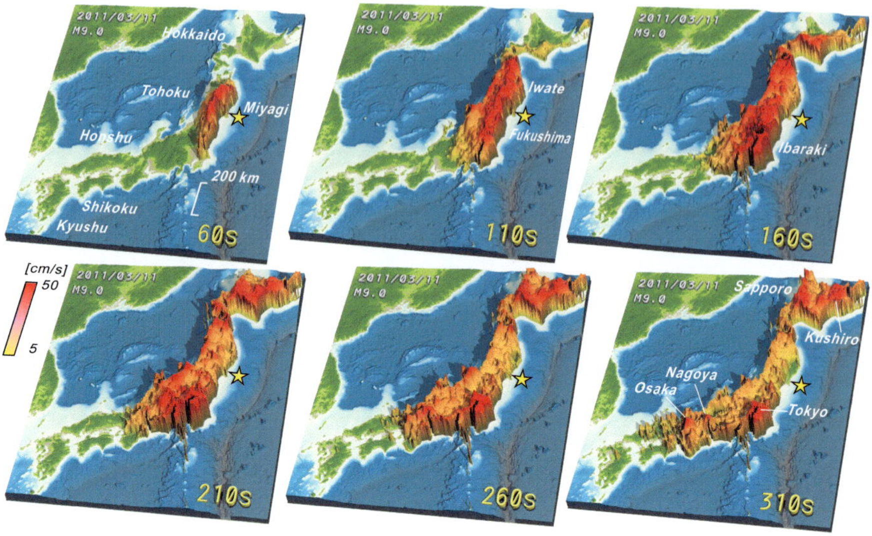

FIGURE 3.2 Propagation of seismic waves generated from the March 11, 2011, Great East Japan Earthquake. The earthquake hypocenter is indicated by the star on the maps. Data were collected by the national K-NET and KiK-net strong ground motion network in near real time. There are more than 1,800 stations across Japan with a station spacing of 20 to 25 km. SOURCE: Furumura et al. (2011). Courtesy of T. Furumura, ERI, University of Tokyo.

tures provided much of the shaking to the south. Ghofrani et al. (2013) used a stochastic model to examine shaking with as many as five sources, which provided the best match with the observed data.

The pseudospectral accelerations derived from the data are particularly useful for engineering studies. The data from KIK-net and (K-NET)[9] were particularly important for these high-frequency studies of shaking (Furumura et al., 2011).

The earthquake was accompanied by large crustal displacements of onshore regions in northern Japan (Figure 3.4). The largest displacements—about 5.3 m to the east and 1.2 m downward—occurred along the coast of

[9] K-NET and KIK-net (http://www.kyoshin.bosai.go.jp) comprise nearly 1,800 strong-motion seismographs (K-NET) and strong-motion seismometer pairs located in boreholes and at the surface (KIK-net) across Japan.

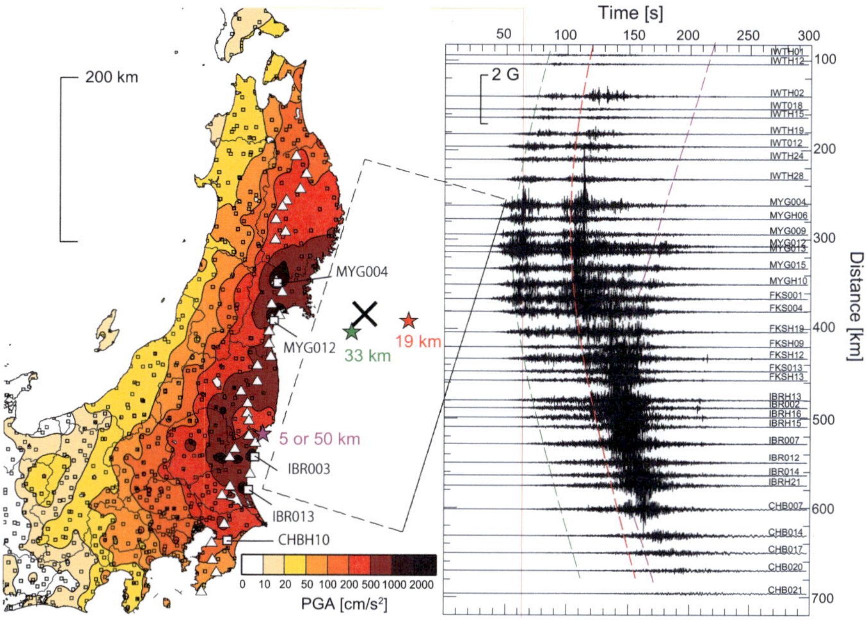

FIGURE 3.3 (Left) Map of northern Japan showing peak ground accelerations (PGA) in cm/s^2 for the Great East Japan Earthquake and tsunami (divide by 100 to obtain accelerations in m/s^2, where g = 9.8 m/s^2). The symbol "X" on the map indicates the location of the earthquake hypocenter and the dashed box indicates the location of the fault plane. (Right) Accelerogram showing the vertical component of ground acceleration as a function of time along an approximate north-south line of onshore seismic stations indicated by the triangles on the map (Left image). There are two and possibly three distinct arrivals of seismic waves indicated by the green, red, and purple lines on the accelerogram. These likely originated from large slips on the fault plane; the hypocenters for these slips are indicated by the colored stars on the map (Left image). SOURCE: Furumura et al. (2011). Courtesy of T. Furumura, ERI, University of Tokyo.

Miyagi Prefecture near the Oshika Peninsula in Ishinomaki City near the Onagawa Nuclear Power Station (see Figure 3.1). The downward displacement of the coast, which was caused by the relaxation of elastic strain at the tectonic plate margin, lowered seawall elevations relative to mean sea level, reducing tsunami protection for coastal communities.

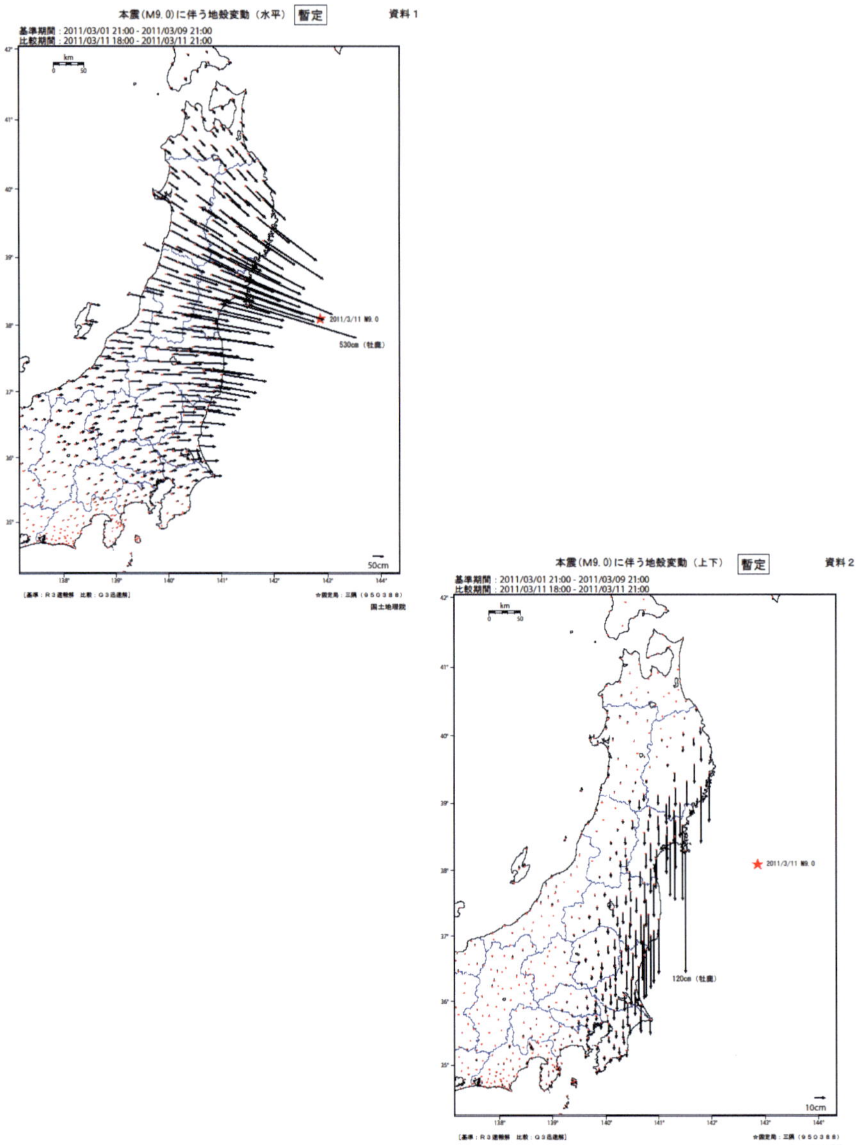

FIGURE 3.4 Horizontal (Left) and vertical (Right) displacements of the crust resulting from the March 11, 2011, Great East Japan Earthquake. The movements shown are relative to a reference point located at Ishinomaki City (Miyagi Prefecture). SOURCE: Geospatial Information Authority of Japan, http://www.gsi.go.jp/chibankansi/chikakukansi40005.html.

3.1.4 Tsunami

Deformation of the seafloor during the earthquake[10] triggered a tsunami that caused substantial damage to coastal regions of northeastern Japan. The first wave struck a 2,000-km-long stretch of the Japan coast starting 20 minutes after the earthquake (Mori et al., 2011, 2012). The first wave arrived at the Fukushima Daiichi Nuclear Power Station about 41 minutes after the earthquake. The second and largest wave arrived at the plant about 9-10 minutes later (TEPCO, 2013). Inundation heights varied along the coast depending on seafloor topography, coastline geometry, and ocean-edge waves. The deepest inundations, in one case approaching 40 m, occurred in Iwate Prefecture (Figure 3.5).

Sugawara et al. (2013) conducted surveys on the Sendai Plain, a low-lying coastal region in east-central Miyagi Prefecture, to assess the extent of inundation from the tsunami. Major inundations (inferred from the type and amount of accumulated debris on land surfaces flooded by the tsunami) were observed at distances of up to 4 km from the coast; minor inundations were observed at distances of up to 5 km. The authors state that the inundation area of the 2011 tsunami is "comparable to that of the 869 Jogan tsunami, although a direct comparison is difficult due to differences in geomorphological contexts between the paleo period and the present" Sugawara et al. (2013, p. 831). The Jogan tsunami was generally considered to be the largest historical tsunami in northeast Japan prior to March 2011.

3.1.5 Discussion

The magnitude of the Great East Japan Earthquake exceeded the then-existing maximum estimated earthquake magnitude for the Tohoku Region (Kagan and Jackson, 2013). The 2005 Seismic Hazard Map, for example, estimates a maximum magnitude of 7.7 to 8.0 for earthquakes offshore from Fukushima and Miyagi Prefectures (HERP, 2005, p. 134). Although seismic hazard mapping is helpful for disaster planning, it has serious scientific limitations and has recently been the subject of controversy (Stein et al., 2012, 2013; Frankel, 2013a).

Seismic magnitudes have been used for decades to compare the sizes of earthquake events. An early measure of magnitude was *body wave magnitude* that could be easily read from a seismogram.[11] Because of the nature of

---

[10] Seafloor displacements of about 50 m horizontal and 7-10 m vertical landward of the Japan Trench were observed from bathymetric surveys (Fujiwara et al., 2011). Deep monuments displayed horizontal seafloor displacements as large as 60-80 m and a pressure gauge recorded a 5-m drop during the tsunami.

[11] It is defined as the logarithm of the amplitude of the first-arriving elastic waves from an earthquake.

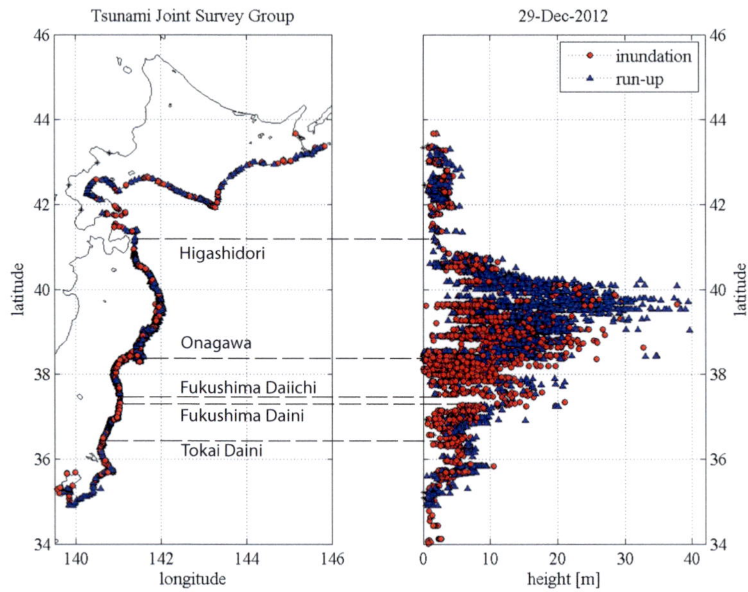

FIGURE 3.5 Tsunami inundation heights (i.e., elevation of maximum water levels taken from watermarks on structures and natural features) and run-up heights (i.e., elevation of the maximum landward extent of debris and seawater marks) in the Tohoku region of Japan. The largest run-ups occurred between latitudes 39° and 40° north, where coastal features focused the tsunami waves. The approximate locations of the five nuclear plants are shown on the figure. SOURCE: 2011 Tohoku Earthquake Tsunami Joint Survey Group. Available at http://www.coastal.jp/tsunami2011/index.php?Field%20survey%20results.

the seismometer and the spectrum of the source, these amplitudes saturate starting around magnitude 6.

A second measure of seismic magnitude is based on surface waves (Rayleigh waves), which arrive later in the seismogram. The formula for this is

$$M = \log_{10}\left(\frac{A}{T}\right) + \sigma(\Delta)$$

where M is the surface wave magnitude, A is the amplitude of the Rayleigh wave, T is the period of the signal (usually 20s), and σ is a distance correction dependent on the distance between the source and receiver, Δ. The logarithmic dependence is clear. Unfortunately, like the body wave magnitude, the surface wave magnitude begins to saturate around magnitude 8.

The currently used measure of seismic magnitude, moment magnitude (M_w), which was described previously in this chapter, does not become saturated at higher magnitudes. However, the most recent (2005) seismic hazard map did not use this metric for predicting future hazards.

Noggerath et al. (2011) noted that construction of the Fukushima plants began in 1967, well before the moment magnitude scale was introduced.[12] At the time, the largest magnitudes were expected not to exceed 8.5, and seismic moment was not yet being used by engineers, regardless of the seismic moment. On the basis of the introduction of the moment magnitude (Kanamori, 1977), we now know that earthquakes larger than magnitude 9 do exist; the largest recorded event was a magnitude 9.5 in Chile in 1960. Earthquakes may be even larger than this but have not occurred since the introduction of modern seismic technologies in the 20th century. Noggerath et al. (2011) note that Professor Hiroo Kanamori had used the original surface wave approach to find that the Tohuku Earthquake (i.e., Great East Japan Earthquake) measured 8.2, versus the 9.0-9.1 based on the moment magnitude scale. Several studies after the 2004 Sumatra earthquake (Stein and Okal, 2007; McCaffrey, 2008) suggested that earthquakes with a moment magnitude of 9 could occur at any subduction zone throughout the world, but these studies do not appear to have been considered by disaster preparedness planners in Japan.

A news report in *Nature* (Cyranoski, 2011) noted that the 2009 seismic hazard map for Japan (March 2009) reported a 30-40 percent chance of a rupture in the region where the Great East Japan Earthquake occurred in the next decade and a 60-70 percent chance in the next 20 years. However, the expected earthquake had a magnitude of only about 7.7. The hazard map segmented northeastern Japan into five seismic zones with probabilities associated with historical data. No consideration was given to the possibility that an earthquake event would involve rupture across zone boundaries. Moreover, the magnitudes may have been biased low by the saturation of early measurements of earthquake magnitude.

Kagan and Jackson (2013) published a paper entitled: "Tohuku Earthquake: A Surprise?" Their conclusion was that the earthquake should not have been considered a surprise. The paper noted that four previous subduction zone earthquakes with magnitudes 9.0 or greater had occurred in the past century and that there was no reason to believe this could not happen near Japan. As noted above, this was also pointed out before the earthquake by Stein and Okal (2007) and McCaffrey (2008).

[12] The scale was introduced in the late 1970s and came into common usage in the 1980s (e.g., Kanamori, 1977).

3.2 IMPACTS ON JAPANESE NUCLEAR PLANTS

Five Japanese nuclear plants were affected directly by the March 11, 2011, earthquake and tsunami. All of these affected plants are located along the northeastern coast of Japan (Figure 3.1; Table 3.1). There are a total of 15 reactors at these plants; 11 reactors were operating when the earthquake occurred and four reactors were shut down for maintenance.

Information on the impacts of the earthquake and tsunami on these plants was obtained from the following sources:

- Government of Japan's reports to the International Atomic Energy Agency (IAEA) (Government of Japan, 2011a,b),
- TEPCO's report on the Fukushima nuclear accident (TEPCO, 2011a,b, 2012b),
- IAEA update logs for the Fukushima accident (see, e.g., http://www.iaea.org/newscenter/news/2011/fukushima070411.html),
- IAEA reports on Japanese nuclear plant inspections (IAEA, 2011, 2012), and
- Briefings to the committee during its meeting in Tokyo (see Appendix B).

Brief descriptions of these impacts based on these reference materials are provided in the following sections.

TABLE 3.1 Japanese Nuclear Power Plants Affected by the Great East Japan Earthquake and Tsunami

Plant	Distance to Earthquake Epicenter (km)	No. of Reactors at Plant	No. of Reactors Operating at Time of Earthquake
Higashidori Nuclear Power Station	300	1	0
Onagawa Nuclear Power Station	80	3	3
Fukushima Daiichi Nuclear Power Station	150	6	3
Fukushima Daini Nuclear Power Station	160	4	4
Tokai Daini Power Station	260	1	1

NOTE: Distances estimated from U.S. Geological Survey–determined location for earthquake epicenter (38.297°N, 142.372°E) and rounded to the nearest 10 km.

3.2.1 Higashidori Nuclear Power Station

The Higashidori Nuclear Power Station is located on the Shimotika Peninsula in northern Aomori Prefecture (Figure 3.1). It has one operating reactor; two other reactors are under construction. The operating reactor was in a maintenance outage when the earthquake occurred and all of its fuel had been offloaded to the spent fuel pool. The earthquake and tsunami did not cause any damage to plant facilities or equipment. However, the earthquake cut all offsite AC power to the plant for part of a day. Emergency diesel generators supplied power to the plant until offsite power was restored.

A 7.1-magnitude aftershock on April 7, 2011, also cut all offsite AC power to the plant. An emergency diesel generator supplied power until offsite AC power could be restored.

3.2.2 Onagawa Nuclear Power Station

The Onagawa Nuclear Power Station is located on the Oshika Peninsula in Ishinomaki City in east central Miyagi Prefecture (Figure 3.1). The plant has three operating reactors (Figure 3.6). Two of the reactors (Units 1 and 3) were operating at full power and one reactor (Unit 2) was in startup operation when the earthquake occurred. The earthquake caused the three reactors at the plant to shut down (i.e., *scram*) automatically. Unit 2 reached cold shutdown[13] about 3 minutes after the tsunami occurred. Units 1 and 3 reached cold shutdown early in the morning on March 12.

Four of the five offsite AC power lines were tripped by the earthquake, cutting most offsite AC power. Emergency diesel generators started up after the loss of offsite power, but two of the five operating generators tripped after the tsunami flooded their cooling-water pumps. The remaining three diesel generators supplied power to the plant until offsite AC power was restored on March 12.

The tsunami arrived at the plant about 43 minutes after the earthquake. The maximum tsunami height was 13 m, which was below the elevation of the main part of the plant (14.8 m[14]) where the reactor and turbine buildings are sited. Consequently, this portion of the site did not experience any flooding.

---

[13] Defined by TEPCO (2012b) to occur when reactor cooling-water temperature falls below 100°C. In the United States, cold shutdown is said to occur when the reactor coolant system is at atmospheric pressure and at a temperature below 200°F (~93°C) following cooldown of a reactor (see http://www.nrc.gov/reading-rm/basic-ref/glossary/full-text.html).

[14] These elevation estimates are relative to the station reference point, Onahama Peil (O.P., Onahama Port Construction Standard Surface). See Table 3.2 notes for definition. As shown in Figure 3.4, co-seismic subsidence in areas near the station exceeded 1 m.

TABLE 3.2 Comparison of Tsunami Heights with Site Elevations at Four Nuclear Power Plants

Plant	Estimated Tsunami Wave Height (m)	Plant Main Elevation (m)	Seawall Elevation (Breakwater Elevation) (m)	Emergency Diesel Generator Elevation (m)	Seawater Pump Elevation (m)
Onagawa	13	14.8	14	14.8	14.8
Fukushima Daiichi	13	10 (Units 1-4) 13 (Units 5-6)	4 (5.5)	2 (Units 1-5)	4
Fukushima Daini	9	12	4 (all units)	3 (Units 1-4)	4
Tokai Daini	5.4	8	6.1	8	

NOTE: Elevations are relative to the Onahama Peil (Onahama Port Construction Standard Surface), abbreviated O.P. for the Onagawa and Fukushima Daiichi and Daini plants; and mean sea level for Tokai Daini plant. Elevations have not been corrected for co-seismic ground subsidence.
SOURCES: Government of Japan (2011b), TEPCO (2012b), Tohoku (2012); TEPCO briefings to committee.

FIGURE 3.6 Oblique aerial photo of the Onagawa Nuclear Power Station prior to the March 11, 2011, Great East Japan Earthquake and tsunami. SOURCE: Courtesy of Tohoku Electric Power Company.

The earthquake and tsunami damaged some plant equipment and structures, but none of the damage affected the structural integrity of the plant. Among the more significant damage reported was an electrical short in a switchgear panel, which caused a fire, the toppling of an oil tank, and flooding in the basement of the Unit 2 reactor building, which submerged a heat exchanger, flooded cooling-water pumps, and resulted in the loss of function of two emergency diesel generators, as noted previously.

3.2.3 Fukushima Daiichi Nuclear Power Station

The Fukushima Daiichi Nuclear Power Station is located in east-central Fukushima Prefecture (Figure 3.1; see also Figure 1.2 in Chapter 1). The plant's six reactors had the following status at the time of the earthquake:

- Units 1, 2, and 3 were operating at licensed power[15] level.
- Unit 4 was in an outage for replacement of the reactor core

[15] That is, the maximum reactor heat output, or maximum power level, that is allowed under the plant license.

shroud.[16] Fuel from the Unit 4 reactor had been relocated to the spent fuel pool in the reactor building.

- Units 5 and 6 were in inspection outages. Fuel remained in their cores and the reactors were being actively cooled. The Unit 5 containment was open and the primary system was undergoing pressure testing.

Reactor Units 1, 2, and 3 automatically shut down when the earthquake occurred. The earthquake cut all offsite AC power to the plant, but emergency diesel generators started up to supply backup power. The plant operator concluded that "major equipment with safety-critical functions maintained its safety functions during and immediately after the earthquake" (TEPCO, 2012b, p. 148). Plant workers were not able to complete an inspection of the plant for earthquake-related damage before the tsunami struck. Subsequent flooding and radioactive contamination prevented inspections for damage in some parts of the plant, particularly the basement levels in some reactor and turbine buildings.

The main tsunami, estimated to be 13 m high (Figure 3.7), flooded areas around Units 1-4, reaching maximum flood depths of up to 5.5 m (Figure 3.8). Floodwaters entered the basement levels of reactor and turbine buildings through building entranceways, emergency diesel generator intake louvers, equipment hatches, and cable and pipe penetrations. The floodwater damaged pumps, electrical distribution panels, backup batteries, and diesel generators, causing loss of most plant power and ultimate heat sink.[17]

In the 72 hours following loss of power, the Unit 1, 2, and 3 reactors melted down, releasing hydrogen and radioactive materials. Hydrogen explosions in the Unit 1, 3, and 4 reactor buildings caused severe structural damage (Figure 3.9). An overview of major events in Units 1-4 is provided in Sidebar 3.1. A more detailed accident time line is presented Chapter 4.

The Japanese government issued a series of evacuation orders for residents around the plant beginning about 5 hours after loss of power and 4 hours after the government declared an Article 15 Emergency Event (loss of emergency core cooling and water injection) on March 11. The initial evacuation order, given at 20:50 on March 11 by the governor of Fukushima Prefecture, was for residents within a 2-km radius of the plant. The Japanese government ordered the evacuation of residents within 3 km of the plant 33 minutes later. That order was further extended to 10 km

---

[16] The core shroud is a stainless steel cylinder that surrounds the fuel assemblies.

[17] Ultimate heat sink was defined in U.S. Atomic Energy Commission (now U.S. Nuclear Regulatory Commission) Regulatory Guide 1.27 (AEC, 1974, p. 1.27-1): "that complex of water sources, including necessary retaining structures (e.g., a pond with its dam, or a river with its dam), and the canals or conduits connecting the sources with, but not including, the cooling water system intake structures for a nuclear power unit." The ultimate heat sink at the Fukushima Daiichi plant is the Pacific Ocean.

FIGURE 3.7 Photo taken looking north toward the Fukushima Daiichi plant on March 11, 2011. The photo shows the tsunami wave as it strikes the seaward side of the plant. Flooding can also be seen in the foreground of the photo. SOURCE: Courtesy of TEPCO (http://photo.tepco.co.jp/library/110409/110409_1f_tsunami_1t.jpg).

FIGURE 3.8 Photo showing flooding at the Fukushima Daiichi plant on the north side of Radiation Waste Treatment Facility on March 11, 2011. This facility is located south of Unit 4 (see Figure 1.2 in Chapter 1). Scale is indicated by the back end of a vehicle propped against the metal door. SOURCE: Courtesy of TEPCO (http://photo.tepco.co.jp/library/110519_2/110519_1_4.jpg).

FIGURE 3.9 March 21, 2011, photo of damaged Unit 3 (middle) and Unit 4 (right) reactor buildings. Unit 2 is shown on the left margin of the photo. SOURCE: Courtesy of TEPCO (http://photo.tepco.co.jp/en/date/2011/201103-e/110321-01e.html).

at 05:44 on March 12. However, cell phone records indicate that residents had largely self-evacuated areas within 10 km of the plant hours before this evacuation order was given (Hayano and Adachi, 2013, especially Fig. 3). The evacuation order was extended to 20 km at 18:25 on March 12, almost 3 hours after the hydrogen explosion in Unit 1.

At Unit 6, one air-cooled emergency diesel generator and its electrical distribution panel were undamaged by tsunami-related flooding. This generator was used to supply power to Units 5 and 6, both of which reached cold shutdown on March 20.

Efforts to restore offsite AC power began immediately after the earthquake. However, because of the extensive damage from the earthquake and tsunami, the first power line to the site was not restored until March 18.

3.2.4 Fukushima Daini Nuclear Power Station

The Fukushima Daini Nuclear Power Station is also located in east-central Fukushima Prefecture, about 12 km south of the Fukushima Daiichi plant (Figure 3.1). The plant has four reactors (Figure 3.10), all of which were operating—and which shut down automatically—when the earthquake occurred. Two of the four offsite AC power lines were lost during the earthquake,[18] but one AC line continued to supply power. A second backup line was restored on March 12 and a third backup was restored on March 13. Because offsite AC power was maintained after the earthquake, control room instrumentation and controls remained available to reactor operators.

The first tsunami struck the plant about 36 minutes following the earthquake. The maximum tsunami height, which occurred at about 41 minutes after the earthquake, flooded areas around reactor units at the southern end of the site (Figure 3.11). Most emergency diesel generators, switchgear for cooling pumps, and seawater pumps were rendered inoperable by the floodwaters. However, Unit 3 had one functional emergency diesel generator and operable reactor heat removal system pumps. All four units also had functional reactor core isolation cooling systems.

During the 36 hours following the tsunami, reactor operators were able to maintain reactor cooling systems while plant personnel replaced damaged pumps and installed nearly 9 km of temporary power cables. Reactor operators also made preparations to vent reactor containments so that low-pressure water could be used to cool the reactors if needed. However, plant personnel were able to replace motors and restore power to the units, so venting was not necessary. The reactors achieved cold shutdown on March 12 (Unit 3), March 14 (Units 1 and 2), and March 15 (Unit 4).

Further discussion of the actions taken at the Fukushima Daini plant to recover from the tsunami is provided in Chapter 4; see particularly Sidebar 4.2.

3.2.5 Tokai Daini Station

The Tokai Daini Station is located in east-central Ibaraki Prefecture (Figure 3.1). It has one reactor, which was operating when the earthquake occurred. It shut down automatically. The earthquake cut all offsite AC power, but emergency diesel generators supplied backup power until offsite power was restored on March 13. The reactor reached cold shutdown early in the morning on March 15.

The first tsunami wave arrived at the site approximately 30 minutes after the main shock of the earthquake. The tsunami flooded the seawater pump for one of the emergency diesel generators, rendering it inoperable,

[18] One of the four offsite AC power lines was shut down for maintenance prior to the earthquake.

SIDEBAR 3.1
Chronology of Key Events for Fukushima Daiichi Nuclear Power Station Accident

March 11, 2011	
14:46	Great East Japan Earthquake (M_w 9.0) occurs.
14:47	Unit 1-3 reactors automatically shut down (scram).
14:48	Offsite AC power is lost; onsite emergency diesel generators automatically start up to provide AC power.
15:27	First tsunami wave arrives at wave height meter[a] (+41 minutes after earthquake).
15:36-15:37[b]	Second (main) tsunami wave (height ~13 m) inundates parts of plant (+50 to +51 minutes after earthquake).
15:37-15:41	Article 10 Notification Event[c] occurs (loss of all plant power).
16:36	Article 15 Emergency Event[d] occurs in Units 1 and 2 (loss of emergency core cooling system water injection sources).
19:03	Japanese government declares a nuclear emergency.
20:50	Fukushima Prefecture governor orders residents within a 2-km radius of the plant to evacuate.
21:23	Japanese government orders evacuation radius to 3 km.
March 12	
05:44	Japanese government increases evacuation radius to 10 km.
15:36	Hydrogen explosion occurs in Unit 1 reactor building.
18:25	Japanese government increases evacuation radius to 20 km.

and a seawater pump for one source of core cooling. However, the main area of the site was not flooded.

3.3 DIFFERENCES IN PLANT IMPACTS

The March 11, 2011, earthquake and tsunami had markedly different impacts on the five nuclear plants located along the northeast coast of Japan. It is instructive to examine the reasons for these differences.

The Higashidori plant and Tokai Daini plant lost all offsite AC power,

March 13

05:10	Article 15 Emergency Event occurs in Unit 3 (loss of emergency core cooling system water injection sources).

March 14

11:01	Hydrogen explosion occurs in Unit 3 reactor building.

March 15

06:14	Hydrogen explosion occurs in Unit 4 reactor building.
08:11	Article 15 Emergency Event occurs in Unit 4 (abnormal release of radioactive materials).

NOTE: All times are local (Japan Standard Time) in 24-hour notation.

[a] The wave height meter is located about 1.5 km offshore of the Fukushima Daiichi plant.

[b] TEPCO (2012b) reported this time as 15:35. TEPCO (2013, Attachment Earthquake-tsunami-1) provides an updated time for wave arrival at the plant based on photographic analysis.

[c] Section 15 of the Japanese Act on Special Measures Concerning Nuclear Emergency Preparedness (Act No. 156, December 17, 1999) specifies conditions under which the Nuclear Emergency Preparedness Manager must notify competent authorities about conditions at the nuclear plant.

[d] Section 15 of the Act on Special Measures Concerning Nuclear Emergency Preparedness (Act No. 156, December 17, 1999) specifies conditions under which a Declaration of a Nuclear Emergency Situation would be made. The Japanese Prime Minister is responsible for notifying prefecture governors, mayors of municipalities, and the public that a nuclear emergency has occurred.

but their emergency diesel generators operated as designed to provide the backup power needed to bring the reactors to cold shutdown.

The Onagawa plant was located closest to the fault that produced the earthquake epicenter. Nevertheless, the plant experienced relatively little damage, and its three operating reactors were able to reach cold shutdown within a half day of the tsunami. Three design features of the plant contributed to this outcome:

FIGURE 3.10 Oblique aerial photo of the Fukushima Daini Nuclear Power Plant. SOURCE: Courtesy of TEPCO.

FIGURE 3.11 Photo showing flooding of the Fukushima Daini plant adjacent to Unit 1 during the March 11, 2011, tsunami. Flooding depth was between 2-3 m. SOURCE: Courtesy of TEPCO (http://photo.tepco.co.jp/library/110411_3/110412_2f_tsunami_6.jpg).

- The site had five offsite power lines, one of which survived the earthquake. Consequently, the site did not lose all offsite AC power.
- The site is elevated (14.8 m above O.P.[19]; see Figure 3.6) and did not flood during the tsunami.
- Seawater pumps are located in pits about 100 m from the plant harbor. The pump motors are elevated within the pits to protect them from flooding.

The positive outcome at the Onagawa plant can be attributed to its good design for earthquake and tsunami hazards.

The Fukushima Daiichi plant also lost all offsite AC power. However, the impacts of the earthquake and tsunami on this plant were markedly different than at the Onagawa plant, primarily because

- Parts of the site, including areas around Units 1-4, were flooded by the tsunami.
- Openings into turbine and service buildings near ground level were not watertight; consequently, the lower levels of those buildings were flooded.
- Seawater pump motors, which are used to move decay heat from the reactors to the Pacific Ocean, were not elevated sufficiently to protect them from flooding.

As noted previously, the tsunami flooded emergency diesel generators, batteries, pumps, and electrical distribution equipment, resulting in a cascade of failures: loss of electrical power, severe damage to the Unit 1-3 reactor cores, and severe structural damage to the Unit 1, 3, and 4 reactor buildings.

The Fukushima Daini plant also had extensive flooding damage from the tsunami. But, unlike Fukushima Daiichi, the reactor operators were able to bring the plant's three reactors to cold shutdown. This was possible because

- One offsite AC power supply to the plant and the battery-powered DC power system survived the earthquake and tsunami. Consequently, critical instruments and control equipment continued to operate.
- Flooding was less severe because the plant site was located at a higher elevation relative to the tsunami. Consequently, some electrical distribution equipment remained operational.
- Plant personnel were able to replace or work around some of

[19] Onahama Peil (Onahama Port Construction Standard Surface). See Table 3.2 notes.

the damaged equipment so that critical reactor safety functions could be restored.

The outcome at Fukushima Daini could have been far worse had the plant lost all offsite AC power as occurred at Fukushima Daiichi.

3.4 TSUNAMI HAZARDS AT JAPANESE NUCLEAR PLANTS

The importance of protecting nuclear plants from large tsunamis was well understood when the Fukushima Daiichi and Daini nuclear plants were designed and constructed.[20] At the time these plants were constructed it was common practice to use records of past tsunamis to estimate expected maximum tsunami wave heights. In the case of Fukushima Daiichi and Daini, TEPCO used the tsunami from the May 22, 1960 earthquake near Valdivia, Chile,[21] as an estimate of the maximum tsunami wave heights that would be expected at these plants. These estimates were O.P. + 3.122 m[22] for Fukushima Daiichi and O.P. + 3.122 m (Unit 1) to O.P. + 3.705 m for Fukushima Daini (Units 3 and 4) (see Table 3.3), where O.P. is the abbreviation of Onahama Peil, which is the tidal level at the Onahama Port (Onahama Port is located to the south of the Fukushima Daiichi Plant).

The maximum wave height from the Great Chilean Earthquake tsunami was also used as the design basis for tsunami protection walls along Japan's eastern coast. These walls typically consisted of large earthen berms with openings for roads. The openings were hardened with concrete frames and heavy doors that could be closed for tsunami protection.

In 2002, the Japan Society of Civil Engineers (JSCE) published a quantitative assessment methodology for estimating maximum tsunami wave heights at nuclear plants in Japan (JSCE, 2006[23]). JSCE used the historical earthquake and tsunami record to develop standard fault models for generating tsunamis. These models were simulated numerically by varying key fault model parameters to identify "design tsunamis" that exceed all recorded and calculated historical tsunami heights.

TEPCO and other nuclear plant operators in Japan used this JSCE

[20] Construction permits for Fukushima Daiichi Units 1-6 were issued between 1966 and 1972. Permits for Fukushima Daini Units 1-4 were issued between 1974 and 1980.

[21] The main shock of this earthquake had a moment magnitude 9.5 and is the largest earthquake ever recorded. The tsunami from this earthquake affected coastal regions throughout the Pacific Rim, including Japan.

[22] The estimates reported by TEPCO in this section and in Table 3.3 give tsunami heights to the nearest millimeter. This precision has no significance for tsunami heights that are measured in meters.

[23] An English-language version of this paper was issued in 2006. See http://committees.jsce.or.jp/ceofnp/system/files/JSCE_Tsunami_060519.pdf.

TABLE 3.3 Tsunami Wave Height Estimates at the Fukushima Daiichi and Fukushima Daini Plants

		Tsunami Height Estimates (meters relative to O.P.[a])	
Date	Estimate Basis	Fukushima Daiichi	Fukushima Daini
1966-1980 (plant permits)	May 22, 1960 earthquake near Valdivia, Chile	3.122 (Units 1-6)	3.122 (Unit 1) 3.690 (Unit 2) 3.705 m (Units 3 and 4)
2002	JSCE (2002)[b] methodology	5.7	5.2
2007	Fukushima Prefecture Disaster Prevention Plan	~5.0	~5.0
2007	Ibaraki Prefecture Disaster Prevention Plan	4.7	4.7
2008	TEPCO trial calculations using fault models for HERP postulated earthquake	8.4-10.2 (Inundation heights: 13.7-15.7)	7.6-8.2 (Inundation height: 15.5)
2008	TEPCO trial calculations for Satake et al. (2008) model for Jogan tsunami	8.7-9.2	7.8-8.0
2009	JSCE (2002) with updated tidal and bathymetric data	6.1	5.2

[a] Onahama Port Construction Standard Surface.
[b] An English-language version of this paper was issued in 2006.
SOURCES: TEPCO (2012b); TEPCO presentation to the committee, September 6, 2012.

methodology to estimate maximum tsunami wave heights at their plant sites. Based on this new methodology, the maximum tsunami wave height at Fukushima Daiichi was estimated to be O.P. + 5.7 m, over 2.5 m higher than the estimate in the plant's original permit. The maximum tsunami wave height at Fukushima Daini was estimated to be O.P. + 5.2 m, over 2 m higher than the estimate in the plant's original permit (Table 3.3).

TEPCO took several countermeasures to protect critical equipment and infrastructure at its plants in response to these new estimates. At Fukushima Daiichi, TEPCO raised the elevations of seawater pumps used to cool emergency diesel generators and feed emergency core cooling systems. At Fukushima Daini, TEPCO made portions of buildings watertight.

Following the December 2004 Indian Ocean earthquake and tsunami[24] there was a renewed effort in Japan to document the 869 Jogan tsunami; that tsunami occurred well before the advent of modern recordkeeping. TEPCO undertook tsunami deposit surveys in the vicinity of the Fukushima Daiichi plant (TEPCO, 2012b, p. 26); it found tsunami deposits at sites north of the plant that were assumed to be from this tsunami. TEPCO reported that those deposits were located at elevations of 0.5 m to 4 m. TEPCO also reported that no tsunami deposits related to the Jogan tsunami were found south of the plant.

In 2006 and 2007 the Fukushima and Ibaraki Prefectures issued disaster prevention plans that contained tsunami wave height estimates ranging from O.P. + 4.7 m to O.P. + 5.0 m (Table 3.3). However, because these estimates were less than estimates based on the JSCE (2006) methodology, TEPCO took no additional countermeasures to protect critical equipment and infrastructure at its plants.

In 2008,[25] TEPCO made what it refers to as "trial calculations" of maximum wave heights based on two information sources: (1) a Headquarters for Earthquake Research Promotion (HERP) conclusion that a magnitude 8.2 earthquake could occur anywhere along the Japan Trench (HERP, 2002); and (2) wave source models for the 869 Jogan tsunami based on the models of Satake et al. (2008). The estimated maximum wave heights from these trial calculations were significantly higher than previous estimates: up to O.P. + 10.2 m at Fukushima Daiichi and up to O.P. + 8.2 m at Fukushima Daini (see last two rows in Table 3.3). The estimated maximum run-up height at the southern portion of the Fukushima Daiichi site was estimated to be 15.7 m (Table 3.3). This is similar to the 15.5-m inundation height at the plant from the March 11, 2011, tsunami.

TEPCO revised its maximum tsunami wave height estimates in 2009 using the JSCE (2006) methodology with updated bathymetric and tidal data. The new estimates were O.P. + 6.1 m for Fukushima Daiichi and O.P. + 5.2 m for Fukushima Daini (Table 3.3). TEPCO raised the elevations of seawater pumps at Fukushima Daiichi based on this new estimate.

In 2009, TEPCO and other nuclear plant operators also requested that JSCE undertake additional reviews of tsunami source fault models and associated methodologies; TEPCO staff reported to the committee that this review was estimated to take about 3 years (2009-2012) with the publication of a new methodology in 2013. Additional time would have been

[24] The tsunami produced 30-m-high waves in the Aceh Province of Indonesia.

[25] TEPCO analysts presented a paper on probabilistic tsunami hazard analysis in 2006. See Sakai et al. (2006).

required to apply this new methodology to obtain updated wave-height estimates and, if necessary, take appropriate countermeasures at Japanese nuclear plants. The new JSCE methodology was not completed prior to the March 11, 2011, earthquake and tsunami.

However, there were indications as early as 2001 that large-scale tsunamis on the east coast of Japan might have recurrence intervals of 800-1,100 years (Minoura et al., 2001). Noggerath et al. (2011) reported on a June 2009 Japanese government committee hearing at which a senior geologist at a government-affiliated research laboratory warned about the risks of large tsunamis based on Jogan tsunami data. They suggest that TEPCO response to the warning was inadequate.

TEPCO staff told the committee that much of the company's thinking about tsunami hazards was based on the location of previous tsunami sources off the east coast of Japan. TEPCO staff also told the committee that there were no records of large earthquakes along the Japan Trench off the coast of Fukushima Prefecture where its plants are located.

TEPCO staff told the committee that the company did not take immediate countermeasures at its plants based on the results of the 2008 trial calculations because (1) there was no record of magnitude 8-level earthquakes off the coast of Fukushima, as noted in the previous paragraph; (2) both JSCE and Japanese government agencies did not consider a large tsunami source to be present off the coast of Fukushima; and (3) the Jogan tsunami source had not been determined and additional tsunami deposit surveys were needed. TEPCO thought it was necessary to further investigate the appropriateness of the tsunami source models used in the 2008 trial calculations.

3.5 DISCUSSION AND FINDING

It should be clear from the descriptions in the preceding section that TEPCO actively implemented countermeasures to protect critical equipment and infrastructure at its Fukushima Daiichi and Daini plants in response to new information about tsunami hazards. Consequently, it is puzzling to the committee why TEPCO appeared to lack a sense of urgency to act after its 2008 trial calculations yielded tsunami wave estimates that were substantially higher than previous estimates (Table 3.3). To the committee's knowledge, TEPCO did not take any steps to implement additional countermeasures at its plants in response to this new information. Instead, TEPCO (as well as other plant operators) called for more studies.

At the same time TEPCO was calling for more studies of tsunami methodologies, it was actively implementing countermeasures in response to new information about earthquake hazards. For example, TEPCO implemented two major countermeasures at the Fukushima Daiichi plant following the 2007 Niigata Chuetsu-Oki Earthquake.[26]

1. TEPCO installed a fire suppression system that could also be used in emergencies to inject water into the plant's reactors.
2. TEPCO also constructed an earthquake-resistant building for the plant's onsite Emergency Response Center.

Both of these countermeasures improved the ability of the Fukushima Daiichi plant's operators to respond to the March 11, 2011, earthquake and tsunami (see Chapter 4).

Implementation of additional countermeasures to protect the Fukushima Daiichi plant from a 10-m tsunami wave estimated from the 2008 trial calculations (Table 3.3) might have required extensive modifications to the harbor front at the plant, for example, augmentation of the existing seawall,[27] construction of a new harbor wall in front of the plant, and/or reconfiguration of the seawater intakes and outfalls. However, other types of countermeasures might have been implemented through less disruptive modifications:

- Moving seawater pumps to protect them from flooding;
- Raising the elevations of emergency diesel generators, batteries, and electrical switching equipment;
- Implementing other means to protect this equipment from flooding, for example, by sealing entryways into buildings against water intrusion; and/or
- Installing additional backup equipment at higher elevations on the plant site.

Of course, even had TEPCO implemented such countermeasures, they might not have been sufficient to protect the plant against flooding from the March 11, 2011, tsunami. Nevertheless, such countermeasures might have reduced flooding damage and provided more time for plant operators to more quickly restore power to critical reactor monitoring and cooling equipment.

TEPCO has been criticized for not being prepared for the March

[26] This 6.6 moment magnitude earthquake occurred on July 16, 2007, in western Japan offshore of Niigata Prefecture. The earthquake affected operations at TEPCO's Kashiwazaki-Kariwa Nuclear Power Station.

[27] The main function of the seawall is to protect the plant harbor from ocean waves during storms. Raising the seawall may not have been sufficient to protect the plant from large tsunamis because there are openings in the seawall to the ocean (see Figure 1.2 in Chapter 1).

11, 2011, earthquake and tsunami at its Fukushima Daiichi plant (e.g., Investigation Committee, 2011, 2012; NAIIC, 2012). This NAS committee was not tasked with assessing whether adequate preparations were taken by TEPCO before or in response to the earthquake or tsunami. Rather, the committee's task is to identify lessons that can be learned from the accident for improving safety and security of U.S. nuclear plants.

FINDING 3.1: The overarching lesson learned from the Fukushima Daiichi accident is that nuclear plant licensees and their regulators must actively seek out and act on new information about hazards that have the potential to affect the safety of nuclear plants. Specifically,

1. **Licensees and their regulators must continually seek out new scientific information about nuclear plant hazards and methodologies for estimating their magnitudes, frequencies, and potential impacts.**
2. **Nuclear plant risk assessments must incorporate new information and methodologies as they become available.**
3. **Plant operators and regulators must take *timely* actions to implement countermeasures when such new information results in substantial changes to risk profiles at nuclear plants.**

The findings and recommendations in Chapter 5 of this report expand on this lesson learned.

Nuclear plants usually operate for many decades. Scientific understanding of hazards, especially hazards arising from natural external events, can advance substantially during such extended periods. The first Fukushima reactors were constructed in the 1960s, when the plate tectonics concept was just coming into acceptance by the scientific community and subduction processes were just beginning to be understood. In the 1960s, knowledge about large earthquakes and tsunamis was based on historical records. Five decades later, advances in scientific understanding of subduction zone processes have enabled the development of new methodologies and technologies for estimating seismic and tsunami hazards. Further substantial advances are likely in the decades ahead.

Japan has expanded its capabilities for monitoring and understanding earthquake and tsunami hazards using extensive geophysical networks installed over the past decade. These networks include

- 816 GEONET stations for GPS geodetic measurements (e.g., Melgar and Bock, 2013),
- 1,800 modern seismic stations (e.g., Furumura et al., 2011),

- Five DART (Deep-Ocean Assessment and Reporting of Tsunami) Buoys,[28]
- Pressure gauges in deep water[29] (e.g., Gusman et al., 2012),
- Nearshore GPS stations operating from buoys, and
- Geodetic monuments in deep water.

Data from these monitoring stations can be used to measure absolute changes in seafloor configurations associated with tsunamis. Geodetic data by themselves are now capable of predicting tsunami run-ups with considerable accuracy. Similar networks are being deployed in other parts of the world as well. For example, the U.S. National Science Foundation has installed new, dense broadband seismic and permanent GPS geodetic stations in coastal Oregon and Washington and is operating seafloor broadband stations near the Cascadia Fault. A new seafloor network, the NSF Ocean Observatory Initiative,[30] will be installed later this year, permitting new geodetic and seismic technologies to be included. A similar network is already operating off Vancouver.[31]

[28] Located offshore of eastern Japan.
[29] Several of the gauges were located near the epicenter of the Great East Japan Earthquake.
[30] http://oceanobservatories.org.
[31] http://oceannetworkds.ca.

4

Fukushima Daiichi Nuclear Accident

The focus of this chapter is on the March 11, 2011, accident at the Fukushima Daiichi nuclear plant: the accident time line, key events during the accident, actions taken to bring the plant's reactors to cold shutdown, and challenges faced in taking those actions. This chapter has two objectives:

1. To address the first charge of the statement of task for this study (see Sidebar 1.1 in Chapter 1) on the "Causes of the Fukushima nuclear accident, particularly with respect to the performance of safety systems and operator response following the earthquake and tsunami."
2. To provide information and analysis to support the committee-identified lessons learned in Chapter 5.

It is not the committee's intention to place blame for the accident or to find fault with how personnel at the Fukushima Daiichi plant responded to the earthquake and tsunami. With the benefit of hindsight, it is easy to second-guess the decisions and actions taken during the accident. In reviewing the accident response, the committee came to appreciate the overwhelming challenges that plant personnel faced in responding to the accident. Some of those challenges are described in the next section of this chapter. Indeed, the conditions at the Fukushima Daiichi plant following the earthquake and tsunami would have challenged any nuclear plant operator.

Many accounts of the Fukushima Daiichi nuclear accident have already been published. These accounts provided the factual information used in this chapter and informed committee judgments about accident causes and lessons learned. The following reports, papers, and presentations were particularly useful for these purposes:

- Post-accident investigation reports by Japanese and U.S. organizations, especially INPO (2011, 2012), Investigation Committee[1] (2011, 2012), NAIIC[2] (2012), TEPCO (2011a,b, 2012b, 2013), ANS (2012), and EPRI (2012a). The Investigation Committee (2011, 2012) and TEPCO (2011a,b, 2012b) reports provide detailed documentation of the decisions and actions taken during the accident as well as key thought processes behind those actions.
- Technical papers on the accident, most notably Gauntt et al. (2012a,b), Levy (2012), Phillips et al. (2012), and EPRI (2013).
- Slides from technical presentations by Japanese researchers at International Atomic Energy Agency conferences in 2012[3] and 2014,[4] technical workshops in Japan, and other international meetings (e.g., 11th International Probabilistic Safety Assessment and Management Conference and The Annual European Safety and Reliability Conference, Helsinki, Finland, June 25-29, 2012).
- Discussions with Japanese technical experts at the committee's November 2012 meeting in Tokyo, Japan.
- Site visits to the Fukushima Daiichi, Fukushima Daini, and Onagawa nuclear plants in November 2012.
- Discussions with U.S. technical experts at the committee's meetings in the United States.

---

[1] The Investigation Committee on the Accident at Fukushima Nuclear Power Stations of Tokyo Electric Power Company was established by the Japanese Government by Cabinet Decision on May 24, 2011. The committee was chaired by Dr. Yotaro Hatamura, professor emeritus of the University of Tokyo and professor at Kogakuin University. The committee published an interim report in 2011 and a final report in 2012.

[2] The Fukushima Nuclear Accident Independent Investigation Commission (NAIIC) was established by the National Diet of Japan on October 30, 2011. The commission was chaired by Dr. Kiyoshi Kurokawa, academic fellow, National Graduate Institute for Policy Studies. The commission published its report in 2012.

[3] International Experts' Meeting on Reactor and Spent Fuel Safety in the Light of the Accident at the Fukushima Daiichi Nuclear Power Plant, March 19-22, 2012. IAEA, Vienna, Austria. Information available at http://www-pub.iaea.org/iaeameetings/43900/International-Experts-Meeting-on-Reactor-and-Spent-Fuel-Safety-in-the-Light-of-the-Accident-at-the-Fukushima-Daiichi-Nuclear-Power-Plant.

[4] International Experts' Meeting on Severe Accident Management in the Light of the Accident at the Fukushima Daiichi Nuclear Plant. March 17-20, 2014. IAEA, Vienna, Austria. Information available at http://www-pub.iaea.org/iaeameetings/46832/International-Experts-Meeting-on-Severe-Accident-Management-in-the-Light-of-the-Accident-at-the-Fukushima-Daiichi-Nuclear-Power-Plant.

Appendix B identifies the technical experts who participated at committee meetings in Japan and the United States.

It is important to acknowledge that there are information gaps and uncertainties about some details of the accident progression. The accident time line presented in this chapter represents the committee's best collective technical judgments informed by the information sources cited above.

This chapter is organized into five sections. The first section provides a time line for the accident. Additional details on the time line are provided in Appendix C. The second section describes some of the challenges in responding to the accident. The third section describes key accident events and responses by plant personnel. The fourth section provides a discussion of six issues that stand out from the committee's analysis of the accident. The fifth and final section provides a committee finding on the causes of the Fukushima Daiichi nuclear accident to address the first charge of the study task.

4.1 TIME LINE FOR FUKUSHIMA DAIICHI ACCIDENT

Table 4.1 provides a committee-constructed summary time line for the accident in Units 1, 2, and 3 at the Fukushima Daiichi nuclear plant. A more detailed description of this time line is provided in Appendix C. A simplified time line of key events is depicted graphically in Figure 4.1.

TABLE 4.1 Time Line of Key Events in Units 1-3 at the Fukushima Daiichi Nuclear Plant

Event/Condition	Unit 1	Unit 2	Unit 3
Prior to earthquake		Operating at rated power level	
Earthquake (3/11/11 @ 14:46)		T = 0 Reactor scram MSIVs close Loss of offsite AC power EDGs start	
Tsunami warnings (Fukushima Prefecture) and estimated wave heights	←	14:49 (+3 min): 3 m 15:15 (+29 min): 6 m 15:30 (+44 min): >10 m	→
Tsunami arrival times (1st/2nd waves)		+41 min/+50 to +51 min (15:27/15:36-15:37)	

TABLE 4.1 Continued

Event/Condition	Unit 1	Unit 2	Unit 3
Loss of onsite AC power (EDGs) and DC power (batteries)	AC lost at +51 min (15:37) DC lost at +60 min (15:46)	AC lost at +55 min (15:41) DC lost at +60 min (15:46)	AC lost at ~+51 min (15:37) DC available until ~+36 h
IC performance	Failed on loss of AC and DC power	NA	NA
RCIC performance	NA	Real-time status uncertain; evidence of ~70 h of running time	~20 h of running time; failed w/o restart at +20 h
HPCI performance	Unavailable due to loss of DC power	Unavailable due to loss of DC power	~16 h of running time beginning at +20 h
RPV depressurization	Depressurized due to assumed RPV failure at +12 h	Depressurized at +75.2 h and +78.3 h	Depressurization occurred at ~+42 h
Time of max containment pressure (Max containment pressure/design pressure)	+11.7 h (0.84 MPa/0.43 MPa)	~+80 h (~0.75 MPa/0.38 MPa)	~+42 h (0.64 MPa/0.38 MPa)
Estimated time of core damage	+4 h to +7 h	+75 h to +85 h	+36 h to +40 h
First indication of offsite release of radioactive materials	←	+8.2 to +14.1 h	→
Containment venting preparation/success	+9.7 h/~+24 h	+26.7 h/not successful	+29.5 h/+42 h
Hydrogen explosion	+24.8 h	None	+68.2 h
Initial injection of freshwater/seawater	+15.0/+28.8 h	None/+77.2 h	+42.6/+46.4 h
Restoration of offsite AC power	March 20	March 20	March 22

NOTES: EDGs = emergency diesel generators, HPCI = high-pressure coolant injection system, IC = isolation condenser, MSIV = main steam isolation valve, RCIC = reactor core isolation cooling system, RPV = reactor pressure vessel.

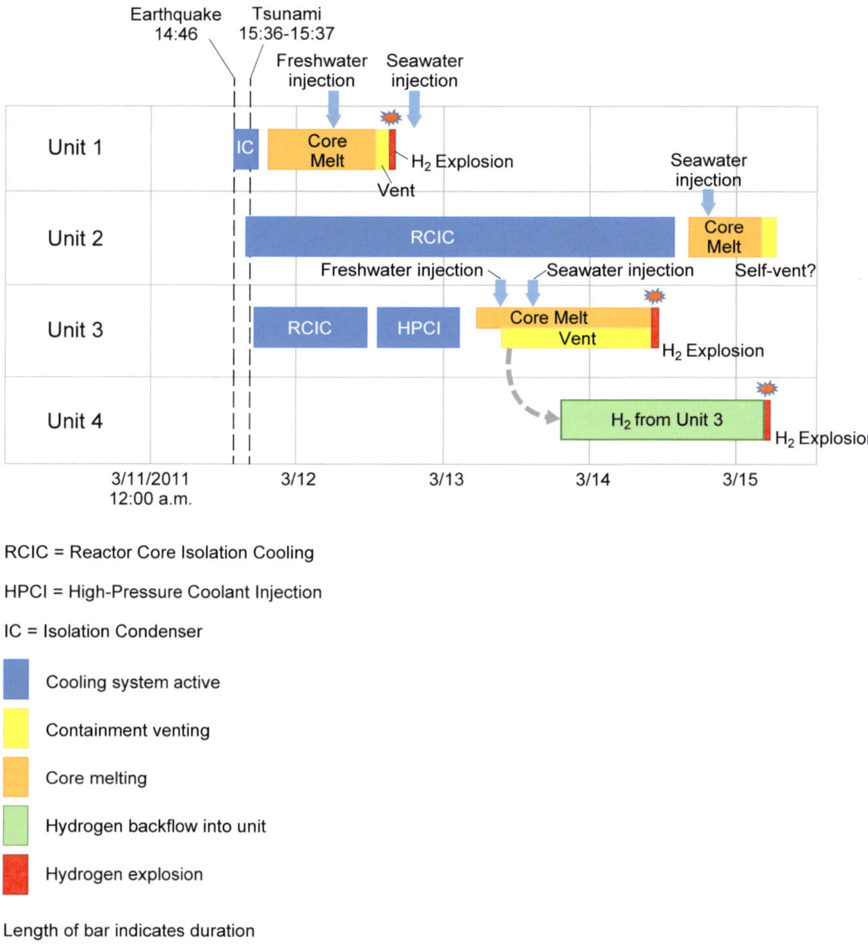

FIGURE 4.1 Graphical depictions of accident time lines for Units 1-4 at the Fukushima Daiichi plant. The key events shown in the time lines are described in the text.

The committee's time line was developed from previously published accident accounts, primarily INPO (2011), Investigation Committee (2011, 2012), TEPCO (2011a,b, 2012b, 2013), and ANS (2012). The committee gathered additional information through discussions with Japanese and U.S. technical experts to better understand some details of the time line.

The zero point of the time line is the afternoon of March 11, 2011, just before the Great East Japan Earthquake struck Japan. Chapter 3 of this report describes the status of the six reactor units at the Fukushima Daiichi nuclear plant at this time:

- Units 1, 2, and 3 were operating at licensed power level.
- Unit 4 was in an outage for replacement of the reactor core shroud. Fuel from the Unit 4 reactor had been relocated to the spent fuel pool in the reactor building.
- Units 5 and 6 were in inspection outages. Fuel remained in their cores and the reactors were being actively cooled. The Unit 5 containment was open and the primary system was undergoing pressure testing; because the reactor was at elevated pressure, it was not strictly in cold shutdown.

The earthquake initiated the following chain of events at the plant (Table 4.1):

- The reactors in Units 1-3 shut down automatically (*scrammed*) as designed when high seismic accelerations (i.e., ground shaking) were detected in the units.
- Offsite AC power to the site was lost because of the collapse of one transmission tower and severe damage to equipment in a substation as a result of ground shaking.
- Following offsite AC power loss, the main steam isolation valves (MSIVs) in Units 1-3 closed automatically to isolate the reactors, limiting the potential loss of coolant, release of radioactivity, and the rate of reactor vessel cooldown.
- Within about a minute of offsite AC power loss, the onsite emergency diesel generators automatically started and were connected to the power distribution system as designed to supply onsite emergency AC power to reactor safety systems.

Normal reactor cooldown and decay-heat removal functions were in place and operating at the plant when the tsunami wave arrived starting about 41 minutes after the earthquake (Table 4.1). The tsunami flooded portions of the plant site (see Chapter 3), damaging pumps, electrical distribution panels, batteries,[5] and emergency diesel generators. Units 1, 2, 3, 4, and 5 lost AC power within 5 minutes after the tsunami and Units 1, 2 and 4 lost DC power shortly thereafter. Unit 3 lost AC power but did not lose DC power immediately after the tsunami because its power distribution panels and backup battery were not damaged by flooding. Once power was

[5] As noted in Chapter 2, nuclear plants have large backup batteries (or banks of batteries) to supply DC power to operate and monitor critical monitoring equipment and safety systems.

lost, the units' control rooms lost lighting, indicators, instrument readouts, and controls.

Although there were intermittent signs of power on some indicators in Units 1 and 2, reliable DC power was only available by connecting arrays of scavenged vehicle batteries to selected systems and instrumentation in the control rooms. Vehicle batteries also had to be employed in Unit 3 to operate critical systems after the installed backup battery was depleted (see Section 4.3.2 for details).

An emergency diesel generator at Unit 6 survived the tsunami because it was air-cooled and was located above flood level. It continued to supply emergency AC power to Unit 6 and was used to supply power to Unit 5 through a crosstie that had been installed during the evening and early morning following the earthquake (see Section 4.3.4 of this chapter for additional details). The crosstie was prepositioned prior to March 11, but installation was not started until after the tsunami and was not completed until 05:00 on March 12.

Three tsunami warnings were issued by the Japan Meteorological Agency following the earthquake[6]:

- Warning 1, indicating a major tsunami with 3-m wave amplitude for Fukushima Prefecture, was issued at +3 min (14:49). This warning was based on an initial analysis of earthquake strong-motion data.
- Warning 2, indicating a major tsunami with 6-m wave amplitude for the Fukushima Prefecture, was issued at +29 minutes (15:15). This warning was based on observed tsunami amplitudes at tsunami meters and tide gauges.
- Warning 3, indicating a major tsunami with 10-m or greater wave amplitude for the Fukushima Prefecture, was issued at +44 min (15:30), again based on observed tsunami amplitudes at tsunami meters and tide gauges.

According to the Investigation Committee (2011), the site superintendent at the Fukushima Daiichi plant (Mr. Masao Yoshida) learned about the first two tsunami warnings from TV news reports. As a result of these warnings, field personnel at the plant were evacuated to the onsite Emergency Response Center (onsite ERC; see Appendix D) or to higher ground. The third tsunami warning came after the first tsunami wave had already arrived at the Fukushima Daiichi plant (see Table 4.1). The tsunami warn-

[6] Information on tsunami warning is from a presentation by Osamu Kamigaichi, Japan Meteorological Agency, at the February 2012 meeting of the Intergovernmental Oceanographic Commission. This presentation is available at http://ioc-tsunami.org/index.php?option=com_oe&task=viewDocumentRecord&docID=8619.

ings affected the site superintendent's thinking about accident management because he was concerned that the tsunami might damage seawater pumps.

Just before the earthquake occurred, there were about 6,400 personnel, including 750 employees of the plant owner-operator (TEPCO), onsite (TEPCO, 2012b, p. 163). Many TEPCO and contractor workers left the plant on their own on March 11. Those who could not leave were evacuated to the seismic isolated building. TEPCO (2012b, p. 166) estimates that an additional 300-400 people were evacuated in buses from March 12 to 14, and some additional unknown number of people self-evacuated during that time. By March 15, there were about 700 people left onsite (TEPCO, 2012b, p. 102). These included people who had no direct role in the emergency response.

Appendix D describes the organization of personnel at the plant at the time of the accident. Ninety-seven personnel were working in the main control rooms at the time of the earthquake. These personnel performed initial actions following the earthquake and tsunami. Additional personnel arrived to support control room staff in the following hours and days.

Staffing reinforcements were dispatched to Fukushima Daiichi by TEPCO following the earthquake and tsunami to support restoration work. They started arriving on March 11 and arrivals continued over the next several days, averaging approximately 400 additional personnel onsite each day. These included "primarily [the] recovery team responsible for restoring power and monitoring instruments, fire brigade units that used fire engines to inject cooling water into reactors, a health physics team that controlled radiation levels within the Fukushima Daiichi NPS [Nuclear Power Station] and its surroundings, and procurement team that provided material support" (TEPCO, 2012b, p. 303). In addition, in accordance with prior agreements, personnel from other utilities arrived to provide support starting on March 13.

Early on March 15, 650 personnel temporarily evacuated to Fukushima Daini following a hydrogen explosion in Unit 4, leaving approximately 70 workers required for station monitoring and restoration activities (TEPCO, 2012b, p. 166). Some of the personnel that had evacuated to the Fukushima Daini plant returned by noon on March 15. These included operators responsible for monitoring data from the main control rooms, the health physics team responsible for performing radiation-level measurements in the field and for access control to the seismic isolated building, and the security guidance team responsible for controlling station access (TEPCO, 2012b, p. 166).

The earthquake and tsunami resulted in three fatalities at TEPCO's plants: two fatalities occurred at Fukushima Daiichi and one at Fukushima Daini.

4.2 CHALLENGES OF RESPONDING TO THE ACCIDENT

The Fukushima Daiichi accident occurred in the midst of a regional disaster involving the largest loss of life and civil disruption in Japan since World War II. The accident is historically unique in this regard. The earthquake and tsunami overwhelmed offsite emergency response efforts (see Chapter 6) and added greatly to the challenges of responding to the accident at the plant.

Japanese investigations of the accident (Investigation Committee, 2011, 2012; NAIIC, 2012) concluded that the Fukushima Daiichi nuclear plant's owner-operator (TEPCO) was not adequately prepared for an earthquake and tsunami of this magnitude. The plant lacked survivable onsite power supply, water pumping, and communications equipment. Moreover, its accident-management emergency operating procedures did not address accident scenarios involving the complete loss of onsite power, instrumentation, and reactor controls; and reactor operators had not been trained to respond to such scenarios. Indeed, the Fukushima Daiichi nuclear accident was "off the map" in terms of preparation, planning, and training for severe nuclear accidents.

Personnel involved in the accident response had to improvise, a fact highlighted by the Investigation Committee (2011, pp. 110-111):

> The shift team[7] used lights with portable batteries and LED flashlights to read the event-based and state-based "Emergency Operating Procedure." However, the content of the material could not be applied directly to the actual events taking place. The team members also checked the "Emergency Operating Procedure" for accident management (AM) to identify the operating procedure necessary to control Units 1 and 2. However, the "Emergency Operating Procedure" for AM contained only internal events as causal events for AM and did not consider external events such as an earthquake or tsunami as causal events. There was no reference taking into account the events where all AC and DC power sources would be lost. In addition, the descriptions of the standards were written on the assumption that the state of the plants can be monitored by the control panel indicators and measuring instruments in the main control room and that the control panel could be manipulated. As a result, the shift team was forced to predict the reactor state according to a limited amount of information and take such procedures [that] operators think best on the spot instead of following the instructions described in the standard manuals.

[7] The shift team comprised the personnel in the control room of each reactor unit. See Appendix D.

Staff in the onsite ERC were stunned to learn of the complete failure of power in three of the reactor units. Their reaction is described by the Investigation Committee (2011, pp. 108-109):

> The NPS [nuclear power station] ERC[8] received reports from the three main control rooms that the nuclear reactors were successively losing their power supplies and Units 1, 2 and 4 in particular had lost all of their power sources. Everyone at the NPS ERC was lost for words at the ongoing unpredictable and devastated state.
>
> . . . .
>
> Site Superintendent Yoshida understood that a situation that far exceeded any expected major accident had actually taken place. He could not think of anything on the spot and so decided to implement the procedure stipulated by the law.

Plant personnel confronted many challenges in responding to the earthquake and tsunami:

- Flooding in the turbine buildings and lower portions of Units 1 and 2 rendered reactor control and safety systems inaccessible or unusable.
- Damage to the site from the tsunami made roads impassable and generally hindered personnel access.
- Loss of instrumentation readouts in the Unit 1 and 2 control rooms and loss of the safety parameter display systems[9] in the Unit 1-3 control rooms and the onsite ERC and offsite center (OFC) made it impossible to obtain timely information about the condition of the Unit 1-3 reactors and Unit 1-4 spent fuel pools. Control room personnel reported basic reactor parameters to the onsite ERC using fixed-line telephones. These data were manually recorded on whiteboards to facilitate the sharing of information within the ERC.
- Loss of lighting made it difficult to work, forcing control room and field personnel to use flashlights.
- Limited means of communication between the control rooms and the onsite ERC and between the control rooms and the field made it difficult to plan and carry out response efforts across the site.
- Hydrogen explosions, radioactive contamination, and high temperatures limited access to some parts of the Unit 1-4 reactor buildings. Field personnel wore standard anti-contamination suits and self-contained

---

[8] This report uses the term "onsite ERC" to refer to this facility.

[9] The safety parameter display system provides detailed real-time plant parameter and component status information.

breathing apparatus, which made their work and communications even more difficult. At one point during the accident the Unit 1 reactor operators had to don full face masks with charcoal filters, anti-contamination coveralls, and at times had to move to the Unit 2 side of the control room and crouch down to avoid excessive radiation exposure.

- The lack of food, working toilets, and relief personnel during the early stages of the accident as well as the extended length of the accident response added greatly to personnel fatigue and distress.

Plant personnel who responded to the accident exhibited a strong degree of self-sacrifice: Many suffered personal losses (homes destroyed or damaged, family members displaced or lost) but continued to work, in some cases for weeks following the tsunami. Personnel volunteered to enter high-radiation zones and many received exposures well over permissible levels.

The OFC, located in Okuma about 5 km southeast of the plant, did not function as intended following the tsunami. It was never fully staffed because of access difficulties owing to transportation system damage and traffic congestion. Additionally, all of its telecommunications circuits except a satellite connection were inoperable.[10] The OFC had to be evacuated on March 14 because of elevated radiation levels following the hydrogen explosion in the Unit 3 reactor building.[11]

The coordination activities that would normally be performed at the OFC were conducted at the TEPCO headquarters ERC, which was located in Tokyo (Appendix D), and at Japanese government offices. This reduced the effectiveness of communications between the onsite ERC, TEPCO, and local and national government agencies (INPO, 2011). According to NAIIC (2012), the loss of telecommunication infrastructures led to the increased involvement of the central government in the response to the accident, partly because the government perceived that it was not receiving accurate and timely information. The Japanese government contacted the headquarters and onsite ERCs directly to get information.

4.3 KEY EVENTS AND RESPONSE ACTIONS

The following sections describe some of the major events during the accident and key response actions by plant personnel. These descriptions are not intended to be comprehensive; rather, they are intended to illumi-

---

[10] Personnel in the OFC were unable to use the videoconferencing system, the Emergency Response Support System, the System for Prediction of Environmental Emergency Dose Information, e-mail, Internet, or ordinary telephone/fax lines.

[11] The OFC was not equipped with filtered ventilation for removing radioactive material even though it was intended for use in nuclear emergencies.

nate the factors that prevented a more successful response to the accident. These factors informed the committee's finding on the causes of the accident (see Section 4.5 in this chapter) and discussions of lessons learned (see Chapter 5). The Investigation Committee (2011, 2012) and TEPCO (2011a,b, 2012b) served as the main sources of information for the descriptions in the following sections.

4.3.1 Unit 1 Reactor

Following the earthquake and scram of the Unit 1 reactor, its two isolation condensers (Figures 4.2 and 4.3) started automatically as designed (see Section 2.2 in Chapter 2). Following established operating procedures, the Unit 1 operators[12] used these isolation condensers to control reactor pressure and cooldown rate. They initially shut down both isolation condensers because reactor cooldown rates were too high; they then cycled one of the isolation condensers (the Train A isolation condenser in Figure 4.3) to maintain reactor pressure and cooldown rates within prescribed specifications.[13]

The Train B isolation condenser was on standby at the time of the tsunami. It was inoperable after the tsunami because the operator had closed off the return line valve (valve MO-3B in Figure 4.3) before the tsunami and was unable to open it afterward due to the lack of AC and DC power (Investigation Committee, 2011, p. 117; TEPCO, 2012b, p. 195).

Subsequently, the tsunami flooded the Unit 1 emergency diesel generators and power panels (Figure 4.2), cutting off all AC and DC power to the unit. With no power for instrumentation or controls, the Unit 1 operators lost the ability to monitor plant indicators from the control room. Most critically, they were unable to check the status of the isolation condenser valves[14] or to actuate them from the control room. Attempts to check the status of the valves in the field were unsuccessful because of access limitations and high radiation fields. Attempts to start up the high-pressure coolant injection system (Figure 4.2) also were unsuccessful due to the loss of DC power.

The loss of AC and DC power in Unit 1 caused its isolation condenser to shut down because of failsafe control logic (this logic is described later in this section). As a consequence, Unit 1 essentially lost all cooling function.

---

[12] The committee uses the following terms to describe TEPCO and contractor staff involved in the response to the accident at the Fukushima Daiichi nuclear plant. The term *operator* refers to personnel stationed in the main control rooms at the plant. The term *ERC staff* refers to personnel stationed in the onsite or headquarters ERCs. The more general term *plant personnel* is used when the locations of personnel at the plant are not specified or important.

[13] That is, maintained reactor pressure between 6 and 7 MPa and a cooldown rate of 55°C (100°F) per hour.

[14] That is, to determine whether the valves were open or closed.

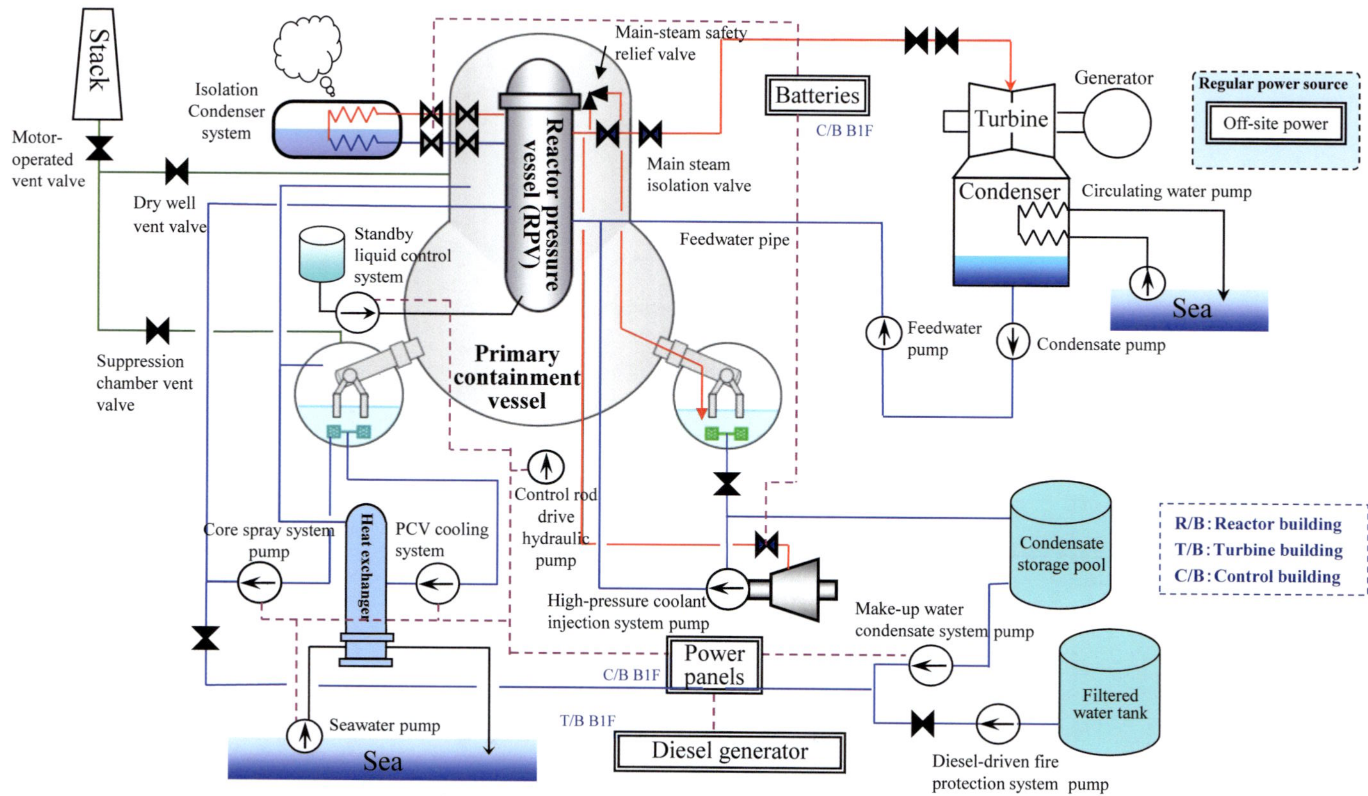

FIGURE 4.2 Schematic illustration of major safety systems in Unit 1 of the Fukushima Daiichi plant. SOURCE: Courtesy of TEPCO.

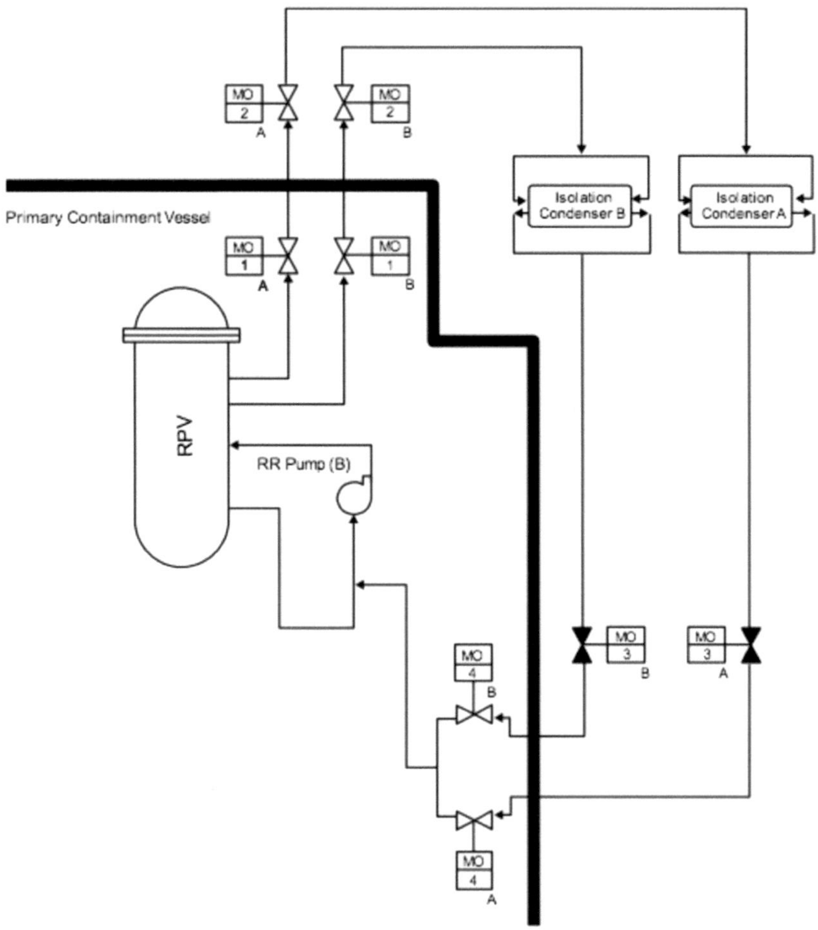

FIGURE 4.3 Schematic of the isolation condenser systems for Fukushima Unit 1. The unit contains two systems, labeled A and B. Motor-operated (MO) valves are indicated by connected triangles. Black indicates valve closed during normal operations; white indicates valve open during normal operation. The valves inside of primary containment are operated by AC power. The valves outside of containment operate with DC power. A fuller description of isolation condenser operation is provided in Chapter 2. SOURCE: Government of Japan, 2011b (Fig. IV-2-4).

However, operators and onsite ERC staff did not understand at first that the isolation condenser had stopped functioning because plant indicators and controls were not available. In fact, the Unit 1 operators initially assumed that the isolation condenser was working.

Because of the failure of the safety parameter display systems and lack of definite information from the Unit 1 operators, the staff in the onsite ERC and the site superintendent could not determine if the isolation condenser was functioning. Site Superintendent Yoshida was sufficiently concerned that he immediately reported to Tokyo that there was a failure of the emergency core cooling systems for Units 1 and 2 (Investigation Committee, 2011, p. 114).

The onsite ERC began to take proactive actions to restore the Unit 1 monitoring systems and establish alternative water injection sources. The site superintendent directed onsite ERC staff to give priority to restoring plant indicators, particularly reactor water level and pressure. At approximately 17:10 on March 11, he instructed onsite ERC staff to begin preparation for two alternative water injection strategies: water injection via the diesel-driven fire protection system (this system is depicted in Figure 4.2), a mitigation strategy specified in the plant's accident management procedures, and water injection through the fire protection system using fire engines, a strategy not specified in those procedures.

Around 18:00 on March 11, some DC power was restored in Unit 1. Operators discovered that the isolation condenser valves outside of containment (i.e., valves MO-2A and MO-3A in isolation condenser A; see Figure 4.3) were closed. The fact that valve MO-2A read closed, when it normally should be open (see Section 2.2 in Chapter 2), caused operators to suspect that all of the isolation condenser valves had closed after loss of AC and DC power. At 18:18, operators decided to open valves MO-2A and MO-3A on the possible chance that the valves in containment (MO-1A and MO-4A) had not fully closed.

At this point the operators inferred that the isolation condenser was functioning; this inference was based on indirect audible (i.e., steam generation was heard) and visual (i.e., a steam plume was observed) cues. The operators informed the onsite ERC that the isolation condenser was functioning. However, operators closed the condensate return valve (valve MO-3A in Figure 4.3) shortly thereafter (at 18:25). The reason for this action is unclear[15] and the onsite ERC was not informed that it had been taken.[16]

[15] The Investigation Committee (2011) and TEPCO (2012b, 2013) discuss possible reasons for this action. The reasons are not relevant to the present discussion, and so they are not described here.

[16] Valve MO-3A was opened again at 21:30.

By around 18:30 on March 11 the Unit 1 operators became convinced that the isolation condenser was not functioning. They recognized then that water injection into the reactor was the only option available to cool it. Preparations for injecting water into the Unit 1 reactor using the diesel-driven fire protection system (Figure 4.2) had already been under way for over an hour; these preparations were completed by 20:50. However, the reactor pressure vessel had to be depressurized first (by opening the safety relief valves; see Figure 4.2) before low-pressure water from the fire protection system could be injected.

The operators asked the onsite ERC to provide batteries so that the safety relief valves could be opened from the control room. However, the ERC team member who received this request did not understand its urgency, possibly because the ERC believed that the isolation condenser was still operating normally. In fact, the onsite ERC did not act on this request for several hours.

Miscommunications, combined with misleading water-level indicators in the reactor pressure vessel (e.g., at 21:19 the water level was shown to be 200 mm above the top of active fuel,[17] which was likely not the case[18]), caused the onsite and headquarters ERCs to continue to believe that the isolation condenser was operating. By about 22:00 on March 11, rising radiation levels were observed in the reactor, drywell, and turbine buildings, suggesting that fuel degradation and core damage were occurring.[19] By 23:50 the site superintendent and other onsite ERC personnel fully understood that the isolation condenser was not operating.

At approximately midnight on March 12, the Unit 1 operators began preparations for venting the containment (Figure 4.2). Operators consulted piping and instrumentation diagrams, valve drawings, and accident management procedures. These procedures assumed that power would be available for remote valve control; consequently, they were not applicable to the then-current situation in Unit 1. The operators needed to develop (in real time) a plan for venting the containment by manual valve operation. This required study of the layout and configuration of the vent valves to determine which valves needed to be opened, their locations, and whether and how they could be opened manually.

Operators confronted a number of additional obstacles for venting containment. These included a need to perform dry runs to keep field work time as short as possible (because of high radiation levels); the need

[17] Top of active fuel, usually denoted TAF, is the uppermost point in a fuel rod that contains uranium fuel. It serves as the reference point for water-level readings in the reactor.

[18] Reactor pressure vessel water-level sensors likely provided misleading values because of sensor degradation.

[19] TEPCO (2013, p. 11) suggests that water levels in the Unit 1 reactor dropped to the top of active fuel at about 18:10 on March 11 and that core damage was initiated at about 18:50.

to gather equipment (fireproof clothing, personal air supply, flashlights, full face masks); and the need to perform the work in shifts (three teams of two people) because the reactor building was pitch dark and radiation levels were high.[20] Team 1 completed its assigned task but teams 2 and 3 had to turn back because of high radiation levels. Venting was eventually performed from the control room after a compressor was procured and installed to enable remote operation of the large air-operated suppression chamber vent valve (see Figure 4.2). Because of these delays venting did not begin until 14:30 on March 12 when containment pressure had reached over 0.75 MPa (110 psig), almost twice the design value of 0.43 MPa (63 psig).

By 02:45 on March 12 the pressure in the reactor pressure vessel was determined to be near containment pressure[21]; freshwater injection was initiated at 05:46.[22] By this time, however, the fuel in the reactor had already been damaged and hydrogen and radioactive materials had likely already leaked into the reactor building. At 15:36 on March 12 a hydrogen explosion occurred on the refueling floor of the Unit 1 reactor building outside of containment. Further discussion of hydrogen generation and the explosion in Unit 1 is provided in Section 4.3.5.

4.3.1.1 Discussion

The isolation condenser in Unit 1 most likely lost its ability to effectively cool the reactor when AC and DC power were lost.[23] However, it was not until approximately 3 hours later (at 18:30) that operators in the Unit 1 control room fully understood that the isolation condenser was not functioning effectively. It took the onsite ERC staff even longer—until about 23:50—to fully understand this fact.

---

[20] The three teams consisted of shift supervisors, deputy managers, and older workers. Younger workers were not permitted to participate because of the danger involved even though they volunteered to do so.

[21] It is not clear whether depressurization occurred because of damage to the reactor pressure vessel, a pipe break, or safety relief valves that had stuck open due to thermal fatigue failure.

[22] Only a fraction of the water injected using the fire truck pumps appears to have reached the reactor. Water may have been lost from leaky fire hoses, open valves, and branches in the piping system that diverted water. See TEPCO (2013, Attachment 1-4) for additional details.

[23] TEPCO has concluded that the valves on the System A isolation condenser did not close fully because some water was lost from the Train A tank; it was measured to be 65 percent full in a post-accident inspection, a decline from the previous, and normal, level of 80 percent. However, as noted by TEPCO (2012b, p. 197), since a substantial amount or water remained in the shell side of the isolation condenser, the amount of heat removal during the accident must have been limited. The Investigation Committee (2011, p. 121) also supports this observation.

In hindsight, shutdown of the isolation condenser was an unanticipated side effect of the design of the failsafe control logic circuit that operates the isolation condenser valves. This circuit is powered by instrumentation DC. If this power is lost, the logic circuit acts as if there were a pipe break in the isolation condenser system and commands all four of its valves to close (see Figure 4.3).

Whether the valves actually close, however, depends on the timing of power loss to three circuits:

- instrumentation DC, which powers the logic circuit,
- 125VDC, which opens and closes the two valves outside containment,[24]
- AC, which opens and closes the two valves inside containment,

as well as the time required to close the valves (20-30 s) once the actuation signals are received by them (Investigation Committee, 2011, p. 118).

The two valves inside containment (i.e., valves MO-1A and MO-4A in Figure 4.3) are of greatest concern for operator control of the isolation condenser because they are not physically accessible. Consequently, once closed, without AC power they cannot be reopened by operators. Based on currently available information (see Footnote 23), it appears that the two valves inside containment received enough AC power to close most of the way, indicating that instrumentation DC power failed first (Craig Sawyer, General Electric (retired), written communication, January 14, 2014). However, the status of the valves inside containment will not be known for certain until they can be inspected, which will require physical entry into containment.

Communications difficulties between operators and onsite ERC may have delayed recovery efforts. As noted previously, they did not communicate effectively about the operation of the isolation condenser. Additionally, the apparent miscommunication between operators and onsite ERC about the urgency of supplying batteries for opening the safety relief valves quite possibly led to delays in depressurizing the reactor pressure vessel.

TEPCO has argued that efforts to vent and set up alternative water sources were initiated in spite of these communication problems. Indeed, the site superintendent and onsite ERC initiated actions to identify alternative water injection means early in the accident.[25] However, the severe conditions at the plant apparently prevented a faster response.

There is some suggestion of lack of clarity in roles and responsibilities

---

[24] If 125VDC is available, it can be routed through inverters to produce AC power to operate the valves. However, such power was not available in this case because of flooding.

[25] The site superintendent directed the onsite ERC staff to develop plans for alternative water injection as early as 17:12 on March 11.

within the onsite ERC, particularly with respect to allocating responsibilities for responding to situations that are not covered by accident management procedures. This led to delays, for example, in developing and implementing the procedure for using fire engines to inject water into the reactor pressure vessel through the fire protection system. Preparations for this procedure (e.g., verifying the availability of fire engines, locating water discharge ports, positioning the fire engines, and laying fire hoses) did not get under way until dawn on March 12.

4.3.2 Unit 3 Reactor

Unit 3 did not lose DC power immediately after the tsunami. Consequently, until its batteries became depleted, operators were able to monitor plant indicators from the control room, including reactor pressure and water levels. They were also able to activate, monitor, and control the reactor core isolation cooling and high-pressure coolant injection systems (Figure 4.4).

Unit 3 operators activated the reactor core isolation cooling system at around 16:00 on March 11. They also cut unnecessary loads on the battery to extend its life. Steam exhaust from the reactor core isolation cooling system to the suppression chamber raised its pressure, prompting operators to initiate suppression chamber spray cooling using the diesel-driven fire protection system (Figure 4.4).

At around 11:36 on March 12, after running for approximately 20 hours, the reactor core isolation cooling system stopped and could not be restarted.[26] The safety relief valves cycled to control reactor pressures; as a result, water levels in the reactor pressure vessel dropped and the high-pressure coolant injection system started automatically at 12:35.

The high-pressure coolant injection system was aligned in full-flow test mode with almost all of the pump flow going back to the suppression pool; only enough flow was directed to the reactor pressure vessel to maintain water levels.[27] This avoided the constant starts and stops that would have occurred had all of the flow been directed to the reactor pressure vessel. However, in this mode of operation, the system turbine consumes enough steam to depressurize the reactor pressure vessel. The turbine had to be shut down when pressures decreased to the point where operators became concerned about the turbine's ability to work without being damaged.

Starting at approximately 20:36 on March 12, operators could no longer monitor water level in the Unit 3 reactor because the 24VDC power

[26] The root cause of the failure of the reactor core isolation cooling system is not known.

[27] This mode of operation was unusual but effective and showed creativity on the part of the operators.

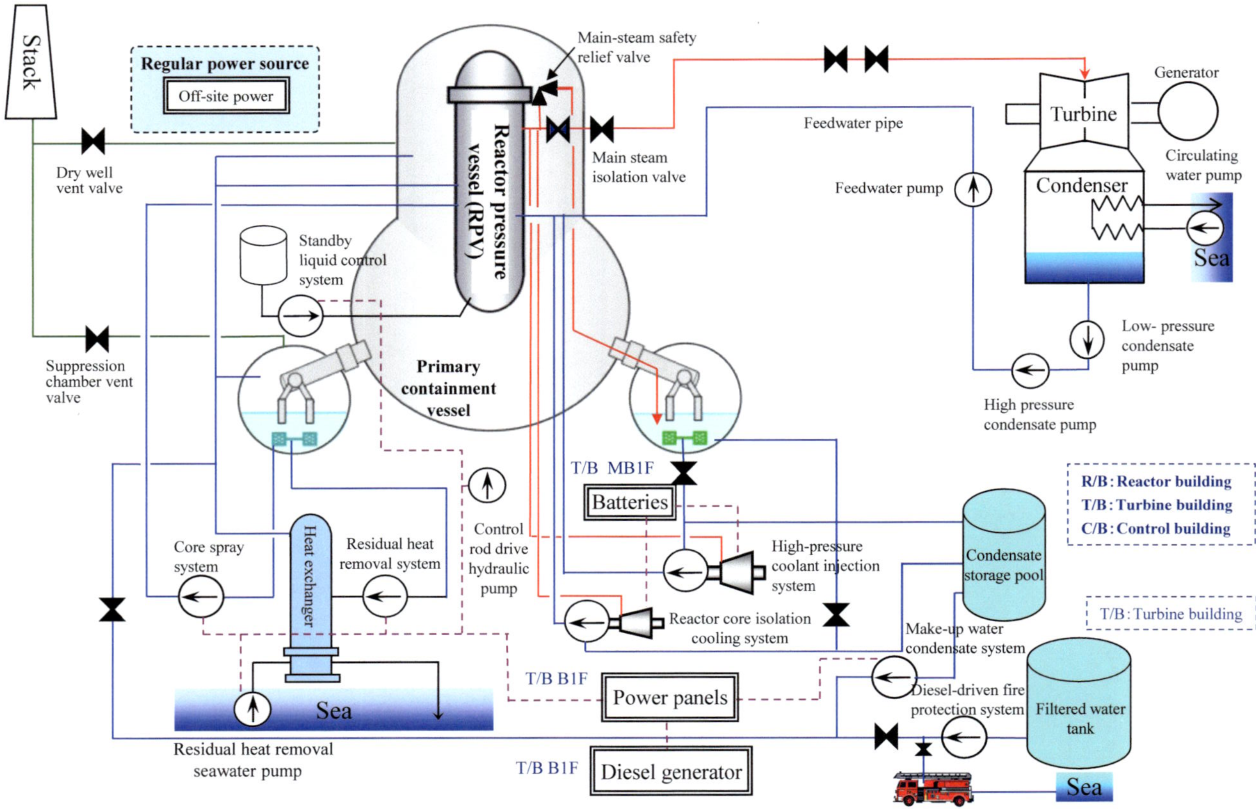

FIGURE 4.4 Schematic illustration of major safety systems in Units 2 and 3 of the Fukushima Daiichi plant. SOURCE: Courtesy of TEPCO.

source for the water-level gauge became depleted. Operators became concerned about the continued availability of the high-pressure coolant injection system,[28] so they developed an alternative plan for water injection into the reactor. This plan involved

- Shutting down the high-pressure coolant injection system;
- Depressurizing the reactor pressure vessel using the safety relief valves, which would vent steam from the reactor pressure vessel to the suppression pool; and
- Injecting water into the reactor pressure vessel using the diesel-driven fire protection system (Figure 4.4).

The Unit 3 operators informed the members of its team in the onsite ERC of their plan. The team members concurred with the plan but failed to communicate it to the site superintendent and other ERC staff.

Operators stopped the high-pressure coolant injection system at 02:45 on March 13 and switched the fire protection system from suppression chamber spray cooling to reactor pressure vessel injection. Operators attempted but were unable to open the safety relief valves, either because the pressure in the reactor pressure vessel was too low[29] or because the batteries were depleted.[30] Attempts to restart the high-pressure coolant injection system also failed probably because the battery was depleted.

The operators informed their team members in the onsite ERC that they had stopped the high-pressure coolant injection system but were unable to open the safety relief valves; this information was not immediately passed on to the site superintendent and other staff in the onsite ERC. It was not until 03:55 on March 13 that the site superintendent and the headquarters ERC learned these facts.[31]

The onsite ERC immediately recognized the need to obtain batteries to operate the safety relief valves and fire engines to inject water into the reac-

---

[28] The operators were concerned specifically about the potential for a steam leak resulting from damage to the high-pressure coolant injection system caused by excessively low speed of the turbine.

[29] The safety relief valves can be manually opened by remote control only if the pressure in the reactor pressure vessel is over 0.686 MPa (gauge); the valves close on their own at 0.35 MPa.

[30] Different trains of battery-powered 125V sources powered the reactor core isolation cooling system, high-pressure coolant injection system, and safety relief valves. The battery may have been depleted from 34 hours of use since the start of the accident.

[31] There was a further miscommunication that led the site superintendent and headquarters ERC to initially believe that the high-pressure coolant injection system had stopped automatically. Because the onsite ERC was so noisy the Japanese word for "manually" was misheard as "automatically." The high-pressure coolant injection system is designed to turn on and off automatically to control pressure and water level. Consequently, automatic stopping would not necessarily be a cause for concern.

tor through the fire protection system (Figure 4.4). Plant personnel salvaged batteries from personal automobiles; it took almost 2 hours to collect them and another hour to connect them to the Unit 3 control panel.[32] Personnel also repaired an onsite road so that the fire brigade could drive a fire truck to the dock by Unit 3.[33]

According to TEPCO (2013, p. 38), reactor pressure reached about 7 MPa (abs) at about 04:30. Reactor pressure then decreased abruptly to below 1 MPa (abs) at about 09:00. The reason for this decrease is not understood at present. Depressurization enabled the injection of freshwater starting at 09:25. By that time, however, the reactor core had been without cooling for over 6 hours[34] and was probably already damaged.

The Unit 3 operators and onsite ERC made preparations for venting the containment through the suppression chamber vent valve (Figure 4.4). This valve is operated with compressed air. These preparations were completed at about 08:41 on March 13. At about 09:24, a drop in drywell pressure was noted, leading the onsite ERC to assume that venting had initiated around 09:20.

At approximately 12:20 on March 13 the store of freshwater used for injection ran out so workers proceeded to hook up a previously constructed seawater injection line. Seawater injection began at 13:12 on March 13 but was interrupted when the Unit 3 reactor building exploded at 11:01 on March 14. Seawater injection was restarted at 15:30. As in the case for Unit 1, only a fraction of the injected water appears to have reached the reactor.

The Unit 3 reactor was cooled by water injection for approximately 30 hours, but cooling was inadequate. There may have been insufficient compressed air pressure[35] and capacity to keep the suppression chamber vent valve open; consequently, pressure in the containment stayed too high to allow the reactor pressure vessel to be depressurized. The high pressure reduced water injection flow rates into the reactor pressure vessel and likely caused hydrogen and fission products to leak from the containment into the reactor building.

A hydrogen explosion occurred in the Unit 3 reactor building at 11:01 on March 14. The explosion caused severe structural damage to the reactor building and destroyed the fire engine and hoses being used to inject water into the reactor. The explosion also prompted the evacuation of field personnel and delayed recovery efforts.

---

[32] Executed while wearing full protective suits and face masks and using flashlights and improvised tools.

[33] Other fire trucks on the plant site were injecting seawater into Unit 1.

[34] The plant superintendent had done a hand calculation at 06:00 on March 13 which showed that the top of active fuel was likely reached at 04:00 on March 13.

[35] First air tanks and then compressors taken from contractor warehouses were used for this purpose.

It is questionable whether the operators' plan for injecting water would have worked even had they succeeded in their initial efforts to open the safety relief valves: the maximum pressure output of the pump in the fire protection system (0.45-0.60 MPa) was likely insufficient to overcome the rapidly climbing pressure in the reactor pressure vessel. The Investigation Committee (2011, p. 221) noted that a

> hasty conclusion should be avoided about whether or not the damage of Unit 3 could have been prevented or mitigated by depressurization and/or earlier alternative water injection because there were many uncertain factors. . . . It could be presumed that, however, if depressurization of Unit 3 had been performed much earlier than it actually had and the alternative method of water injection using fire engines had been conducted smoothly, the progress of core damage might have been slower, radiation dose in the RPV [reactor pressure vessel] would have been less and subsequent work might have been easier.

4.3.2.1 Discussion

Several factors contributed to the severity of the accident in Unit 3:

1. The Unit 3 operators apparently did not assess the viability of their alternative water injection plan before turning off the high-pressure coolant injection system. Once the system was stopped, the operators would have had less than 2 hours to initiate water injection into the reactor before initiation of core damage. The operators turned off the system before working water injection sources and means for depressurizing the reactor pressure vessel and venting the containment were in place.
2. The operators informed team members in the onsite ERC of their plan. However, the team members did not pass this information on to the site superintendent and other ERC staff. Consequently, the onsite ERC was unable to check the plan's viability.
3. The reason for stopping the high-pressure coolant injection system was miscommunicated to the onsite ERC. The onsite ERC did not learn that the Unit 3 operators had manually stopped the system until about an hour later.
4. Work on constructing an alternative water injection line was not started until the onsite ERC learned that the high-pressure coolant injection system had been stopped.[36]

[36] According to the Investigation Committee (2011, p. 218), a major contributor to delays in establishing water injection via fire engines was the fact that the staff organizational structure within the onsite ERC did not effectively support activities that were not explicitly called out in accident management procedures.

5. The hydrogen explosion in Unit 1 further complicated accident management in Unit 3. TEPCO (2012b) reported that the onsite ERC became fearful when this event, which was not initially understood to be an explosion, occurred. The explosion damaged cables that were prepared to recover power in Unit 3.

4.3.3 Unit 2 Reactor

The tsunami flooded the Unit 2 emergency diesel generators, power panels, and batteries, cutting off all AC and DC power to the unit.[37] Control room lighting, alarms, plant indicators, and controls were lost as a result. Fortunately, the Unit 2 operators had started the reactor core isolation cooling system (Figure 4.4) just before power was lost. As a consequence, the system's isolation valves failed in the "as is" position (i.e., open) thereby allowing it to continue to operate.[38] The system functioned for almost 3 days, although operators could not monitor or control it.[39]

Operators were unable to verify visually that the system was operating because it was not physically accessible.[40] They were also unsure of the water level in the reactor pressure vessel. As a consequence, the onsite ERC staff and site superintendent initially doubted that the system was operating and believed that Unit 2 was in more serious difficulty than Unit 1.

The tsunami also flooded the diesel-driven pump for the fire protection system.[41] The onsite ERC requested that the Unit 2 operators locate the fire protection system connections, which are accessible from outside the reactor building, so that water from the makeup-water condensate system (Figure 4.4) could be injected into the reactor pressure vessel using fire trucks. The operators also had to line up the system valves manually (they are normally motor operated) by entering the reactor building; they had great difficulty accomplishing this operation because of a lack of knowledge about the location of the valves, missing keys for locked doors, and the

---

[37] Flooding completely submerged the large 120V battery system that supplied DC power to the high-pressure coolant injection system. Seawater pumps used for reactor heat removal and containment cooling were also unavailable.

[38] It appears that flooding by the tsunami caused the loss of actuation power in the reactor core isolation cooling system's isolation valves before the activation of the interlocking logic circuit function. Evidently, all of the isolation valves remained open so the system was able to maintain its cooling capability after AC and DC power was lost. More details on the reactor core isolation cooling system and failsafe control logic are provided in Section 2.2.3.2 in Chapter 2.

[39] Specifically, operators could not monitor or control the rate at which the system delivered water to the reactor pressure vessel.

[40] The system is located in a basement room in the reactor building. Flood waters prevented access to this room.

[41] The pump was located in the flooded basement of the turbine building.

physical effort required to turn the valve wheels. Nevertheless, this operation was completed in Unit 2 late on March 11 after it was first performed in Unit 1. The early timing of this operation was fortuitous because the reactor buildings later became too contaminated for extended entry.

At about 22:00 on March 11, workers entered the Unit 2 reactor building to manually read the reactor pressure vessel water level; they found it was 3,400 mm above the top of active fuel. At this point the onsite ERC staff realized that the reactor core isolation cooling system must be functioning; the ERC's focus then shifted away from providing emergency injection water for Unit 2. By 23:35, operators obtained further indirect confirmation that the reactor core isolation cooling system was functioning when they were able to connect emergency power to a drywell pressure gauge; the gauge reading was 0.14 MPa (abs), as expected for normal operation of the reactor core isolation cooling system.[42]

During the early hours of March 12, workers visited the equipment room in the basement of the Unit 2 reactor building to check on the status of the system.[43] They were equipped with self-contained air breathing sets, small flashlights, and rubber boots. Their attempt to confirm system operation was unsuccessful. However, at approximately 02:55 on March 12 the shift supervisor reported to the onsite ERC that he believed the system was operating because the system pump discharge was higher than the pressure in the reactor. At this point the site superintendent decided to give priority to Unit 1 containment venting.[44]

Operators noticed a decrease in water level of the condensate storage tank (Figure 4.4), which was then being used as the water source for the reactor core isolation cooling system. Operators switched the system water supply from the tank to the suppression pool at around 04:00 on March 12. Operators again checked on the operation of the system during this changeover. However, the operation of the system was not checked again until about 2 days later (at 04:30 on March 14). By that time the pressure

[42] The exhaust from the reactor core isolation cooling system turbine is discharged below water level in the suppression pool. After the steam condensing capacity of the water in the suppression pool is partially lost, pressure in the upper space of the suppression chamber would start to rise and in turn cause the drywell pressure to increase.

[43] TEPCO has not confirmed whether there were any attempts to check the status of the system before about 02:00 on March 12.

[44] TEPCO upper management was given information that led it to believe that the Unit 1 isolation condenser was working and that the Unit 2 reactor core isolation cooling system was not. This created substantial misunderstandings between the plant regulator (Nuclear and Industrial Safety Agency [NISA]), which understood that the Unit 2 system was functional, and TEPCO upper management. These misunderstandings were manifested in announcements to the press about venting on the morning of March 12. This is one of many miscommunications during the response to the accident.

in containment had reached 0.4 MPa (abs), approaching its design pressure of 0.48 MPa (abs).

Somewhat surprisingly, the pressure in Unit 2 increased much less rapidly than in Unit 3, suggesting that there was either a leak in the Unit 2 containment or an unusually effective cooling mode, such as external cooling of the suppression pool due to flooding of the torus room. Analyses using MAAP and MELCOR (see Section 4.3.5) by groups in the United States and Japan have been used to support both hypotheses, alone or in combination.

The reactor core isolation cooling system in Unit 2 continued to operate until about 13:30 on March 14.[45] After the system stopped, the safety relief valves operated mechanically to vent steam from the reactor pressure vessel to the suppression pool. Steam loss from the reactor pressure vessel caused its water levels to drop continuously for the next 5 hours.

At 17:45 on March 14, a safety relief valve was manually actuated to depressurize the reactor pressure vessel. This reduced pressure in the reactor pressure vessel from 7.8 MPa to 0.7 MPa within 45 minutes. Depressurization also discharged a substantial fraction of the reactor pressure vessel's water inventory to the suppression pool, uncovering the reactor core. After the valve was closed the reactor immediately started to repressurize. The safety relief valves were manually actuated several times over the next 10 hours[46] to keep reactor pressure sufficiently low for water injection to be effective.

Preparations for water injection had begun on March 11. Valves in the fire protection system had been aligned to make it possible to inject water into the reactor pressure vessel using fire trucks (Figure 4.4) once the reactor had been depressurized. However, fire trucks were not hooked up to the Unit 2 fire protection system until after 23:00 on March 14. Prior to that time, all available fire trucks were being used at Units 1 and 3. Moreover, there was not sufficient space in the valve backwash pit (from which seawater was being pumped) to place another suction line.

Additional fire trucks and staff arrived at the site at 05:00 on March 14, and a pumping path from a valve backwash pit was established. However,

---

[45] The reactor core isolation cooling system can, under ideal circumstances, inject at full reactor pressure, but it was unable to do so after noon on the March 14. There is evidence (Investigation Committee, 2011, pp. 258-259) that pressure increased above the discharge pressure of the reactor isolation cooling system pump at about noon on March 14. At this point the flow of cooling water into the reactor would have stopped.

[46] The core remained either completely or partially uncovered during this time according to reactor accident simulations. Because of uncertainties in injected water flow rates, the amount of reactor cooling during this time period is poorly understood.

the fire trucks were placed on standby while seawater in the valve backwash pit was being used to supply Unit 3.[47]

Operators and onsite ERC staff also began making preparations on the evening of March 11 for depressurizing the Unit 2 containment. They had previously vented the containment in Unit 1 through its air-operated suppression chamber vent valve (Figure 4.2). They knew that a powerful compressor and DC batteries would be needed to open this vent in Unit 2. The site superintendent instructed staff to complete preparations for venting Unit 2 by 17:30 on March 12.

Workers located and connected air cylinders that could be used to operate the suppression chamber vent valve. An emergency generator in the control room was used to energize the air solenoid. By 23:00 on March 13 all the valves had been prepared for actuation or actuated; only the rupture disk needed to be broken to vent the unit. The rupture disk was set to break at 0.528 MPa (abs). According to the drywell pressure readings, which continued to increase, it was presumed that the valves failed to stay open. Venting on March 13 or 14 was apparently never successfully accomplished from either the suppression chamber or drywell in Unit 2 (Investigation Committee, 2011, p. 266).

The Prime Minister's office, Nuclear Safety Commission, and TEPCO upper management became concerned about the delays in venting the containment in Unit 2. TEPCO's president ordered the site superintendent to depressurize the reactor pressure vessel without waiting to vent containment. The site superintendent accepted the president's directive and gave instructions to start venting and water injection into the Unit 2 reactor pressure vessel while concurrently continuing preparations for containment venting.[48]

Operators struggled to depressurize the reactor pressure vessel. They had trouble opening the safety relief valves and keeping them open once they were actuated. Additionally, steam exhausted to the suppression pool

---

[47] In one of many miscommunications, NISA staff in Tokyo became impatient because Unit 2 was not being cooled with seawater as ordered by METI Minister Kaieda. NISA staff had not realized that the pumps were unable to take water directly from the ocean (pump suction was not adequate to lift water directly from the ocean, which was more than 10 m below the fire engine inlet) and that the supply of seawater was a limitation (Investigation Committee, 2011, p. 230).

[48] Site Superintendent Yoshida's reaction is apparently visible on the video tapes of teleconferences that have been released to the media (*Asahi Shimbun*, August 12, 2012): "The videos did show Yoshida on that date frustrated at one point with questioning and advice from TEPCO officials and asked them to let him have his own way to vent the core in the No. 2 reactor to reduce mounting pressure." According to an August 7, 2012, article in *Asahi Simbun*, Yoshida told TEPCO headquarters during the video conference "Don't ask us any questions," he says. "Don't disturb us, because we are now in the middle of trying to open the vent for the containment vessel."

from the reactor did not condense efficiently because pool temperatures were high. Water injection became possible around 19:00 when the reactor pressure was 0.630 MPa (gauge).[49] However, the fuel in the reactor had likely been uncovered for some time prior to water injection.

Although the pressure gauges were probably unreliable, the drywell pressure at one point was indicated to be as much as 0.85 MPa (abs), more than twice its design pressure. Reactor pressure was in excess of the maximum fire truck pump head (about 1 MPa; TEPCO, 2013, Attachment 1-4), for substantial periods of time, preventing seawater injection from taking place and placing the containment vessel under significant thermal and pressure stresses.

The operators and onsite ERC struggled through the evening of March 14 and early morning of March 15 to vent containment. Their attempts to open both small and large vent valves in the suppression chamber were unsuccessful.[50] Attempts to vent the drywell (Figure 4.4) were also unsuccessful.

Pressure in containment remained high until early on March 15 (06:14) when the explosion in the Unit 4 reactor building occurred.[51] Because of the evacuation and the confusion following the explosion, it took some time to understand what had happened. There were apparently two events that occurred on the morning of March 15: (1) a hydrogen explosion in Unit 4 and (2) and a loud noise accompanied by an apparent drop in the suppression chamber pressure in Unit 2. This noise was subsequently determined by TEPCO to be associated with the hydrogen explosion in Unit 4. The drop in suppression chamber pressure appears to be an instrument malfunction (the drywell pressure did not drop and remained above atmospheric for some time). The actual condition of the Unit 2 containment system is still unknown.

Once the Unit 2 reactor core isolation cooling system shut down, measures to cool the Unit 2 reactor were ineffective. The Unit 2 containment was never deliberately vented and it is unclear how or when depressurization occurred. Some fraction of the core is almost certainly highly degraded but the amount and location of core material is unknown. Also unknown are the timing, mechanisms, and magnitude of releases of fission products from Unit 2, although it appears likely that fission products did leak from containment during the period of time that it exceeded its design pressure.

---

[49] TEPCO (2013, p. 21) suggests that not all of the injected water reached the reactor but instead went to other systems and equipment.

[50] High radiation levels prevented workers from entering the reactor building to hook up alternative air sources to the vent valve, for example.

[51] At that point about 600 staff evacuated to Fukushima Daini (50 staff remained at the Fukushima Daiichi plant); the staff that evacuated did not return until the morning of March 16.

4.3.3.1 Discussion

Several factors contributed to the severity of the accident in Unit 2:

- The onsite ERC staff were trying to manage responses at multiple reactor units, which taxed their ability to maintain awareness of the rapidly changing conditions at Unit 2 and appropriately prioritize and direct response activities. ERC staff were occupied with Unit 1 through the morning and afternoon of March 12. Staff attention then focused on the Unit 3 reactor. The hydrogen explosion in the Unit 3 reactor building on March 14 caused extensive damage to the site and temporarily halted response activities at Unit 2.
- The Unit 2 operators had to depressurize the reactor pressure vessel and vent containment to enable injection of low-pressure cooling water.[52] Venting of the Unit 2 containment was difficult to implement on an ad hoc basis: emergency air supplies were inadequate; the torus room environment was too hot, humid, and contaminated for the staff to manually operate the suppression chamber vent valves; and the rupture disks were designed to operate at higher containment overpressures than were achieved and could not be bypassed. The hydrogen explosion in the Unit 3 reactor building further impeded efforts to vent the Unit 2 containment.
- Water injection methods and alternative water supplies were limited. Water injection into the Unit 2 reactor came too late to prevent core damage.
- Miscommunication between the onsite ERC, headquarters ERC, and NISA contributed to misunderstandings and lack of confidence by the Prime Minister's office in TEPCO's ability to manage the accident.

The transition from installed cooling equipment (e.g., the reactor core isolation cooling system and high-pressure coolant injection system) to ad hoc cooling measures (i.e., injection of low-pressure water) was not carefully orchestrated in Unit 2. This transition requires

- Timely depressurization of the reactor pressure vessel and venting of containment,
- Maintenance of the reactor and containment at low pressures after initial depressurization, and
- Provision of an adequate and reliable water supply and sufficiently high injection pressures.

---

[52] The fire truck pump pressure heads were not high enough to inject water into the reactor pressure vessel without first depressurizing it. This also required venting of the containment. The pressure in the reactor pressure vessel and contanment had to be less than the pressure of the fire truck pump minus the pressure losses in the water piping system. See also footnote 22.

Coordination of depressurization and low-pressure water injection proved impossible to accomplish under the conditions at the plant following the tsunami, even with advance planning and some on-the-ground experience with depressurization of the Unit 1 and Unit 3 reactors. Only a few hours separated success (i.e., timely depressurization and water injection) from failure (core damage due to the rapid boiloff of the water once cooling systems stop). More time was available to achieve success in preventing the release of fission product aerosols from containment; however, this success was only partial because the delays in venting containments allowed them to spend long periods of time above their design pressures, causing substantial releases into the reactor buildings and the environment. *Indeed, the events at the Fukushima Daiichi plant demonstrate the extraordinary difficulty of executing a successful response to accidents involving multiple reactor units under the difficult conditions that existed at the site.*

4.3.4 Unit 5 and 6 Reactors

Both Units 5 and 6 were shut down at the time of the earthquake and remained so thereafter. The units experienced thermal transients following the earthquake when active cooling was temporarily lost,[53] as described below. The reactors were eventually brought to cold shutdown without damage to the fuel in the core or spent fuel pool. The events at these reactors would likely have been a major story in the annals of nuclear safety had they not occurred in the shadow of the accidents in Units 1-4. In fact, the events at Unit 5 have important implications for safety risks during reactor shutdown conditions.

Unit 5 was undergoing pressure testing[54] on the day the earthquake occurred. The unit lost all AC power due to flooding of seawater pumps and power panels. As a consequence, lighting and AC-powered instrumentation on the Unit 5 side of the common control room were inoperable; the Unit 5 side of the control room went dark after the batteries for emergency lighting were depleted.

The containment in Unit 5 was open so that visual inspections for leaks could be carried out during pressure testing. The MSIV apparently was also open. A number of safety systems were either unavailable (i.e., reactor core isolation cooling system, high-pressure coolant injection system; residual heat removal system) and some other safety features (e.g., the automatic depressurization system) had been disabled for pressure testing.

[53] Although both reactors were shut down, they still required active cooling to remove decay heat from the fuel in their cores; see Chapter 2.

[54] The Unit 5 reactor vessel was being pressure tested to 7 MPa, but the temperature was only 90°C.

The safety relief valves had been deactivated by pulling circuit breakers and depressurizing nitrogen lines. Additionally, as part of the pressure test, the low-setpoint safety relief valves had the manual operation locked out by inserting tools into the mechanisms. However, the high-setpoint safety relief valves could still be actuated once nitrogen and power were restored.

The shift supervisor for Unit 5 appears to have taken a strong role in managing the response to the flooding and loss of power because the onsite ERC was occupied with other units at the plant. Nevertheless, the Unit 5 operators and onsite ERC worked together to identify a strategy for depressurizing the reactor. This strategy was identified by 05:00 on March 12 and implemented about an hour later. It involved brute-force prying open of the nitrogen supply line to the vent valve on top of the reactor head from outside the containment.[55]

Workers apparently had to enter the reactor building or containment to connect an ad hoc nitrogen supply line that could be used to activate the safety relief valve that was ultimately used to maintain the reactor pressure at desired levels (Investigation Committee, 2012, p. 110).

A 125V battery that was being used to power some monitoring equipment was depleted at 01:00 on March 12.[56] The operators used another gauge with independent power to read the reactor water level until the AC crossties to Unit 6 (described below) were completed. A 250V battery was depleted at about 17:00 on March 12, shutting down additional monitoring equipment including the process computer.

An air-cooled emergency diesel generator in Unit 6 survived the tsunami. It was cross-tied to Unit 5 using a cable that had been prepared as part of the unit's accident management strategy. The generator was then able to supply emergency AC power to both units. This was a key element of the successful outcome for these two units relative to Units 1-3. (This was also key evidence that advanced preparation, including advanced positioning of portable equipment, enabled a more effective response than was possible for Units 1-3.) Power was restored to Unit 5 at around 05:00 on March 12, allowing reactor parameters to be read in the main control room.

The Unit 5 operators had decided to use the makeup water condensate system as an alternative cooling method but needed AC power to operate its pump. The laying of the power cables for the pump required cooperation between the electrical contractors and onsite ERC recovery teams. There was also cooperation in laying power cables for instrumentation. The

---

[55] The shift supervisor judged that it was too dangerous for workers to enter the containment vessel because of aftershocks and lack of lighting (Investigation Committee, 2012, p. 107).

[56] Roughly in agreement with 8-hour coping time that is possible with DC power.

Investigation Committee (2012, p. 112) noted that "Since Units 5 and 6 were undergoing routine inspections, a number of cables were stored in the warehouse of contractors and were used for this task. Regarding the task of interconnecting the power, four members of the ERC Recovery Team laid and connected about 70 m length of cables."

Some personnel from Units 5 and 6 had been sent to other units to support the accident response. They were recalled on March 16 to restore the residual heat removal system by installing submersible seawater pumps and using portable diesel-driven generators to supply electric power. Cold shutdown of the Unit 5 reactor was achieved on March 20.

The response at Unit 6 was more straightforward given the continued availability of AC power following the earthquake and tsunami. The Unit 6 containment was closed but its vent line was open. Water was supplied to the reactor pressure vessel and the spent fuel pool on a reliable basis from March 13 onward. Cold shutdown was achieved on March 20.

4.3.4.1 Discussion

TEPCO attributed the successful achievement of cold shutdown in Unit 5 to close cooperation between the onsite ERC and operators, the early restoration of monitoring instruments, reactor depressurization, coolant supply via seawater, and the restoration of AC power. TEPCO noted that the response utilized concepts learned via training and work experience. That is, TEPCO's accident management abilities were sufficient for the challenges at these units.

In fact, it took several ad hoc measures (such as brute-force opening of the nitrogen line and laying of cables) and fortuitous conditions (such as the working emergency diesel generator in Unit 6 and low decay heat in the Unit 5 reactor) to bring Unit 5 to cold shutdown. Although the report by TEPCO (2012b) minimizes or does not mention many of these measures, it is clear that a successful response might not have been mounted without them. It is also clear that emergencies in other units at the plant impacted the timeliness of response in Unit 5.

The response in Unit 5 points out the need once again for specialized training and appropriate prepositioned equipment (i.e., power cables in this case) to carry out ad hoc solutions that require going beyond prepared measures. The challenges encountered during the Unit 5 response demonstrate the importance of developing severe accident management guidelines (SAMG; see Chapter 5 and Appendix H) specifically for reactors in shutdown or maintenance conditions, that is, SSAMG.

4.3.5 Hydrogen Explosions

Perhaps the most conspicuous and dramatic aspects of the Fukushima Daiichi accident were the hydrogen explosions in the Unit 1, 3, and 4 reactor buildings. These explosions took place at +24.8 hours, +68.2 hours, and about +87.2 hours, respectively (Table 4.1). A hydrogen explosion did not occur in Unit 2, although there were precursor conditions similar to those in Units 1 and 3.

There is a good general understanding of the basic chemical processes that generated the hydrogen that led to these explosions (see Sidebar 4.1); however, there is substantial uncertainty about how hydrogen leaked from reactor containments into the buildings and the specific conditions that led to its ignition. Although no measurements of hydrogen concentrations in Units 1-3 were made, it is likely that large quantities of hydrogen were produced because the reactor cores in these units were not actively cooled for long periods: about 14 hours in Unit 1, 6 hours in Unit 3, and between 5 and 15 hours in Unit 2 (Phillips et al., 2012). Further forensic studies and analyses of hydrogen distribution, combustion, and structural damage to the units are needed to improve the understanding of these explosion events.

Video recordings of the explosions in Units 1 and 3 and the visible damage in Units 1, 3, and 4 indicate that significant overpressures consistent with hydrogen combustion occurred. At the present time it is not known if the combustion events were deflagration, detonation (see Sidebar 4.1 for definitions of these terms), or more complex events. The explosions caused extensive damage to the reactor buildings (Figure 4.5), opening up an easier path for the direct release of radioactive materials to the atmosphere and spreading contaminated debris inside and around the units.

The explosions also had a significant impact on the accident response: they injured workers; destroyed equipment and temporary water-line and power cables; prompted evacuations to onsite buildings or offsite facilities that slowed and in some cases halted recovery work; and created a general atmosphere of fear at the plant, throughout Japan, and in other parts of the world. The extensive damage and contamination were totally unexpected by the operators at the Fukushima Daiichi plant and in the view of two U.S. safety experts was a "game changing" event[57] in the accident.

The current understanding of the accident progression in the Fukushima Daiichi reactors depends substantially on computer simulations of the accident. These simulations are used to make predictions about the state

---

[57] The event was "game changing" because it made already difficult access to reactor buildings essentially impossible and inhibited travel throughout the plant, hampering accident recovery and mitigation measures (Jeff Gabor and Doug True, Erin Engineering, communications with the committee on April 23, 2013, and February 11, 2014).

SIDEBAR 4.1
Hydrogen Generation and Combustion

Hydrogen is generated in a reactor when zirconium in the fuel cladding reacts with steam at elevated temperatures:

$$2H_2O + Zr \rightarrow 2H_2 + ZrO_2.$$

This reaction is highly exothermic, releasing 5.6 megajoules per kilogram of zirconium. The reaction heat increases zirconium temperatures, accelerating the reaction and generation rate of hydrogen (Lee and McCormick, 2011, pp. 266-267). This reaction can become self-sustaining at high enough temperatures. The USNRC limit of 1,204°C (2,200°F) during accident conditions was established (in 10 CFR § 50.46) in part to address concerns of runaway oxidation above that temperature (Hache and Chung, 2001).

Once a significant amount of hydrogen is released, a risk of explosion exists because hydrogen–air–steam mixtures are flammable over a wide range of compositions (Camp et al., 1983) and are easily ignited by sparks and hot surfaces (Gelfand et al., 2012). Combustion of hydrogen and air,

$$H_2 + \tfrac{1}{2}O_2 \rightarrow H_2O,$$

releases 120 megajoules per kilogram of H_2 burned and produces hot steam. Combustion of hydrogen in confined spaces can generate severe overpressures leading to structural failures of confining structures. This failure process was observed in Units 1, 3, and 4 at the Fukushima Daiichi nuclear plant.

A hydrogen–air–steam mixture will ignite and burn when its composition is within a critical range, illustrated in Figure S4.1. If there is too much or too little hydrogen or too much steam, there will not be enough energy to sustain combustion; in such cases the mixture is said to be *nonflammable.* As the amount of steam decreases, the mixture enters the *flammable* range and combustion will occur if a sufficiently strong ignition source is available.

Combustion starts at the ignition source and propagates through the mixture as a chemical reaction wave. If this wave moves through the mixture at less than the speed of sound then combustion is said to be a *deflagration.* If the wave moves through the mixture faster than the speed of sound, then combustion is said to be a *detonation.* The more general term *explosion* encompasses both deflagration and detonation combustion.

Hydrogen deflagration results in much lower pressures and less structural damage to confining structures than detonations; the latter produce very damaging overpressures and are likely to result in structural failures. A deflagration can, under certain circumstances (discussed in NEA, 2000), accelerate and transition to a detonation wave. Regardless of combustion speed, hydrogen explosions can be very destructive when large volumes of combustible gas within confining structures are involved, as was the case for Units 1, 3, and 4 at the Fukushima Daiichi nuclear plant.

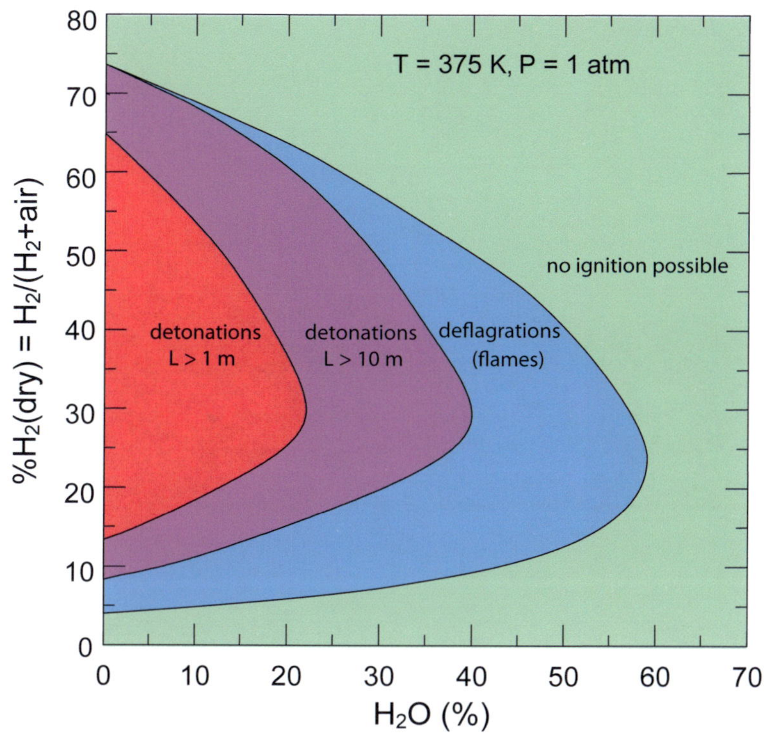

FIGURE S4.1 Flammability diagram for a nominal temperature of 100°C and a pressure of 1 bar. Values on the *x* axis represent the steam concentration as a percentage of the hydrogen–air–steam mixture and the values on the *y* axis represent the hydrogen concentration as a percentage of the hydrogen–air mixture. Steam concentration can be as high as 100 percent at 100°C. The distances L are the characteristic dimensions of a compartment or room in the reactor building or containment that will allow detonation to occur. Note that the boundaries in the diagram are guidelines intended only to indicate possible outcomes of ignition of a mixture. A wide range of behaviors—including no explosion, deflagration, high-speed flames, and detonation—can be observed in the flammable region depending on the strength and location of ignition sources as well as the spatial distribution of hydrogen and steam (see NEA [2000] for further discussion). For large volumes, such as the refueling areas on the upper floors of the Fukushima Daiichi reactor buildings, there may be potential for transition to detonation for some mixtures that are within the blue-shaded region marked "deflagration" (see Chapter 3 of NEA (2000) for further discussion).

SOURCE: Adapted from NEA (2000, Fig. 7.1.1-3). Based on data from Figure 7.1.1-3 from Breitung et al. (2000).

FIGURE 4.5 Photos showing damage to the reactor buildings at the Fukushima Daiichi plant from the Unit 1, 3, and 4 hydrogen explosions. (A) Unit 1 exterior; (B) Unit 3 exterior; (C) Unit 3 interior at the operating floor level (i.e., the upper floor of the building where the spent fuel pool is located); (D) Unit 4 exterior; (E) Unit 4 interior at the operating floor level. SOURCE: Courtesy of TEPCO.

and location of the reactor core and generation of hydrogen during the accident. They are based on detailed physical models of the reactor units (e.g., models of the reactor core, reactor pressure vessel, containment, and reactor building) and information about important operational events that occurred during the accident (e.g., operation of various safety systems and the timing and rates of water injection).

Two simulation tools have been used for this purpose: MELCOR, developed for the U.S. Nuclear Regulatory Commission (USNRC) by Sandia National Laboratories; and MAAP, developed for industry by the Electric Power Research Institute. These models have been extensively compared against a wide range of experiments with simulated accident conditions as well as analysis of the 1979 Three Mile Island accident.[58] Simulations of the accident using these tools have been carried out to date by Gauntt et al. (2012a,b), TEPCO (2012c), Yamanaka (2012), and EPRI (2013). Some of the key results of these simulations for Units 1-3 are described below.

There is extensive experience with modeling severe accidents in Mark I containments[59] for the relatively simple accident scenario that occurred in Unit 1: the reactor pressure vessel was isolated except for mechanical venting into the suppression chamber through automatic operation of the safety relief valves, and there were no active cooling measures for at least 14 hours after the earthquake. Consequently, the MELCOR and MAAP simulations for Unit 1 likely have better fidelity to reality than the simulations for the other units.

The results of simulations for Unit 1 are summarized in Table 4.2. They agree reasonably well even though the groups doing the simulations used different assumptions about operator actions, equipment behavior and effectiveness, cooling-water flow rates, and other key events. These simulations are being refined as new information is emerging from TEPCO's continuing investigations into the accident (e.g., Kawabe, 2012; TEPCO, 2013).

The simulations suggest the following sequence of events for the accident: With the isolation condenser valves closed and no cooling to the reactor after the loss of all electrical power (see Section 4.3.1 of this chapter), decay heat generated in the fuel boiled the water in the reactor pressure vessel, increasing its pressure and causing the safety relief valves to open. This allowed steam to exit the reactor pressure vessel, dropping its liquid water level. The steam was exhausted to the suppression pool. Condensa-

[58] The accident involved the partial meltdown of the Unit 2 reactor core.

[59] U.S. severe reactor accident consequence studies (USNRC, 1990, 2013c,d) have modeled core-melt accidents at the Peach Bottom Atomic Power Station, which also has a Mark I containment.

TABLE 4.2 Key Results for Accident Progression Simulations in Unit 1

Event	Time After Earthquake (+ h)	Notes
Core exposure (TAF)	+2.5 to +3	
Core damage begins	+4	Core damage timing is nominal and based on Sandia MELCOR analysis (Gauntt et al., 2012a)
Core fully uncovered	+4.5 to +5	
MSL ruptures	+6.5	Considered by Sandia MELCOR analysis only (Gauntt et al., 2012a)
RPV damage	+9 to +11	
RPV melt-through	+14	Probably occurred at +13 h, could have been as late as +16 h
Containment leaks	+3 to +6	Depends strongly on assumed failure modes
Hydrogen generated (kg)		900 kg; amount depends on extent of core–concrete interaction
Containment venting	+23.7	Known from actions of operators and pressure records
Explosion	+24.8	Known from both seismic and video recordings

NOTES: MSL = main steam line; RPV = reactor pressure vessel; TAF = top of active fuel.
SOURCE: Estimates based on MELCOR and MAAP simulations by Gauntt et al. (2012a), TEPCO (2012c), Yamanaka (2012), and EPRI (2013).

tion of steam in the pool raised its temperature and also increased pressure within the suppression chamber.

Continued depressurization lowered water levels in the reactor pressure vessel. At about +2.5 to +3 hours, all simulations predict that the water level in the Unit 1 reactor pressure vessel dropped enough to expose the active portion of the fuel in the reactor core; within +4.5 to +5 hours, the liquid level dropped below the bottom of the active portion of the fuel. Uncovering of the fuel in the reactor core likely initiated the following sequence of events:

- The temperature of the fuel cladding increased quickly, which accelerated the highly exothermic oxidation reaction between the cladding and steam in the reactor. This reaction generated hydrogen (see Sidebar 4.1) and released heat.

- As temperature continued to increase, pressures inside the fuel rods (from buildup of gaseous fission products) also increased and cladding strength decreased. This caused the cladding to balloon and fail, releasing fission products.
- As temperature increased further, the fuel cladding (and eventually the fuel itself as well as the cladding, control rods, and reactor internals) began to melt, forming a molten mixture referred to as *corium*.
- The corium flowed downward onto the lower head of the reactor pressure vessel causing it to melt and fail.
- The molten mass flowed onto the concrete floor of the containment. (The simulations estimate that about 139 tonnes of molten material were released into containment.)
- The molten material attacked the concrete and was further oxidized by water that was injected into the reactor starting at about +15 hours.

MELCOR simulations predict that about 900 kg of hydrogen was generated in Unit 1. Production of hydrogen likely started 12-15 hours before the explosion in the unit. The hydrogen and fission product aerosols (particularly iodine and cesium) probably leaked into the reactor building over a similar period of time. Sandia (Gauntt et al., 2012b) proposed that the extended period of high pressure within the containment caused stretching of the bolts fastening the containment head, opening a gap and allowing gas and fission products to flow directly from the containment into the reactor building. TEPCO (2012b, p. 340) suggests a number of potential leakage pathways in addition to the containment head seal, including equipment and personnel access hatches, and/or electrical cable penetrations; or through the standby gas treatment system when the containment was deliberately vented.[60]

The hydrogen explosion in Unit 1 occurred at 15:36 on March 12. The steel siding covering the upper portion of the building (above the refueling floor) was blown off (see Figure 4.5A); five workers were injured and preparations for connecting water injection and power were disrupted. The ignition source for the hydrogen explosion is unknown but could have been

[60] TEPCO (2012b, pp. 351-352) notes that the standby gas treatment system may not have been isolated from the stack at the time of venting.

a hot surface, an electric arc from damaged electrical wiring, or a spark from exposed contacts on a motor.[61]

The explosion was quite disturbing to the plant personnel, who did not initially understand what had happened. Once they recognized that a hydrogen explosion had occurred and realized (Investigation Committee, 2011, pp. 244-247) that explosions could occur in the other units, they began to fear more explosions could take place. Methods were considered for venting the reactor buildings such as removing blowout panels. However, it was not feasible to send personnel into the buildings to remove the panels because of high radiation levels and explosion hazards. Plant personnel considered using a water jet to cut holes in the building, but this type of complex operation was not feasible under the working conditions at the plant.

Videos of the explosions were captured on a camera set up by a local television station and replayed over the Internet and evening news. This contributed substantially to the public's anxiety, particularly in Japan. The videos were, however, useful to the site superintendent and the staff in determining what had happened to the unit.

Simulations similar to those carried out for Unit 1 have been performed for Units 2 and 3. They are not described in detail here in the interests of brevity.

As noted previously, there was no hydrogen explosion in Unit 2. The reasons for this are unclear. Based on the amount of time that the core was uncovered and the estimated hydrogen generation rates, a substantial amount of hydrogen is likely to have been generated. The containment pressure in Unit 2 reached the extreme values (0.75 MPa) between +80 and +90 hours after the reactor core isolation cooling system stopped (at +70 hours) and seawater injection was initiated. This pressure is similar to that reached in Unit 1, and so similar venting of hydrogen from the containment to the reactor building might have occurred.

There is speculation (Investigation Committee, 2012, p. 70; TEPCO, 2012b, p. 342) that a hydrogen explosion was prevented in Unit 2 because a building panel was blown out[62] from the upper level, just above the refueling deck. The presence of an opening in the building may have created

---

[61] An extensive discussion of possible ignition sources is given by the Investigation Committee (2012, pp. 65-70). Just prior to the explosion, efforts were nearing completion to reconnect power to the standby liquid control system. It is possible that an electrical fault in equipment attached to the buses being energized caused the ignition (2012, pp. 68-69).

[62] The panel was apparently knocked out by the pressure wave from the explosion in Unit 1 and was hanging by its restraining chains. The pressure differential required to cause this is about 3.5 kPa (0.5 psi). Following the explosion in Unit 3, the chains were observed to be broken and the panel had dropped to the roof of the turbine building (Investigation Committee, 2012, p. 70).

a pathway for hydrogen to flow outside the building, thereby preventing a buildup of an explosive atmosphere. Although plausible, this scenario needs more careful analysis to understand the effectiveness of this vent. Further information about hydrogen generation and transport will likely be obtained as the Unit 2 reactor is dismantled and studied.

The vent effectiveness will depend on the location, rate, and duration of the gas release from the containment, the hydrogen content, and the gas motion within the upper volume of the reactor building. It is unclear if the panel opening (approximately 4.3 × 6 m; see Investigation Committee, 2012, p. 70) would be effective in preventing a flammable atmosphere from being formed. The panel was near the refueling floor, and a substantial portion of the fifth-floor volume was above the upper edge of the opening.

There may be other reasons that an explosion did not occur in Unit 2: for example, insufficient hydrogen may have been generated and released within the Unit 2 reactor building. Alternatively, there may have been a special set of circumstances that resulted in an inert atmosphere or the lack of an ignition source.

The hydrogen explosion in Unit 3 occurred at 11:01 on March 14. It destroyed the upper portion of the reactor building (floors 3 and 4 were heavily damaged and floor 5 was demolished; see Figure 4.5B,C) and injured 11 workers. As noted previously in the chapter (see Section 4.3.2), the debris from this explosion damaged equipment and spread radioactive debris. It also forced field workers to retreat to the onsite ERC, further delaying the accident response.

The hydrogen explosion in Unit 4 (see Figure 4.5D,E) caused much concern at the time of the accident because the reactor did not contain any fuel; it had been offloaded to the spent fuel pool in the reactor building. There was initial concern that spent fuel in the pool had become uncovered and had reacted with the steam to form hydrogen.[63] This would have likely resulted in large uncontained fission product releases.

In fact, it now appears that the hydrogen in Unit 4 reactor building came from Unit 3, through an unexpected path. There was substantial destruction on the fourth and fifth floors of Unit 4. The pattern suggests that the hydrogen reached the building by flowing back through the ventilation system for the standby gas treatment system.

4.3.6 Spent Fuel Pools

As noted in Chapter 1, the committee has deferred its analysis of spent fuel safety and security to a subsequent report. Consequently, this section

[63] Currently, there is no evidence of fuel damage in the Unit 4 spent fuel pool, but this was not immediately known at the time of the accident.

TABLE 4.3 Spent Fuel Storage at the Fukushima Daiichi Nuclear Plant on March 11, 2011

Storage Location	Spent Fuel (Assemblies[a])	Fresh Fuel (Assemblies)
Unit 1	292	100
Unit 2	587	28
Unit 3	514	52
Unit 4	1,331	204
Unit 5	946	48
Unit 6	876	64
Common pool	6,375	0
Cask storage building	408	0

[a] A boiling water reactor fuel assembly contains about 170-185 kg of uranium.
SOURCE: TEPCO (2012b, p. 299).

provides only a brief discussion of events in the spent fuel pools at the Fukushima Daiichi plant.

A large amount of spent fuel was in storage in pools in the Unit 1-6 reactor buildings at the time of the earthquake and tsunami (Table 4.3). The September 2011 supplemental report by the Japanese government to the International Atomic Energy Agency (Government of Japan, 2011a) concluded that it is a highly likely that spent fuel was not exposed to air in the Unit 1-3 spent fuel pools and that mass damage did not occur in the Unit 4 pool.[64] Power was restored to the Unit 5 and Unit 6 pools and the common spent fuel pool before their temperatures increased significantly.

Investigations of the Unit 1-4 spent fuel pools have to date uncovered no evidence to contradict the Japanese Government's initial conclusions. Efforts are now under way by TEPCO to move spent fuel from the Unit 4 pool into the common pool.

Nevertheless, the events at the Fukushima Daiichi plant highlight concerns about the vulnerability of spent fuel pools to severe accidents. These pools are located outside plant containment; consequently, accidents involving loaded spent fuel pools have the potential to produce substantial radioactive material releases.

[64] TEPCO was transferring spent fuel from the Unit 4 pool to the common spent fuel pool when the present report was being completed. There have been no indications of damage to the fuel to date.

4.4 DISCUSSION

In the committee's judgment, the severity of the accident at the Fukushima Daiichi nuclear plant following the March 11, 2011, earthquake and tsunami were the result of six factors:

1. The loss of all AC/DC power in Units 1-4 narrowed options for responding to the accident.
2. Operators lacked resources, procedures, and training to promptly reestablish reactor cooling and to vent containments using alternative methods for accidents involving loss of all AC/DC power.
3. Multiunit interactions complicated operator responses.
4. Communication failures hindered responses to the accident.
5. Confusion about ERC roles and responsibilities delayed and in some cases prevented effective responses to the accident.
6. Staffing levels were insufficient for responding to the accident.

These factors are discussed briefly in the following sections.

4.4.1 Loss of Power

The Fukushima Daiichi accident significantly exceeded the beyond-design-basis events that TEPCO postulated and resulted in different conditions than those assumed when accident management strategies were developed (TEPCO, 2012b, p. 51). Neither the plant's engineered systems nor severe accident management procedures were sufficient to handle the situation; in fact, a majority of the "preplanned" response options as embodied in the TEPCO accident management procedures were not applicable to the situations that the plant's operators confronted.

For example, procedures to cool the reactors using various installed emergency core cooling systems (e.g., isolation condenser system, reactor core isolation cooling system, high-pressure coolant injection system) were specified in accident management procedures. Other means of emergency cooling via control-rod-drive hydraulic pressure systems, makeup water condensate system, and the fire protection system were also identified options in these procedures. However, once power was lost, all of the motor-operated systems became inoperable.

Provisions were in place in the plant's accident management procedures to handle loss-of-power incidents. These included multiple emergency diesel generators in each unit to cope with loss of offsite power. In addition, plans were in place to allow high- and low-voltage AC power supply to be fed from adjacent units. This was intended to cope with delays in AC power restoration or unavailability of DC power at one of the units. In spite of these provisions, AC and DC power could not be restored to some plant units for several days.

Loss of AC and DC power also had unanticipated systems effects. The best example is the isolation condenser system and its complex interlocks described in Section 4.3.1.1. In fact, nuclear plants have numerous systems containing complex interlock and failsafe logic that are not readily apparent from user interfaces. It can be a challenge to anticipate the effects of power losses on such systems. Experience in the aviation industry has led to development of principles and guidelines for how to design automated systems so that their behavior can be more readily anticipated (e.g., Norman, 1990; Sarter et al., 1997). This experience base can be leveraged in design of next-generation nuclear plant control rooms, as well as control room upgrades to existing plants, to enable operators to maintain better situational awareness of the status of automated systems and how systems are likely to be affected by events such as loss of power. The nuclear industry's FLEX initiative (NEI, 2012) and the USNRC's station blackout order (USNRC, 2012b) are intended to address potential systems effects in current-generation nuclear plants. (The FLEX initiative and USNRC actions are discussed in Chapter 5 and Appendix F.)

4.4.2 Resources, Procedures, and Training

The majority of the "preplanned" response options embodied in the TEPCO accident management procedures were not applicable to the situations that operating staff confronted following the earthquake and tsunami. Although operators underwent extensive training, that training did not cover the accident scenarios that unfolded at the plant following the March 11 tsunami. For example, although there were procedures and training for venting, these procedures assumed that power would be available to operate the venting valves from the control room. Procedures and training also assumed that plant indicators would be available in the control room. Onsite ERC staff training assumed that the safety display parameter system and communication lines with control rooms would provide good situational awareness of plant state and operator actions.

Operators could not take critical control actions from the control room; instead, they had to take manual actions in the field. Radiation releases in the plant and limited access to personnel dosimeters[65] hampered the ability of personnel to perform their duties, both in the control room

[65] TEPCO (2012b, Attachment 2, pp. 13-14) notes that approximately 5,000 personnel dosimeters that were stored at the plant were rendered inoperable by the tsunami. TEPCO was able to recover about 320 dosimeters from various sources at the plant by the night of March 12.

(see Section 4.2) and in the field. Some field activities required multiple teams because of difficult onsite conditions. Flooding, debris, and other hazards caused by the tsunami challenged the field response; hydrogen explosions further set back response activities. The operators encountered situations that went well beyond their training for responding to off-normal conditions.

Unit operators and the onsite ERC staff had to fall back on "first principles" reasoning and problem solving to respond to the rapidly unfolding events at the plant.[66] This required active diagnosis and tracking of plant conditions, goal identification and prioritization, adaptive problem solving, and development and rehearsal of ad hoc response plans. Plant personnel displayed creativity in responding to the accident. For example, plant operators

- Restored some control room lighting, instrumentation, and control systems using batteries from employee automobiles and portable generators from contractor warehouses;
- Used fire engines to inject cooling water into the reactor, an option not specified in accident management procedures;
- Injected seawater into the reactors when freshwater supplies became unavailable; and
- Developed and implemented a plan to vent containment without power.

Some of these response actions are similar to the accident response actions required under Section B.5.b of the Order for Interim Safeguards and Security Compensatory Measures (described in Chapter 2). These include the use of fire engines for water injection and batteries to restore water-level gauges and operate steam safety relief valves (TEPCO, 2012b, p. 54). However, TEPCO (2012b, p. 54) notes that B.5.b information was not available to private electrical utilities in Japan.[67]

Accidents frequently involve a confluence of interacting faults resulting in situations that have not been previously anticipated, placing a premium on the ingenuity and adaptability of plant personnel. *In the committee's judgment, the personnel at the Fukushima Daiichi nuclear plant showed*

[66] This type of "on-the-spot" reasoning and problem solving is referred to as "knowledge-based" performance in the human factors literature.

[67] As noted in Chapter 2, Section B.5.b was designated by the USNRC as Safeguards Information and so it was exempt from public release. Consequently, TEPCO would not have had direct access to this information. However, as discussed in Chapter 7, the USNRC shared some B.5.b information with Japanese government authorities. Moreover, as noted in Appendix H, information about B.5.b requirements was released to the public as a result of a 2009 rulemaking by the USNRC.

SIDEBAR 4.2
Accident Response at Fukushima Daini

TEPCO's Fukushima Daini nuclear plant (see Chapter 3) sustained severe damage from the March 11, 2011, earthquake and tsunami. However, operators were able to bring the plant's four reactors to cold shutdown by the morning of March 15. Their actions illustrate the successful application of emergency operating and accident management procedures in response to a severe external event.

The earthquake shut down two of the three available offsite AC power lines to the plant (another line was shut down for inspection at the time of the earthquake). Flooding from the tsunami damaged power distribution systems and pumps for the emergency core cooling and residual heat removal systems in the Unit 1, 2, and 4 reactors. However, AC power from one offsite power line and onsite DC power remained available following the earthquake and tsunami. Consequently, operators were able to maintain instrument and control room command over critical plant systems.

Operators used safety relief valves and reactor core isolation cooling systems to lower reactor pressures in Units 1, 2, and 4 following the tsunami; reactor pressures were less than 1 MPa 8 hours after the tsunami. Cooling was then transitioned seamlessly to low-pressure water injection with an alternative water supply (the makeup water condensate system) by midnight of March 11. The water levels in the reactors were maintained at or near the "L8" level, over 5 m above the top of active fuel, during the cooldown phase. Drywell and suppression chamber sprays were used to control containment pressures to less than 0.4 MPa until power was restored to the residual heat removal systems on the morning of March 14.

Operators were able to quickly and successfully execute several critical tasks that operators at Fukushima Daiichi attempted but could not complete. These included lining up vent valves, arranging alternative water supplies, controlling reactor core isolation cooling systems, and, most important for recovering the residual heat removal system, laying and connecting alternative power cables and replacing damaged motors, all carried out by hand or by using crane trucks. Operators took some actions (e.g., lining up vent valves) in anticipation that the accident

courage and resilience in responding to the March 11, 2011, accident under extraordinarily difficult conditions. Their actions potentially prevented even more severe outcomes at the plant.

The response of operators at the Fukushima Daini plant to the earthquake and tsunami (Sidebar 4.2) demonstrate the successful application of accident management/emergency operating procedures and operator training to extreme accident scenarios. The response at Fukushima Daini was a huge success story in its own right (albeit because some power sources survived the tsunami). However, this success was overshadowed by events at the Fukushima Daiichi plant.

might become more severe; however, existing emergency operating procedures were adequate for bringing the reactors to cold shutdown. Only one ad hoc measure suggested by the onsite ERC—water injection into the suppression chamber using an alternative water source—was employed (TEPCO, 2012b, p. 54).

Although operators at Fukushima Daini faced some of the same challenges as those at Fukushima Daiichi—most notably onsite access difficulties due to tsunami-related flooding and damage and earthquake aftershocks—there were some key differences: flooding at the Fukushima Daini plant was not as severe, AC and DC power were continuously available in functioning control rooms, and onsite response efforts were not hindered by debris and radioactive contamination from hydrogen explosions. Operators also did not have to enter dark and contaminated reactor buildings to mount a response, but could monitor and control reactors from their control rooms. The communications and command structure functioned properly: the onsite ERC had a functional safety parameter display system and continuous communication with the control rooms.

According to TEPCO (2012b, p. 55):

> During the accident, the decision-making procedure where the Shift Supervisor made determinations and the ERC at the power station made verifications was generally adhered to. This allowed operational manipulations to be implemented in a timely manner according to plant conditions and also was effective in allowing the ERC at the power station to fulfill its function of keeping a big-picture perspective to maintain oversight of response strategies and to manage equipment restoration activities.

Comparing the responses at the Fukushima Daiichi and Daini plants, where operators presumably received the same levels of training, it is clear that the loss of all AC and DC power at Fukushima Daiichi precipitated a series of cascading failures that simply overwhelmed operators. In a sense, the events at Fukushima Daiichi represent a "cliff edge" in accident management capabilities.

TEPCO anticipated and trained its operators for the situations they encountered at Fukushima Daini and the response was effective. TEPCO never anticipated or trained its operators for the events at Fukushima Daiichi; the response was ineffective and the consequences were disastrous.

In discussing the difficulties experienced at Fukushima Daiichi, the Investigation Committee (2011, p. 141) noted:

> [T]hey did not assume that a situation in which multiple nuclear reactors losing all power sources almost simultaneously would occur and thus did not provide the training and education necessary to implement measures to control such a serious situation.

Koichio Kitazawa (Chairman of the Rebuild Japan Foundation Investigation Commission; see RJIF, 2014) put it more succinctly (NPR interview,

March 9, 2012): "You can't adequately prepare for a disaster that you don't admit can ever happen."

4.4.3 Multiunit Interactions

The colocation and close spacing of Units 1-4 and the extensive sitewide impacts from the tsunami and earthquake also hindered the accident response. In particular, harbor-side tsunami damage, earthquake damage to cisterns and water-supply piping, displacement of road surfaces, landslides, and blockage of roads and building access by debris are examples of damage common to Units 1-4 at the site. This damage impeded efforts to establish alternative cooling water, power, and compressed air sources.

Control rooms at the Fukushima Daiichi plant are shared between pairs of reactor units (see Appendix D). The ventilation systems in Units 1 and 2 and 3 and 4 are also paired together. This pairing apparently allowed hydrogen generated in the damaged Unit 3 reactor to flow into the Unit 4 reactor building. Hydrogen explosions in the Unit 1, 3, and 4 buildings scattered debris and caused substantial ground contamination around the buildings, damaged temporary installations for water injection and electric power, and injured workers. The hydrogen explosions in the Unit 3 and Unit 4 buildings also affected the management of the accident at all units because personnel at the site were reduced to a bare minimum for a time and recovery operations at the reactor units were halted.

The units also competed for physical resources and attention and/or services of the onsite ERC staff. Some of these resource competitions were described previously in the chapter: the competition for fire trucks to pump water into the Unit 1-3 reactors and the limited space in a valve backwash pit to siphon water. These limitations made it impossible to supply seawater to both Unit 3 and Unit 2 simultaneously.

Interviews with onsite ERC staff as reported by the Investigation Committee (2011) suggest that at different points in time, the onsite ERC focused attention on one unit at the possible expense of others. For example, the delay in recognizing that the isolation condenser was shut in Unit 1 was partly explained by the fact that ERC was initially focused on Unit 2 because it could not confirm that its reactor cooling isolation system was functioning.

In contrast, colocation had great value for the accident response at Units 5 and 6 at the Fukushima Daiichi plant and at the Fukushima Daini plant despite the sitewide earthquake and tsunami damage. Colocation enabled power to be cross-connected between units and also enabled mutual aid for the timely recovery of cooling and reactor pressure vessel depressurization, thereby preventing reactor damage.

4.4.4 Communications

Failures to transmit information and instructions in an accurate and timely manner played an important role in shaping actions at certain points during the accident response. These include the inability of operators to get appropriate attention and clear instructions from the onsite ERC. Lack of cooperation between operators and contractors' "partner companies" was also a challenge. The overwhelming nature of the accident and the lack of training to cope with its many challenges may have played key roles in these communication and coordination failures.

The earthquake and tsunami damaged physical communications (voice and data transmission) systems and hampered travel on the site. As noted previously, the failure of instrumentation within the control rooms and lack of operational safety parameter display systems required that data be relayed verbally over a single telephone line between the control rooms and onsite ERC. Data had to be written on whiteboards in the ERC and reported by video or telephone conference to Tokyo. Radios and cell phones had limited functionality in many parts of the site, requiring plant personnel to traverse the debris-strewn site to report findings, coordinate activities with unit operators, and obtain instructions from the onsite ERC.

As the accident progressed the site became progressively contaminated due to the spread of radioactive materials. This further hampered communications, particularly when site personnel were temporarily evacuated following the hydrogen explosion in Unit 3. These same factors also made it difficult for site personnel to communicate with the outside world, including with their families. (A notable aspect of the accident was the fact that the plant personnel remained onsite and worked diligently without news about their families.)

Some notable examples of communication failures were mentioned previously in this chapter:

- Miscommunications about operations of valves and status of the isolation condenser in Unit 1;
- Miscommunications about the need for batteries to operate the safety relief valves in Unit 1;
- Lack of coordination between shift team and firefighters because neither understood the responsibility given to them by the site superintendent for hooking up the fire truck pump to the Unit 1 fire protection system;
- Incorrect battery types (2V instead of 12V) were supplied to depressurize the safety relief valves in Unit 2; plant personnel had to scavenge their automobile batteries instead;
- Portable generators were delivered with incorrect voltage and connectors;

- Miscommunications about why the high-pressure coolant injection system was halted in Unit 3 and the need for an alternative water injection supply; and
- Breakdown in communications among the shift teams, onsite ERC, offsite ERC, NISA, and the prime minister's office about the situation inside and outside of the plant.

4.4.5 ERC Roles and Responsibilities

The lack of clarity about roles and responsibility within the onsite ERC as well as between the onsite ERC and the headquarters ERC in Tokyo proved to be a source of distraction for members of the onsite ERC and may have contributed to response delays. TEPCO personnel, in their presentations to this committee, indicated that there was confusion in the chain-of-command structure due to the complexity and multiunit nature of the accident. In particular, the organizational structure within the onsite ERC (which defined 12 function teams) was effective for situations that were explicitly covered by the accident management procedures, but they proved to be inadequate for the performance of tasks that fell outside the procedures. In particular, defining roles and responsibilities for tasks that were not covered by the procedures (e.g., water injection using fire engines) proved challenging and resulted in substantial response delays.

4.4.6 Staffing Plan

A presentation from TEPCO (Kawano, 2012) and discussions with TEPCO personnel during the committee's November 2012 meeting in Tokyo indicate that staffing levels were inadequate for managing the accident (see also TEPCO, 2012b, pp. 472, 474). TEPCO had not anticipated accident scenarios where multiple units were impacted simultaneously. The staffing plan assumed that operators from one unit could cover for another unit. Consequently, the plan did not cover accidents that involved multiple units.

The staffing plan also did not anticipate accident situations that extended over multiple days. TEPCO indicated to the committee that there were no handoffs during the initial days of the Fukushima accident: The first shift was present for the next three shifts, and the shift manager was present for 2-3 days straight during the accident. TEPCO recognized that this situation was not sustainable.

TEPCO also indicated to the committee that one of the lessons learned was the need to provide additional shifts and shift handovers in a prolonged emergency. TEPCO's plan for the future is to have two shifts in the ERC. They also plan to increase shift team staffing because there were so many things that needed to happen quickly. TEPCO noted to the committee that had the accident occurred at night or on a weekend, the response could have been worse because fewer people would have been present onsite. TEPCO indicated that it plans to increase the number of operators in the night shift from six to eight.

TEPCO also indicated to the committee (Kawano, 2012) that it is taking measures to strengthen the organizational structure for handling simultaneous and compound accidents at multiple units. This includes increasing the number of technical support personnel at the onsite ERC and establishing two technical support rooms in the headquarters ERC to handle the simultaneous occurrence of a nuclear accident and a natural disaster. In a presentation to the committee at its November 2012 meeting in Tokyo, TEPCO General Manager Mr. Arika Kawano specifically noted that TEPCO would

- Designate personnel responsible for individual units to the operation support team and restoration team in the Emergency Response Organization;
- Enhance the night duty structure to strengthen the functions of collecting plant information and external communications immediately after a disaster strikes; and
- Have personnel on standby at each plant 24 hours a day to quickly restore emergency power or inject cooling water.

4.5 FINDING

The committee developed one finding to address the first charge of the statement of task (see Sidebar 1.1 in Chapter 1) on causes of the Fukushima nuclear accident.

FINDING 4.1: The accident at the Fukushima Daiichi nuclear plant was initiated by the March 11, 2011, Great East Japan Earthquake and tsunami. The earthquake knocked out offsite AC power to the plant and the tsunami inundated portions of the plant site. Flooding of critical plant equipment resulted in the extended loss of onsite AC and DC power with the consequent loss of reactor monitoring, control, and cooling functions in multiple units. Three reactors sustained severe core damage (Units 1, 2, and 3); three reactor buildings were damaged by hydrogen explosions (Units 1, 3, and 4); and offsite releases of radioactive materials contaminated land in Fukushima and several neighboring prefectures. The accident prompted widespread evacuations of local populations and distress of the Japanese citizenry, large economic losses, and the eventual shutdown of all nuclear power plants in Japan.

Personnel at the Fukushima Daiichi plant responded with courage and resilience during the accident in the face of harsh circumstances; their actions likely reduced the severity of the accident and the magnitude of offsite radioactive material releases. Several factors prevented plant personnel from achieving greater success—in particular, averting reactor core damage—and contributed to the overall severity of the accident:

1. Failure of the plant owner (Tokyo Electric Power Company) and the principal regulator (Nuclear and Industrial Safety Agency) to protect critical safety equipment at the plant from flooding in spite of mounting evidence that the plant's current design basis for tsunamis was inadequate.[68]

2. The loss of nearly all onsite AC and DC power at the plant—with the consequent loss of real-time information for monitoring critical thermodynamic parameters in reactors, containments, and spent fuel pools and for sensing and actuating critical valves and equipment—greatly narrowed options for responding to the accident.

3. As a result of (1) and (2), the Unit 1, 2 and 3 reactors were effectively isolated from their ultimate heat sink (the Pacific Ocean) for a period of time far in excess of the heat capacity of the suppression pools or the coping time of the plant to station blackout.

4. Multiunit interactions complicated the accident response. Unit operators competed for physical resources and the attention and services of staff in the onsite emergency response center.

5. Operators and onsite emergency response center staff lacked adequate procedures and training for accidents involving extended loss of all onsite AC and DC power, particularly procedures and training for managing water levels and pressures in reactors and their containments and hydrogen generated during reactor core degradation.

6. Failures to transmit information and instructions in an accurate and timely manner hindered responses to the accident. These failures resulted partly from the loss of communication systems and the challenging operating environments throughout the plant.

7. The lack of clarity of roles and responsibilities within the onsite emergency response center and between the onsite and headquarters emergency response centers may have contributed to response delays.

8. Staffing levels at the plant were inadequate for managing the accident because of its scope (affecting several reactor units) and long duration.

---

[68] See Chapter 3 for a discussion. NAIIC (2012) criticized TEPCO for the lack of adequate tsunami countermeasures at the Fukushima Daiichi plant.

5

Lessons Learned: Plant Operations and Safety Regulations

The final three chapters of this report are intended to address the third and fourth charges of the study task (see Sidebar 1.1 in Chapter 1):

- Lessons that can be learned from the accident to improve commercial nuclear plant safety and security systems and operations;
- Lessons that can be learned from the accident to improve commercial nuclear plant safety and security regulations, including processes for identifying and applying design-basis events for accidents and terrorist attacks to existing nuclear plants.

The focus of this chapter is on nuclear plant safety systems, operations, and regulations. Chapter 6 focuses on offsite nuclear emergency planning and emergency management, whereas Chapter 7 focuses on the nuclear safety culture. As noted in Chapter 1, a discussion of spent fuel and related security issues will be addressed in a subsequent report.

This NAS study is one of many investigations/assessments initiated in the wake of the Fukushima Daiichi nuclear accident (see Table 1.1 in Chapter 1). The reports from these other studies have been invaluable for informing the committee's thinking about potential lessons learned. The committee has provided a tabular summary of key recommendations from selected reports in Appendix E.

The committee presents three findings and five recommendations in this chapter. These findings and recommendations are organized into two major sections:

1. Nuclear plant systems, procedures, and training; and
2. Nuclear plant safety risks.

Additional supporting information is provided in Appendixes E through L.

These findings and recommendations are directed primarily at the U.S. nuclear power industry and its regulator (U.S. Nuclear Regulatory Commission [USNRC]). However, the committee anticipates that they will also have value for nuclear power industries and regulators in other countries.

5.1 NUCLEAR PLANT SYSTEMS, PROCEDURES, AND TRAINING

FINDING 5.1: Nuclear plant operators and regulators in the United States and other countries have identified and are taking useful actions to upgrade nuclear plant systems, operating procedures, and operator training in response to the Fukushima Daiichi accident. In the United States, these actions include the nuclear industry's FLEX (diverse and flexible coping strategies) initiative as well as regulatory changes proposed by the U.S. Nuclear Regulatory Commission's Near-Term Task Force. Implementation of these actions is still under way; consequently, it is too soon to evaluate their comprehensiveness, effectiveness, or status in the regulatory framework.

In the weeks following the Fukushima nuclear accident, many national governments and international bodies initiated reviews of nuclear power plant performance and current safety measures (see Table 1.1 in Chapter 1). Some of the outputs of these efforts are described in Appendix E.

In the United States, two major initiatives were begun:

- The USNRC appointed a six-member task force headed by Dr. Charles Miller, the Near-Term Task Force. Its charge was to perform a "systematic and methodological review of the U.S. Nuclear Regulatory Commission processes and regulations to determine whether the agency should make additional improvements to its regulatory system and to make recommendations to the Commission for its policy direction, in light of the accident at the Fukushima Dai-ichi Nuclear Power Plant" (USNRC NTTF, 2011, p. vii).
- At about the same time, the U.S. nuclear industry, led by the Institute of Nuclear Power Operations (INPO), Nuclear Energy Institute (NEI), and Electric Power Research Institute (EPRI), initiated a voluntary effort to "integrate and coordinate the U.S. nuclear industry's response to events at the Fukushima Daiichi nuclear energy facility. This will ensure that lessons learned are identified and well understood, and that response actions are

effectively coordinated and implemented throughout the industry" (NEI et al., 2012, p. 1).

Brief discussions of these initiatives and key results to date are provided in Appendix F.

The results from these initiatives that have been documented to date have been helpful to the committee in informing its thinking about potential lessons learned. However, these initiatives were still in progress when the present report was completed; many decisions have yet to be made or fully implemented. Moreover, the committee had neither the time nor resources to carry out in-depth reviews of these initiatives, which in some cases would have required plant-by-plant examinations.

5.1.1 Nuclear Plant Systems

RECOMMENDATION 5.1A: As the nuclear industry and its regulator implement the actions referenced in Finding 5.1, they should give specific attention to improving plant systems in order to enable effective responses to beyond-design-basis events, including, when necessary, developing and implementing ad hoc[1] responses to deal with unanticipated complexities. Attention to availability, reliability, redundancy, and diversity of plant systems and equipment is specifically needed for

- **DC power for instrumentation and safety system control;**
- **Tools for estimating real-time plant status during loss of power;**
- **Decay-heat removal and reactor depressurization and containment venting systems and protocols;**
- **Instrumentation for monitoring critical thermodynamic parameters in reactors, containments, and spent fuel pools;**
- **Hydrogen monitoring (including monitoring in reactor buildings) and mitigation;**
- **Instrumentation for both onsite and offsite radiation and security monitoring; and**
- **Communications and real-time information systems to support**

[1] The term "ad hoc" in this finding refers to responses that are not planned and trained on in advance but rather are developed on the spot—operators' use of car batteries at the Fukushima Daiichi plant (see Chapter 4) is an example of an ad hoc response. This type of on-the-spot reasoning and problem solving is referred to as "knowledge-based" performance in the human factors literature. Knowledge-based performance is necessary when a situation is novel or not fully covered by the available procedural guidance. In these situations individuals need to have a deeper level of understanding of how a system works (e.g., the physical laws and principles that apply) to be able to correctly assess the situation, establish appropriate response goals, and formulate a plan of action to achieve those goals (Rasmussen, 1983; Mumaw et al., 1994).

communication and coordination between control rooms and technical support centers, between control rooms and the field, and between onsite and offsite support facilities.

The quality and completeness of the changes that result from this recommendation should be adequately peer reviewed.

5.1.1.1 DC Power for Instrumentation and Safety System Control

As noted in Chapter 4, the loss of DC power at the Fukushima Daiichi plant severely impacted operators' ability to monitor the status of reactor pressure, temperature, and water level and operate critical safety equipment. A lesson that emerges from this accident is that *high priority must be given to protecting DC batteries and power distribution systems at nuclear plants so that they remain functional during beyond-design-basis events.*

Both the USNRC and industry are taking useful steps to improve the ability of nuclear plants to cope during extended loss of power (see Appendix F). The USNRC issued a Mitigation Strategies Order requiring U.S. nuclear plant licensees to implement strategies for coping without permanent electrical power sources for an indefinite period of time. This order is being followed by a formal rulemaking. The industry's FLEX initiative (Appendix F) is intended to address this USNRC order using installed and portable equipment. The specific strategies to be used will be different for each nuclear plant.

Neither the USNRC order nor FLEX specifically addresses the need to protect station DC batteries and power distribution systems so that they remain functional during beyond-design-basis events. The baseline FLEX strategy for the Peach Bottom plant, for example, simply assumes that station DC batteries and power systems would be available during a beyond-design-basis external[2] event and that emergency portable power would be needed only for battery charging (Exelon Generation Company, 2013). However, the functional requirements in NEI (2012) provide for capabilities that can be effective in responding to the loss of DC power. This includes the ability to operate the reactor core isolation cooling system, the capability to read certain instruments, and the capability to depressurize the reactor pressure vessel without DC power.

The Fukushima Daiichi accident demonstrates that without AC or DC power, operators would have a few hours at most to restore critical reactor

[2] FLEX was developed specifically to address external events. See Appendix F.

monitoring and cooling functions to prevent core damage.[3] If station DC batteries or power distribution systems are destroyed or damaged, there may not be enough time to install backup DC power even if the necessary equipment is available onsite.

Existing battery rooms and associated power distribution systems at U.S. nuclear plants might need to be retrofitted and/or relocated to protect them during beyond-design-basis events. The specific actions required, if any, will be plant specific. That is, it will depend on both the design of the plant as well as the specific event scenarios that emerge from plant risk evaluations.

5.1.1.2 Tools for Estimating Real-Time Plant Status During Loss of Power

During abnormal transients or accident conditions in nuclear reactors, key thermodynamic parameters (e.g., temperature, pressure, and water level in the reactor vessel; temperature, pressure, and radiation level in the containment; and water level and temperature in spent fuel pools) must be known to facilitate appropriate operator actions. Indeed, the reliability of information gained from the instruments is a key to decision making and action taking by operators. *Another lesson that emerges from the Fukushima Daiichi accident is that alternative means for estimating these parameters is needed during loss-of-power situations.*

Under certain severe accident conditions and with disruption in power supplies, instruments may give faulty information. Although the committee is recommending that critical instruments be upgraded to cope with events that may severely impact their reliability (see Section 5.1.1.4), alternative means are still needed to guide the operators in coping with accident situations in which power is unavailable or unreliable.

Operators and technical support center staff should be provided with upgraded simulation tools and knowledge-based reasoning aids for both training and operation: for example, system-level analysis software installed on independent computers (e.g., laptops with extended battery life) to aid the operators and technical support center staff with the diagnosis of the plant state and appropriate actions under conditions of incomplete or

---

[3] The time limitation has been known since the early days of reactor engineering and can be estimated from basic engineering principles of heat transfer and thermodynamics for a given accident sequence. Some of the earliest estimates of the time to uncover the core and time to core melt are documented in the Reactor Safety Study published in 1975 (USNRC, 1975, Appendixes III and IV). The engineering models and examples of estimates for boiling water reactors are given in Appendix VIII-A of that study (USNRC, 1975), and more recent results are given in the State-of-the-Art Reactor Consequence Analyses Study draft report published in 2012 (Borchardt, 2012b; USNRC, 2013c,d).

confusing information. Such software needs to execute rapidly to provide operators with immediate feedback in crisis situations; have a modern, intuitive graphical interface; and carry out simplified mass and energy balances to give realistic estimates of plant states, particularly critical reactor and containment parameters. The software needs to have an inference engine that uses both operator inputs and a knowledge base of plant systems, including failsafe control logic, and provides prioritized recommendations on diagnostic and corrective actions.

It is also important to provide technical support center staff with similar or greater capabilities, which could include enhancement of simulators to include accident scenarios involving core damage. Currently, operators perform only tabletop exercises for severe accidents because presently available simulators cannot handle core-damage events.

These new capabilities should be integrated with existing procedures, guidance, computational aids, and software tools. Any future changes to procedures, guidance, aids, and software tools also need to be reflected in these capabilities.

The committee recognizes that the real-time decision support tools and aids called for above will require some developmental efforts; the committee judges that the potential benefits of these tools and aids warrant the necessary investments in such efforts. The shortfalls in real-time situation assessment that were exhibited by control room and emergency response center (ERC) staff at the Fukushima Daiichi plant underscore the value of providing real-time decision support tools and aids for plant status assessment and response planning, both for control room and technical support center staff. The committee further judges that the existing thermal-hydraulics knowledge base can be leveraged to create aids for generating real-time estimates of key thermodynamic parameters and liquid level in the reactor pressure vessel and provide real-time support for response planning.

5.1.1.3 Decay-Heat Removal, Reactor Depressurization, and Containment Venting Systems

The loss of AC and DC power at the Fukushima Daiichi plant severely impacted operators' ability to remove decay heat from the Unit 1-3 reactors and depressurize reactor press174ure vessels and vent containments, both to restore cooling to the core and to prevent leakage of fission products. Another lesson that emerges from the accident is that *strategies and capabilities must be in place for removing decay heat from reactors, depressurizing reactor pressure vessels, and venting containments under loss of AC and DC power conditions.*

Reactors continue to generate decay heat even after shutdown (see Chapter 2). This decay heat must be removed reliably over a long period of time to avoid damage to the integrity of the reactor core. Boiling water reactors (BWRs) have a number of core cooling systems that can be used to remove decay heat (see Chapter 2):

- Low-pressure cooling systems (low-pressure coolant injection system) require power to operate pumps and actuate valves.
- High-pressure cooling systems (isolation cooling,[4] reactor core isolation cooling, and high-pressure coolant injection systems) require power to actuate valves.
- "Ad hoc" cooling systems (e.g., injection of water from the fire protection system using diesel-driven fire pumps or fire truck pumps) can be utilized only when reactor pressure vessels are at low pressure (see Chapter 4).

The Fukushima Daiichi accident revealed two problems with the operation of these cooling systems under loss-of-power conditions (Chapter 4):

1. The isolation condenser system in Unit 1 did not function after AC and DC power were lost, apparently because the valves inside containment were closed.
2. Ad hoc low-pressure water injection systems were not effective for cooling the Unit 1-3 reactors because of difficulties in depressurizing reactor pressure vessels and venting containments.

The subtle failsafe logic of the DC electrical system impacted the ability of the isolation condenser system of Unit 1 to function following loss of AC and DC power. This same logic system was also operative in the reactor core isolation cooling system in Unit 2 (however, because of the fortunate time sequencing of the loss of AC power, the Unit 2 system was able to operate for many hours).

There may well be other safety-critical plant control systems and subsystems that could be similarly affected by the near-simultaneous loss of AC and DC power. The design bases for these systems need to be better understood and appropriately reflected in plant operating procedures. Alternatively, such systems need to be redesigned to reduce the subtleties of the interactions.

Section 4.3.3.1 in Chapter 4 describes the careful orchestration required to depressurize a reactor pressure vessel and begin injection of low-pressure water. Depressurization removes heat from the reactor core through steam

[4] Isolation condensers can provide cooling for an indefinite period of time as long as water is available on the secondary (shell) side of the heat exchanger and system valves are open. See Chapters 2 and 4.

flashing, which provides time to bring external cooling-water injection systems online. However, steam flashing can also result in the loss of a significant fraction of a reactor pressure vessel's water inventory. Core damage can occur if low-pressure injection does not restore water levels in a timely fashion.[5] Consequently, reactor operators must have well-defined strategies and capabilities for depressurizing reactor pressure vessels and venting containments in a timely manner under loss-of-power conditions. Additionally, there must be a low-pressure heat removal capability that is independent of electrical power.

The use of ad hoc water sources for cooling reactors is not addressed in standard design-basis accidents involving loss of reactor coolant. Moreover, the use of ad hoc water sources requires the availability of portable pumps, not installed core cooling systems. To the committee's knowledge, the only analysis relevant to the type of scenario that occurred in Unit 1 at Fukushima Daiichi is a rudimentary discussion in EPRI (2012b, Volume 2, Appendix AA).

The U.S. nuclear industry has already identified depressurization as an issue and recognizes that there is a trade-off between lowering pressure and operating steam-driven cooling systems (i.e., reactor core isolation cooling and high-pressure coolant injection systems). Williamson et al. (2013) reported on the BWR Owners' Group revisions to Emergency Procedures Guidelines. The guidance on depressurization places core cooling as the highest priority: if depressurization of the reactor pressure vessel results in the loss of systems needed for core cooling, then the guidelines specify that operators: (1) terminate depressurization and (2) maintain reactor pressure vessel pressure as low as possible. This guidance applies during all depressurization steps.

The revised guidelines instruct operators of reactors with reactor core isolation cooling systems to lower reactor pressure to about 200 psi (~1.4 MPa) during an extended loss-of-AC-power event. This will enable a more timely response and less loss of water inventory when transitioning to low-pressure cooling sources such as might be provided through FLEX, thereby helping prevent the core from becoming uncovered.

The FLEX guidance (NEI, 2012) also addresses depressurization:

> Regardless of installed coping capability, all plants will include the ability to use portable pumps to provide RPV/RCS/SG [reactor pressure vessel/reactor coolant system/steam generator] makeup as a means to provide a diverse capability beyond installed equipment. The use of portable pumps to provide RPV/RCS/SG makeup requires a transition and interaction with installed systems. For example, transitioning from RCIC [reactor

[5] The time window could be established through a fuel cladding heatup analysis.

> core isolation cooling] to a portable FLEX pump as the source for RPV makeup requires appropriate controls on the depressurization of the RPV and injection rates to avoid extended core uncovery.

There is a specification in this guidance for providing an indefinite capability to depressurize reactor and supply water to the reactor pressure vessel under loss-of-power conditions.[6] However, the details of how this strategy will be implemented are left up to each plant.

Moreover, if FLEX is not initially successful and core degradation occurs, radiation levels may impede access to locations where FLEX water and power connections are made—just as radiation levels hindered workers' responses at the Fukushima Daiichi plant. FLEX would be greatly enhanced if it focused on preventing core damage as well as on mitigating damage severity should it occur.

5.1.1.4 Instrumentation for Monitoring Critical Thermodynamic Parameters

The loss of AC and DC power in Units 1 and 2 at the Fukushima Daiichi plant shut down key monitoring instrumentation for the reactor pressure vessel, drywell, and suppression chamber (see Chapter 4). The DC-powered monitoring instrumentation in Unit 3 shut down when that unit's batteries were depleted nearly a day and a half later. The validity of readings from working instruments was difficult to ascertain after power was restored. Thermocouples on the exterior surfaces of reactor pressure vessels had been exposed to temperatures above their operating ranges and therefore were likely unreliable. Water-level gauges were likely affected by pressure transients and seawater use for cooling. Some pressure gauges also gave erroneous readings.[7] A lesson that emerges from these observations is that *robust and diverse monitoring instrumentation that can withstand severe accident conditions is essential for diagnosing problems, selecting and implementing accident mitigation strategies, and monitoring their effectiveness.*

The availability and adequacy of monitoring instrumentation were identified as important issues following the Three Mile Island accident in 1979 (see Rempe et al., 2012). In the 1990s, U.S. nuclear power plant licensees and the USNRC addressed this issue through a systematic needs analysis. This analysis involved the identification of (1) sensor information required to monitor key plant functions; (2) locations and operating ranges

---

[6] The indefinite coping time requirement is designed to address outages such as the 72-hour power loss experienced by some units at the Fukushima Daiichi plant. See Chapter 4.

[7] See Gauntt et al. (2012b) for further discussion of data reliability during the accident.

of sensors that provide such information, and (3) environmental conditions that these sensors must withstand during the accident sequences that dominate risks. Additional monitoring instrumentation was added to U.S. nuclear plants as a result of this analysis: for example, reactor pressure indications, a wider range of reactor core temperature indications, and more robust temperature sensors.

The Fukushima Daiichi accident demonstrates the need to further harden essential reactor, containment, and spent fuel pool monitoring instrumentation to better withstand severe-accident conditions. The U.S. nuclear industry and the USNRC have already recognized the need for enhanced reactor and containment monitoring instrumentation, in particular with respect to monitoring spent fuel pool water levels (see Appendix F). The committee judges that further work is needed to evaluate the adequacy and reliability of existing reactor, containment, and spent fuel pool monitoring instrumentation for the risk-dominant accident sequences that emerge from the committee's recommended plant-specific risk evaluations (see Recommendation 5.2A later in this chapter).

The USNRC issued an order[8] in March 2012 requiring that all U.S. nuclear power plants install additional water-level instrumentation in their spent fuel pools (see Appendix F). The order required that this instrumentation provide at least three distinct water levels (the following material is quoted from p. 35 of the Order):

> 1. level that is adequate to support operation of the normal fuel pool cooling system,
> 2. level that is adequate to provide substantial radiation shielding for a person standing on the spent fuel pool operating deck, and
> 3. level where fuel remains covered and actions to implement make-up water addition should no longer be deferred.

The USNRC staff provided interim guidance on implementing this order (USNRC, 2012c).

The USNRC's Advisory Committee on Reactor Safeguards[9] (ACRS) commented on the sufficiency of this monitoring instrumentation:

> [Water-level monitoring] instrumentation should be capable of detecting unexpected changes in SFP [spent fuel pool] level and provide appropriate alarms to alert the operations staff. Emphasis should be on the ability to detect water level reductions early during the event. The system should

---

[8] Order Modifying Licenses with Regard to Reliable Spent Fuel Pool Instrumentation. Available at http://pbadupws.nrc.gov/docs/ML1205/ML12056A044.pdf.

[9] Committee member Dr. Michael Corradini is a member of the Advisory Committee on Reactor Safeguards.

> also have the capability to track and display changes in the SFP water level. This capability would provide the operations staff with the ability to know whether the rate of water level reduction was accelerating, slowing, or remaining constant. (ACRS, 2012, p. 5)

Additionally,

> The [interim staff guidance] should be modified to specify direct measurement of temperature in the SFP. Operators should know, as early as possible, if pool cooling is degrading. Information about SFP temperature provides operators with defense-in-depth information about the status of spent fuel cooling. Temperature information about the approach to boiling may also affect decisions regarding local personnel actions in the vicinity of the SFP. The temperature instrumentation should be simple, capable of being monitored continuously, and displayed in the main control room. (ACRS, 2012, p. 5)

As a result of the systematic evaluation recommended here, nuclear plant licensees and the USNRC might conclude that additional temperature sensors should be placed in pools to provide confirmatory information about the thermodynamic state of water inventories.[10]

5.1.1.5 Hydrogen Control

Based on what has been known about hydrogen behavior since 1980 (see Appendix G), the explosions and damage to reactor buildings at the Fukushima Daiichi plant should not have been that surprising. They illustrate in dramatic fashion the importance of hydrogen control in severe reactor accidents. Hydrogen explosions in Units 1, 3, and 4 at the Fukushima Daiichi plant caused severe structural damage to reactor buildings, created pathways for radioactive material releases to the environment, and greatly impeded onsite accident responses (see Chapter 4). The explosions also caused damage to fuel handling equipment and cooling systems for these units' spent fuel pools. Large additional releases of radioactive materials to the environment might have occurred had the integrity of the spent fuel pools in Units 1, 3, and 4 been compromised. The accident *highlighted the need to examine the adequacy of current hydrogen mitigation measures in some types of reactor containments.*

Nuclear plants with Mark I and Mark II containments worldwide are equipped with nitrogen inerting systems to maintain reduced oxygen concentrations in containment (see Appendix G). Igniters are also used in

---

[10] Water-level sensors provide no information about the thermodynamic state of the pool water until water levels begin to decrease due to boil-off.

boiling water reactors with Mark III–type containments (see Chapter 2) and pressurized water reactors with ice condenser–type containments[11] to prevent the buildup of hydrogen.

The Fukushima Daiichi accident demonstrated in dramatic fashion that inerting the containment is inadequate for preventing hydrogen explosions if the containment fails. This emphasizes the key importance of managing thermal and pressure loads inside containment in order to maintain containment integrity. Being able to safely vent the containment in timely fashion with a minimum release of fission products is a key accident management step that must be available to operators (see Sidebar 2.2 in Chapter 2 for a discussion of venting). Preventing accidental releases of hydrogen into a reactor building even though containment is inerted is important—the large volume of hydrogen generation during a severe accident can overwhelm the inert gas when a hot hydrogen–nitrogen–steam mixture is released into a reactor building. When this mixture leaks into confined spaces outside of containment (i.e., into a reactor building) the steam will condense and a flammable mixture can be formed if the concentration of hydrogen is sufficiently high.

Following the Fukushima Daiichi accident, the USNRC issued orders requiring installation of reliable venting systems in reactors with Mark I and Mark II containments. In June 2013 the USNRC modified this order to require severe-accident-capable venting systems (see Appendix F). These vents should help to reduce hydrogen explosion hazards during severe accidents.

However, the Fukushima Daiichi accident demonstrated that the mere presence of containment vents[12] does not eliminate hydrogen explosion hazards during severe accidents. Indeed, the effectiveness of these vents in limiting hydrogen releases in the buildings will depend on their operability under severe accident conditions (e.g., under loss of DC power and compressed air, as happened at Fukushima Daiichi), as well as the interaction of the vents with building ventilation systems.

The committee judges that reexamination is needed of the potential hazards of hydrogen explosions within the secondary containment (i.e., reactor buildings) of Mark I and Mark II plants. Mitigation strategies such as deliberate ignition, passive autocatalytic recombiners, and post-accident inerting that have been previously examined for large dry containments (NRC, 1987) could be reexamined for secondary containments.

---

[11] Plants with ice condenser containments utilize water ice to condense steam generated during an accident. Plants of this design generally have smaller-volume containments than pressurized water reactors with dry containments.

[12] All of the units at the Fukushima Daiichi plant had containment vents (see Section 2.5.2 in Chapter 2).

Such efforts are in progress in Japan and other countries with Mark I and II BWR plants. The USNRC has identified hydrogen control as an important safety issue but has designated it as a TIER III issue (see Appendix F) to be addressed at some later time.

Flames propagating in spaces filled with equipment and piping or within a building complex generate turbulence that results in substantial increases in flame speed, accelerating flames from low to high speeds and substantially increasing the pressure loading on structures. The severity of the explosions at the Fukushima Daiichi plant also suggests that the deliberate ignition strategies currently in use in Mark III and ice condensers reactors should be reexamined to determine if they will be adequate for accidents involving severe core damage under loss-of-power conditions.

5.1.1.6 Instrumentation for Onsite Radiation and Security Monitoring

The loss of AC and DC power shut down the Fukushima Daiichi plant's onsite radiation monitoring and security systems. The loss of the plant's radiation monitoring systems impeded efforts to monitor radioactive material releases from the Unit 1, 2, and 3 reactors and estimate the timing and magnitude of offsite releases (see Chapter 6).

The loss of onsite security monitoring systems reduced physical protection of the plant grounds and critical plant infrastructure. The reduction of physical protection at the plant increases its vulnerability to attacks from external forces or determined insiders. Additionally, the voluminous amount of information published about the accident provides potential adversaries with data about critical plant systems, their interdependencies, and key personnel; this information could be used to plan and carry out attacks on other nuclear plants. The committee intends to discuss security issues in its second report (see Chapter 1).

A clear lesson learned from the accident is that *onsite radiation and security monitoring systems need to be hardened so that they continue to function during severe accidents.* Alarm annunciation and communication equipment at U.S. nuclear plants are currently required to have a secondary power supply such as an emergency diesel generator. Additionally, intrusion detection and assessment equipment at the protected area perimeter of the plant is required to have an uninterruptible power supply so that it remains operable in the event of the loss of normal power. This equipment may need to be hardened to protect it against severe accidents. The need for and approaches to hardening should be based on plant-specific risk evaluations recommended elsewhere in this chapter (see Recommendation 5.2A).

5.1.1.7 Communication and Real-Time Information Systems

The Fukushima Daiichi accident highlighted the need for reliable communication links between control rooms and technical support centers, between control rooms and the field, and between onsite and offsite support facilities during severe accidents. The limited means of communication during the Fukushima Daiichi accident degraded the ability of plant personnel to plan and coordinate their response actions. The loss of the offsite emergency response center disrupted lines of communication with local and national government agencies. The loss of communication infrastructure contributed to the central government's concerns that it was not receiving timely and accurate information about the status of plant.

The USNRC's Near-Term Task Force (USNRC NTTF, 2011) report highlighted the need for reliable communications equipment (e.g., hard-wired telephones, cellular telephones, satellite telephones, radios, and pagers) for communicating onsite and offsite, including during events that may involve extended loss of AC power and/or damage to external telecommunication infrastructure (e.g., phone switches and cell towers). The committee concurs with this assessment.

The committee suggests that there is also a need to ensure the reliability of data communications, both onsite (e.g., between the control room and the technical support center[13]) and offsite (e.g., between the plant and offsite government and regulatory agencies), particularly during extended AC power loss. The Fukushima Daiichi accident highlighted the importance of real-time information systems (e.g., Safety Parameter Display Systems) for enabling personnel to maintain situational awareness of plant conditions. In discussions with the committee, TEPCO personnel commented that the lack of availability of this system in the control rooms and ERCs contributed to delays in diagnosing plant conditions (see Chapter 4).

The committee also concurs with the Near-Term Task Force recommendation on developing reliable and secure data pathways between U.S. nuclear plants and USNRC headquarters to enable direct and automatic electronic transmission of critical plant parameters during emergencies. The task force report notes that

> Having data provided directly from automated sources at the site also gives confidence to government authorities and the public that the plant operator is not filtering the details of an evolving accident. (USNRC NTTF, 2011, p. 55)

---

[13] Technical support centers at U.S. nuclear plants carry out many of the same functions as ERCs at Japanese plants. More information about this and related facilities is provided elsewhere in this chapter.

It is particularly important that these data pathways be functional during extended loss-of-AC power events, multiunit events, and events that affect multiple plants simultaneously.

5.1.1.8 Peer Review

The committee's call for adequate peer review is intended to increase the quality and completeness of the changes resulting from its recommendations and thereby enhance nuclear plant safety. The committee judges that peer review will also enhance the transparency, credibility, and public confidence in actions taken by industry and its regulator (USNRC) to implement lessons learned from the Fukushima nuclear accident. Peer review has the following characteristics: expert (including national and international perspectives), independent, external, and technical (NRC, 1998, p. 2; see also USNRC, 1998) and is transparent to audiences external to the industry and its regulator.

The nuclear industry and its regulator already carry out a large number of technical and operational reviews. Industry reviews are carried out, for example, by reactor owners' groups, the Institute of Nuclear Power Operations (see Sidebar 7.2 in Chapter 7), and plant-specific safety review committees. The industry regulator obtains peer reviews from an independent advisory committee, the Advisory Committee on Reactor Safeguards.

The committee acknowledges the importance of these review groups and their continuing engagement in the process of ensuring the adequacy of the U.S. response to the lessons learned from Fukushima. At the same time, it is essential that the regulator and industry be vigilant to the one key lesson from Fukushima, which is the value of independent, informed perspectives that are outside the immediate community of decision makers as provided by peer review in the broad sense described above.

5.1.2 Procedures and Training

RECOMMENDATION 5.1B: As the nuclear industry and its regulator implement the actions referenced in Finding 5.1, they should give specific attention to improving resource availability and operator training to enable effective responses to beyond-design-basis events, including, when necessary, developing and implementing ad hoc responses to deal with unanticipated complexities. Attention to the following is specifically needed:

1. Staffing levels for emergencies involving multiple reactors at a site, that last for extended durations, and/or that involve stranded-plant conditions.[14]

2. Strengthening and better integrating emergency procedures, extensive damage mitigation guidelines, and severe accident management guidelines, in particular for

- **Coping with the complete loss of AC and DC power for extended periods,**
- **Depressurizing reactor pressure vessels and venting containments when DC power and installed plant air supplies (i.e., compressed air and gas) are unavailable,**
- **Injecting low-pressure water when plant power is unavailable,**
- **Transitioning between reactor pressure vessel depressurization and low-pressure water injection while maintaining sufficient water levels to protect the core from damage,**
- **Preventing and mitigating the effects of large hydrogen explosions on cooling systems and containments, and**
- **Maintaining cold shutdown in reactors that are undergoing maintenance outages when critical safety systems have been disabled.**

3. Training of operators and plant emergency response organizations, in particular,

- **Specific training on the use of ad hoc responses for bringing reactors to safe shutdown during extreme beyond-design-basis events, and**
- **More general training to reinforce understanding of nuclear plant system design and operation and enhance operators' capabilities for managing emergency situations.**

The quality and completeness of the changes that result from this recommendation should be adequately peer reviewed (see Section 5.1.1.8).

5.1.2.1 Staffing Levels

Staffing levels at the Fukushima Daiichi plant were inadequate for managing the accident response (see Chapter 4) because the accident extended over multiple days and involved multiple reactor units. A clear lesson from this accident is that *staffing levels and responsibilities at nuclear plants need to be reassessed to ensure that they are adequate for managing complex emergencies.*

During an emergency at a nuclear plant in the United States several onsite and offsite emergency response facilities are activated to provide

[14] That is, when the plant is cut off from outside supply of materials and personnel.

technical and management support: technical support centers, which provide management and technical support to control room personnel; operational support centers, which are used as an assembly area for damage repair teams; and emergency operations facilities, which provide information about the emergency to federal, state, tribal, and local authorities (USNRC NTTF, 2011, p. 53).

Staffing numbers, roles, and responsibilities at U.S. nuclear plants and these associated emergency response facilities need to be reassessed to ensure that critical personnel functions, including communication and coordination functions, can be supported in complex emergencies, particularly emergencies involving multiple reactor units (at multiunit sites), and/or require 24-hour operations with shift turnovers. The reassessment should ensure that the support facilities are organized and staffed to have high reliability, appropriate levels of authority, and appropriate mixes of knowledge and experience to develop and orchestrate response plans in real time.

The analysis of staffing needs should also take into account any additional functions or workloads arising from the industry's FLEX initiative to establish regional centers as a common source of emergency equipment (see Appendix F). Although the regional centers can provide equipment and resources that can aid onsite staff in responding to an accident, they are also likely to impose additional burdens on onsite staff to handle communications, coordination, and logistics.

The analysis of staffing needs should also consider stranded-plant conditions—that is, when the plant is cut off from outside supply of materials and personnel. U.S. plants have stranded-plant procedures that address staffing levels if natural disasters restrict access to plants. These should be reviewed and augmented as necessary to ensure the availability of personnel and resources during severe accidents.

5.1.2.2 Emergency Procedures and Guidance

Reactor operators and emergency response center personnel at the Fukushima Daiichi plant lacked written guidance for bringing the plant's reactors to cold shutdown under loss-of-power conditions. TEPCO (2012b, p. 52) described the situation this way:

> [I]n this accident, due to the tsunami impact, which was far beyond the previous estimations, almost all equipment and power sources expected to operate to respond to the accident lost their functions, resulting in a situation that was outside of the assumptions that were made to plan accident response.

An important lesson from this accident is that *the written guidance used at nuclear plants to guide operator actions during off-normal events needs to be strengthened and better integrated to address loss-of-power conditions in operating and shutdown reactors.*

Nuclear plant operators have written aids to guide them in responding to off-normal events at nuclear plants; these include emergency operating procedures (EOPs), severe accident management guidelines (SAMG), and extensive damage mitigation guidelines[15] (EDMGs). Information about development and use of EOPs, SAMG, and EDMGs in the United States is provided in Appendix H.

Off-normal events involving the loss-of-offsite AC power are within the design basis for nuclear plants. Operators are trained to respond to such events using EOPs and other plant procedures such as abnormal operating procedures and alarm response procedures. EOPs typically apply as long as reactor pressure and water level can be monitored and remain within acceptable ranges. The shift supervisor, who is stationed in the control room, and the plant manager have command-and-control responsibilities for implementing EOPs. (Both individuals possess senior reactor operator licenses.)

Operators would transition to SAMG or EDMGs when an off-normal event progresses beyond conditions covered by EOPs. The decision about which of these procedures to use would be based on plant conditions:

- Transition to SAMG would take place when core damage was determined to be imminent. The technical support center director would have command-and-control responsibilities for implementing SAMG. However, the control room could begin implementation if the technical support center was not yet staffed. The goals of SAMG are to stabilize the reactor core, maintain containment, and minimize the release of the core's radioactive materials after fuel damage has occurred.
- Operators may elect to implement EDMGs when large fires or explosions damage large areas of a plant or disable the plant's command-and-control structure. Responsibility for implementing EDMGs could reside in the control room, technical support center, or emergency operations facility (see USNRC NTTF, 2011, pp. 46-49). EDMGs provide for the use of portable equipment (e.g., generators, pumps) to restore basic plant monitoring and safety functions.

---

[15] EDMGs provide strategies for maintaining or restoring core cooling and containment (and spent fuel pool cooling) in emergencies involving the loss of large areas of the plant as a result of fires and explosions. These guidelines were developed after the September 11, 2001, terrorist attacks.

The Fukushima Daiichi plant did not have EDMGs, and the SAMG in effect at the plant at the time of the accident proved to be inadequate because it did not anticipate complete loss-of-power conditions. SAMG in place in the United States at the time of the accident also did not anticipate such conditions.

The Fukushima Daiichi accident exposed plant operators to complex conditions and competing demands. Had this accident occurred in the United States, it would have taken plant operators out of EOPs and into EDMGs or SAMG, depending on plant conditions. Arguably, operators in the United States may have been able to use guidelines and equipment available via EDMGs to prevent or delay damage to fuel in the reactor core. If core damage had occurred, then operators could have used SAMG to stabilize the core and maintain containment. However, it is not at all clear that U.S. operators could have prevented core damage given the severity of the accident; to the committee's knowledge, there is no experience in this regime in the U.S. nuclear industry.

Although most emergency response drills involve scenarios that include core damage, operator training does not routinely exercise the range of SAMG response options and does not involve multiple-unit scenarios. Examination of the factors that drive human responses under such conditions is essential for integrating EOPs, SAMG, and EDMGs.

Recommendation 8 of the USNRC's Near-Term Task Force (USNRC NTTF, 2011) called for strengthening and integrating EOPs, EDMGs, and SAMG. This recommendation will be implemented through rulemaking, perhaps leading to a final rule in 2016. The enhanced capabilities available through the U.S. nuclear industry's FLEX initiative will undoubtedly be considered during the rulemaking process.

The committee concurs with this USNRC NTTF assessment and recommends that the following issues be specifically examined and relevant guidance developed where appropriate:

1. Coping with the complete loss of AC and DC power for extended periods (e.g., up to 72 hours during the Fukushima Daiichi accident), not just for standard "station blackout conditions" involving loss of AC power for a limited (4- to 8-hour) duration;
2. Depressurizing reactor pressure vessels and containments when DC power and installed plant air supplies are unavailable;
3. Injecting low-pressure water when plant power is unavailable;
4. Transitioning between reactor pressure vessel depressurization (point 2) and low-pressure water injection (point 3) while maintaining sufficient water levels to protect the core from damage;

5. Mitigating the effects of large hydrogen explosions on cooling systems and containments; and

6. Maintaining cold shutdown in reactors that are undergoing maintenance outages when critical safety systems have been disabled.

With respect to point 6, SAMG are needed not only for operating reactors but also for reactors that are in "cold shutdown," as was the case for Unit 5 at the Fukushima Daiichi plant (see Chapter 4). Unit 5 was in a maintenance outage when the earthquake and tsunami occurred; its containment was open for inspection, some safety equipment had been disabled, and the reactor pressure vessel was pressurized for leak testing. There was a substantial loss of water inventory in the reactor pressure vessel after cooling system functions were lost. It is not unusual to have one or more reactors in maintenance outages at multiunit plants.

The foregoing underscores the importance of understanding and coping with risks during shutdown conditions. It was increasingly recognized in the global nuclear safety community by the 1990s that core-damage risk at shutdown could be comparable to that at power operation (e.g., IAEA, 1994; Jo et al., 1995). While various plant and operational improvements have been considered since then, it is important to continue to recognize that severe events need to be considered for shutdown conditions.

The ACRS considered the issue of overlapping guidelines and procedures in connection with plant fire response procedures and commented on the need for better integration:

> These procedures provide operator guidance for coping with fires that are beyond a plant's original design basis. Some plant-specific fire response procedures instruct operators to manually de-energize major electrical buses and realign fluid systems in configurations that may not be consistent with the guidance or expectations in the EOPs. Experience from actual fire events has shown that parallel execution of fire procedures, Abnormal Operating Procedures (AOPs), and EOPs can be difficult and can introduce operational complexity. Therefore, these procedures should also be included in the comprehensive efforts to better coordinate and integrate operator responses during challenging plant conditions. (ACRS, 2011, p. 8)

The integration of EOPs, EDMGs, and SAMG will be a complex effort that requires substantial interactions among several parties: plant operations, engineering, and management personnel; reactor owners' groups; EPRI and INPO; technical experts; and regulators. Extensive testing of the integrated procedures will also be required at each nuclear plant.

The nuclear industry could develop accident management advisory tools to assist in the development of SAMG, assess their effectiveness and completeness, and better inform operator actions for accident management. The usefulness of such tools will depend on the availability of accurate data for key plant operating parameters, the ability to model accidents that progress beyond the design basis, and the ability to model the potential range of operator actions. It is important that the regulator (the USNRC) have the ability to evaluate the technical accuracy and utility of these tools.

FLEX strategies at individual nuclear plants might need to be augmented to provide the resources required to implement revised SAMG. For example:

- Coping with power loss will likely require the availability of portable batteries, emergency generators, and prepared power cables.
- Depressurizing reactor pressure vessels and containments might require the availability of portable power supplies and compressed gas (air or nitrogen).
- Low-pressure water injection might require the availability of self-powered portable pumps that can generate sufficiently high pressures to overcome a partially depressurized reactor vessel or partially vented containment.

Work is already under way by industry to address some of these issues. For example, the BWR Owners' Group Emergency Procedures Committee issued revisions to guidelines affecting emergency procedures and severe accidents for boiling water reactors in the United States. Special emphasis in the revised guidance has been given to loss-of-onsite-power scenarios (Williamson et al., 2013). New generic procedures are being implemented through workshops being held throughout the international community, including Japan.

Consideration should also be given to explicitly ensuring that emergency procedures, severe accident management guidelines, and support documents (e.g., blueprints and calculations) are available to workers during loss-of-power events. Initial and continuing training of operators and other workers should include exercising their ability to diagnose plant conditions and implement necessary actions without relying on computer systems that might not be unavailable during such events.

Peer reviews of these procedures, guidance, and strategies will be needed to ensure that they are based on appropriate sets of plant damage states and do not contain unidentified "cliff-edge" effects.[16]

[16] That is, plant damage states that would prevent an adequate response using FLEX and revised SAMG.

5.1.2.3 Training of Operators and Plant Emergency Response Organizations

The Fukushima Daiichi accident demonstrated that extreme beyond-design-basis events pose multiple challenges to human performance, including challenges to situation assessment, planning, decision making, communication, coordination, and task execution. Because events of this complexity had not been anticipated, the training received by the Fukushima Daiichi operational and ERC staff did not sufficiently prepare them for these extreme challenges. Given this experience, U.S. nuclear power plant training for responding to extreme beyond-design-basis events should be reviewed to ensure that it is sufficiently effective. This includes training for control room operators, technical support center personnel, and other plant personnel who would be involved in decision making and response to severe accidents.[17]

The committee judges that two types of training are important:

1. Specific training on the use of strategies for bringing reactors to safe shutdown during extreme beyond-design-basis events. This includes training on

- Operation of reactor heat removal systems, including failsafe logic of control systems and manual-control backup options, with special attention being paid to scenarios where AC and DC power sources fail in different sequences;
- Depressurizing reactors while tracking and controlling reactor water level, pressures, and temperatures;
- Means for injecting low-pressure water from various plant sources; and
- Recognizing instrument failures and degradations and developing alternative means to obtain critical monitoring data, especially with respect to reactor water level and pressure, containment pressure, and temperatures.

The plant-specific risk evaluations recommended by the committee (see Recommendation 5.2A) will be important sources of scenarios for this training. The training should also account for conditions that are likely to be encountered during these scenarios: poor lighting, flooding, high radiation, fires, and other plant damage. The role of other support systems, for example, instrument air, should also be considered in the training.

[17] At present, U.S. nuclear plants have accredited training programs that are conducted annually for a range of maintenance, engineering, technical personnel, and operators.

2. More general training intended to support effective performance of the broader emergency response organization. This includes training not only for control room operators but also the Shift Technical Advisor resident in the control room, and technical support staff operating out of the technical support center and emergency operations facility. It is important that this training (i) reinforce fundamental understanding of nuclear plant system design and operation—this includes having a full grasp of the capabilities of all plant equipment (not just so-called "safety critical" equipment) and how it can be marshaled in emergency situations; and (ii) enhance capabilities for managing emergency situations including, for example, capabilities for the following:

- Reasoning with missing, conflicting, and misleading data (e.g., from degraded sensors);
- Reasoning that requires understanding complex system interlocking, automated system behavior, and failsafe operation;
- Reasoning under data overload conditions;
- Managing competing demands on attention;
- Prioritizing and making goal trade-offs;
- Developing and implementing mitigation plans that are not fully covered by available procedures and guidance;
- Communicating and coordinating activities within and across physical locations and shifts; and
- Establishing and exercising clear roles, responsibilities, and lines of authority within and across the various control centers (e.g., control room, technical support center, emergency operations facility), particularly in situations where roles and responsibilities have to be dynamically redefined in response to evolving situations.

The objective of this training is to develop capacities to respond adaptively in the face of unforeseen situations. These training activities would help build the kinds of problem-solving, decision-making and communication skills that were demonstrated to be critically important in the Fukushima Daiichi accident.

There is extensive literature that can be drawn on for training techniques to improve cognitive skills for responding adaptively under high-stress conditions. This includes training for decision making under stress (Cannon-Bowers and Salas, 1998); training for emergency responders (Wall et al., 2004); and training for coping with complex severe accident conditions (Mumaw et al., 1994).

Training to improve cognitive skills is intended as an adjunct to, rather than a substitute for, development of robust preplanned procedures and decision support tools for guiding performance. In general, availability of effective decision support is the preferred solution because cognitive performance is prone to error under high-stress, time-pressured conditions. However, as the Fukushima Daiichi accident illustrates, reliance on cognitive skills can become critically important when ad hoc responses are required for coping with unanticipated situations that are not well handled by the available procedural guidance and decision support.

The Fukushima Daiichi accident also highlighted the importance of training to enable efficient planning and execution of manual actions that may need to be performed under harsh time-pressured conditions (e.g., lack of lighting and high radiation levels). This includes manual actions that may be needed when remote control capabilities are lost (e.g., planning and execution of manual valve operation for containment venting); and movement and activation of portable auxiliary equipment (e.g., portable pumps) that might be called for as part of severe accident response strategies such as FLEX.

5.1.3 Discussion

Many of the committee-identified lessons learned for nuclear plant systems, procedures, and training (Sections 5.1.1 and 5.1.2) have been anticipated in previous analyses, some over three decades ago. A 1981 Oak Ridge National Laboratory (ORNL) report (Cook et al., 1981), for example, examines the consequences of an unmitigated station blackout at Browns Ferry, a BWR/4 Mark I plant in Alabama. Among the insights gained from that study are the following:

- Neither existing training nor emergency operating procedures adequately prepared operators for an unmitigated station blackout accident.
- The plant could cope with loss of offsite AC power as long as onsite AC power and/or station DC batteries were available. Station battery lifetime was a primary determinant of accident sequence progression timing.
- Plant instrumentation (sensors, detectors, indicators, and annunciators) would not be functional and/or provide reliable information once DC power was lost and core damage was initiated. Operators would be "flying blind" during the most critical phases of the accident.

ORNL also published several papers on the role of BWR reactor buildings in severe accidents (Greene, 1986, 1987, 1988, 1990; Greene and Hodge, 1986). The analyses in those papers suggested that

- Intact reactor buildings could play a significant role in mitigating the consequences of severe accidents in BWRs.

- Hydrogen explosion-induced differential pressures in BWR Mark I reactor buildings could exceed their design differential pressures by a factor of four. Consequently, hydrogen explosions present a real potential for reactor building failure and secondary containment bypass.

Greene (2014) provides an interesting historical perspective on severe accident initiation, progression, and mitigation in BWRs. The author posits that many lessons learned from this work have been forgotten or ignored. Indeed, he observes that

> Based on historical BWR station blackout studies, and given the hybrid short-/long-term station blackout sequence that occurred at Fukushima Daiichi, we have little reason to be surprised about the course and timing of events that occurred in Fukushima Daiichi Units 1-3.

The committee agrees that the Fukushima accident was not a technical surprise and was in fact anticipated by previous severe reactor accident analyses. Indeed, there is a well-documented and logical progression of knowledge regarding severe reactor accidents, beginning with WASH-740 (AEC, 1957) and continuing through to the present-day State-of-the-Art Reactor Consequence Analyses (USNRC, 2013c,d). There is a continuing stream of technical reports, papers, conferences, and books that sustain and augment the knowledge base. See Sidebar 5.1 for a brief description of the history of severe accident analysis.

5.2 NUCLEAR PLANT SAFETY RISKS

FINDING 5.2: Beyond-design-basis events—particularly low-frequency, high-magnitude (i.e., extreme) events—can produce severe accidents at nuclear plants that damage reactor cores and stored spent fuel. Such accidents can result in the generation and combustion of hydrogen within the plant and release of radioactive material to the offsite environment. There is a need to better understand the safety risks[18] that arise from such events and take appropriate countermeasures to reduce them.

RECOMMENDATION 5.2A: The U.S. nuclear industry and the U.S. Nuclear Regulatory Commission should strengthen their capabilities for identifying, evaluating, and managing the risks from beyond-design-basis events. Particular attention is needed to improve the identification of such events; better account for plant system interactions and the per-

[18] Risk is defined and discussed in Appendix I.

SIDEBAR 5.1
Brief History of Severe Accident Analyses

The U.S. Atomic Energy Commission sponsored the first major study of the theoretical consequences of severe accidents at large nuclear power plants in the mid-1950s. This study was performed by Brookhaven National Laboratory and resulted in the WASH-740 report (AEC, 1957). The subsequent Reactor Safety Study, which was issued as the WASH-1400 report in 1975 (USNRC, 1975), concluded that a severe accident was "the only way that potentially large amounts of radioactivity could be released by melting the fuel in the reactor core." All risk studies performed subsequent to WASH-1400 have found this to be the case.

Industry also advanced the state of the art of severe accident analysis in the early 1980s as a result of the full-scope PRAs performed for the Indian Point, Zion (Commonwealth Edison, 1981; Consolidated Edison, 1982), and Limerick nuclear plants (Philadelphia Electric Company, 1981). These PRAs made major advancements to severe accident analysis, particularly with respect to containment response analysis and to radiological source-term analysis.

A substantial research program on severe accident phenomenology was planned and initiated by the USNRC following the Three Mile Island Unit 2 accident in 1979. This program included experimental and analytical studies of accident phenomenology (i.e., the physical, chemical, and radiological processes that occur during a severe accident). In 1980, the USNRC issued a *Federal Register* Notice for a proposed rulemaking on severe accident design criteria (USNRC, 1980d). In parallel with this regulatory effort, the nuclear industry sponsored the Industry Degraded Core Rulemaking (IDCOR) program. This program, which was active during 1981-1984, also involved experiments and analytical studies. The USNRC later withdrew the proposed rulemaking and issued a severe accident policy statement in 1985 (USNRC, 1985) which set the regulatory course for ad-

formance of plant operators and other critical personnel in responding to such events; and better estimate the broad range of offsite health, environmental, economic, and social consequences that can result from such events.

RECOMMENDATION 5.2B: The U.S. Nuclear Regulatory Commission should support industry's efforts to strengthen its capabilities by providing guidance on approaches and by overseeing independent review by technical peers (i.e., peer review).

RECOMMENDATION 5.2C: As the U.S. nuclear industry and the U.S. Nuclear Regulatory Commission carry out the actions in Recommendation 5.2A, they should pay particular attention to the risks from

dressing severe accidents. The USNRC also issued a policy statement on safety goals (USNRC, 1986).

By the mid-1980s, new computational models of severe accident phenomenology had been developed and subjected to peer review. Studies of reactor severe accidents and their public health consequences were being carried out throughout the 1980s in many countries with light-water reactor programs. Many conferences and symposia took place, and papers and reports were widely disseminated. In the United States, a major update of the WASH-1400 report was issued (USNRC, 1990). It evaluated severe accident risks at five nuclear plants.

Beginning in 1988, the U.S. nuclear industry performed assessments of severe accident vulnerabilities for each licensed nuclear power plant. These individual plant examinations were done for both internal and external event initiators and were essentially PRAs. The USNRC issued its perspectives documents starting in the late 1990s (USNRC, 1997a, 2002c) which summarized the plant vulnerabilities and proposed modifications for each plant.

At present, severe accident studies are continuing in most countries with light-water reactors. Many international conferences and symposia feature studies on refinement of tools and confirmation of theoretical models based on experiments. Most university programs on nuclear engineering cover severe accidents in their curricula, and the topic is covered in contemporary textbooks and monographs on reactor safety. Short courses on severe accidents are typically offered in conjunction with conferences on PRA. Severe accident management guidelines have been developed and refined based on insights from phenomenological studies.

The most recent risk study that uses current severe accident information is the State-of-the-Art Reactor Consequence Analyses Project (USNRC, 2013c,d). The USNRC is also performing a Level 3 risk analysis of a pressurized water reactor, which will be completed in the next few years.

beyond-design-basis events that have the potential to affect large geographic regions and multiple nuclear plants. These include earthquakes, tsunamis and other geographically extensive floods, and geomagnetic disturbances.

A *design-basis event* is a postulated event that a nuclear plant system, including its structures and components, must be designed and constructed to withstand without a loss of functions necessary to ensure public health and safety. Such events can include malfunctions of plant structures or components due to manufacturing defects or wear or failures caused by outside agents, for example, natural hazards. An event that is *beyond design basis* has characteristics that could challenge the design of plant structures and components and lead to a loss of critical safety functions. (See Sidebar 1.2 in Chapter 1 for additional discussion of severe accident terminology.)

Extreme external events—that is, large-magnitude environmental events such as earthquakes and floods that have recurrence intervals of centuries to millennia—have the potential to cause failures in multiple plant operating and safety systems, resulting in core-damage accidents that involve the release of explosive hydrogen within the plant and release of radioactive materials to offsite environments. The Fukushima Daiichi accident demonstrated that the health[19] (including mental well-being), environmental, economic, and social consequences of such accidents can be considerable.

The accident at the Fukushima Daiichi nuclear plant has prompted the U.S. nuclear industry and its regulator, the USNRC, to take several actions (Appendix F) to better understand and mitigate the risks from extreme external events. Of relevance to the present discussion are the following three actions (see Appendix F, especially Table F.1, for details):

1. The USNRC requested that nuclear plant licensees perform detailed inspections (referred to as "walkdowns") of currently installed seismic and flooding protection features at U.S. nuclear plants and identify, correct, and report any degraded conditions.
2. The USNRC also requested that nuclear plant licensees use present-day information to reevaluate seismic and external flooding effects and hazards that could impact plants to determine if plant structures, systems, and/or components need to be updated.
3. The USNRC ordered nuclear plant licensees to implement strategies for coping without permanent electrical power sources for an indefinite amount of time.

These initiatives are important and necessary steps to evaluate and mitigate the risks associated with beyond-design-basis external events. However, as currently organized, they are one-time efforts directed at two specific external hazards (i.e., seismic and flooding hazards). In the committee's judgment, there is a need for a broader examination of extreme external hazards that can affect nuclear plant safety. The committee's recommendation that the U.S. nuclear industry and USNRC strengthen their capabilities for identifying and evaluating the risks associated with beyond-design-basis external events is intended to address this broader need.

There are several approaches that could be used to address the committee's recommendations. The choice of approaches involves nontechnical policy considerations and, for regulatory actions, would also require input

[19] See Section 6.1.1 in Chapter 6 for more details on health effects from the Fukushima Daiichi accident.

from potentially affected stakeholders. Whatever approaches are adopted, however, the committee suggests that they

- Be implemented by the nuclear industry with oversight from regulators,
- Use established and credible risk evaluation tools and criteria,
- Use peer review to assess the quality and completeness of the risk evaluations, and
- Be updated as new information about extreme external hazards becomes available.

The nuclear industry in the United States and many other countries already uses a risk evaluation method that could be used to examine risks from beyond-design-basis external events such as occurred at the Fukushima Daiichi plant: probabilistic risk assessment (PRA). PRA is used routinely in the United States and many other countries for designing and operating nuclear plants. Information about the development and use of PRA can be found, for example, in Bley et al. (1992), Keller and Modarres (2005), and Garrick (2008).

PRAs are required for new nuclear plants in the United States but not for existing plants. Nevertheless, they exist in various forms for all existing plants and are used extensively in decision making about plant operations. Appendix I defines PRA and examines its applications in Japan and the United States.

PRAs in use at existing U.S. nuclear plants would need to be enhanced to make them useful for assessing beyond-design-basis external events such as occurred at the Fukushima Daiichi plant; in particular, they would need to

1. Integrate external events, including extreme events: earthquakes; floods (including tsunamis); and other offsite events that can disrupt electrical power, damage the electrical grid, and make it difficult to resupply equipment, fuel, communication systems, and personnel. Such "other" offsite events could include, for example, regional failures of the electric power grid as a result of equipment malfunctions, human error, terrorism (not discussed in this report), or geomagnetic disturbances (see Sidebar 5.2). Such regional-scale events could simultaneously affect multiple nuclear plants.
2. Account for potential interactions among plant components (e.g.,

SIDEBAR 5.2
Coronal Mass Ejections

Coronal mass ejections (CMEs) are massive bursts of charged plasmas from the surface of the Sun that travel through space at hundreds of kilometers per second. They can produce severe geomagnetic disturbances (e.g., terawatt-scale oscillating electrical currents) if they encounter Earth's magnetosphere, which in turn can induce quasi-DC currents in electrical transmission lines. These currents can enter and exit power systems at transformer grounds, disrupting power system operations and damaging equipment (EPRI and NEC, 2011).

Large CME-induced geomagnetic disturbances have affected the electrical and communications infrastructure in North America during recent history:

- The "Carrington Event" in September 1859 produced aurorae that could be seen as far south as Cuba and Hawaii. This event induced currents in telegraph lines causing large-scale failures of telegraph systems; some systems continued to operate even after they were disconnected from their power sources (Carlowicz and Lopez, 2002).
- In May 1921, the largest CME of the 20th century, the "Great Storm," disabled most telegraph service in the United States and damaged underwater trans-Atlantic cables.
- A CME in March 1989 collapsed the Hydro-Québec power grid and nearly toppled the U.S. grid. The net cost of the grid failure was estimated to be $13.2 million; some damaged transmission-system equipment was not returned to service for several months (Bolduc, 2002, p. 1794).

multiple power sources, actuators, and control circuits) that can cause nuclear plant systems to fail in unexpected ways[20] (Sidebar 5.3).

3. Account for potential interactions among reactors at multiunit plants.[21]

4. Account for situations that could hamper plant recovery efforts (e.g., blocked roads or a damaged electrical grid) and slow offsite assistance.

5. Account for the performance of plant operators and other critical personnel. This includes a consideration of situational challenges (e.g., unavailability of or misleading sensor indications and lack of relevant procedural guidance) that are likely to arise in severe accidents, and the individual, team, and organizational decision-making processes that are

[20] The isolation condenser failure in Unit 1 at the Fukushima Daiichi plant is an example of such an interaction. See Chapter 4.

[21] PRAs for existing plants generally address risk on a unit-by-unit basis. Furthermore, essentially all existing risk-informed rules and regulations are based on single-unit analyses.

Riley (2012, p. 1) notes that "By virtue of their rarity, extreme space weather events [e.g., geomagnetic disturbances], such as the Carrington event of 1859, are difficult to study, their rates of occurrence are difficult to estimate, and prediction of a specific future event is virtually impossible." Nevertheless, Riley (2012) and Kappenman (2010, 2012) suggest that such events have occurrence frequencies on the order of one or more per century; Kappenman (2012) also suggests that extreme geomagnetic disturbances can cause severe damage to the electrical grid. A 2011 JASON report (MITRE, 2011) questions the plausibility of Kappenman's worst-case scenario for damage to the electrical grid from an extreme geomagnetic disturbance but also calls for a study of the vulnerability of the U.S. grid.

The potential impacts of CME-induced geomagnetic disturbances on the electrical grid are well recognized (e.g., CENTRA, 2011; EPRI and NEC, 2011). Measures can be taken to protect the grid from damage from such disturbances, as was done, for example, by Hydro-Québec following the 1989 CME (see Bolduc, 2002). In 2013, the Federal Energy Regulatory Commission ordered the development of electrical grid reliability standards for geomagnetic disturbances (FERC, 2013). The standards are to be developed over a 2-year period and implemented thereafter. It could be several more years before a plan is developed and executed to implement those standards.

The USNRC has initiated a phased rulemaking to ensure long-term cooling and unattended water makeup of spent fuel pools that could be affected by prolonged disruptions to the electrical grid resulting from geomagnetic disturbances (USNRC, 2012e). This action was initiated in response to a petition asserting that prolonged outages of the North American power grid caused by geomagnetic disturbances could result in diesel generator fuel depletion and failure of resupply.

likely to influence performance under time-pressured, high-stress conditions. Additional discussion of human performance during severe accidents is provided in Appendix J.

6. Consider[22] the offsite health, environmental, economic, and social consequences that can result from severe nuclear accidents:

- Health: death and injury (including mental distress) resulting from evacuations and exposures to radioactive releases.
- Environmental: contamination of air, water, and land; waste remediation and disposal costs.

[22] Some offsite consequences of severe nuclear accidents are difficult to quantify, especially some types of health and social consequences. Nevertheless, the Fukushima Daiichi nuclear accident demonstrated that such consequences can be substantial (see Chapter 6). Silva et al. (2013) describe a methodology for assessing some health, economic, social, and environmental consequences from severe nuclear accidents.

SIDEBAR 5.3
Race and Corner Conditions

The response of engineering safety systems in Units 1 and 2 at the Fukushima Daiichi plant to the loss of AC and DC power revealed a subtle but significant vulnerability of control systems; this vulnerability has important implications for risk analysis. As noted in Chapter 4, the power in both the AC and DC circuits was lost nearly simultaneously, resulting in a "race" between DC logic circuits commanding the failsafe closure of the isolation valves and the loss of AC power to the valve motors. This race had different outcomes in Unit 1 and Unit 2: In Unit 1, the isolation condenser's AC-operated valves inside containment were effectively closed before the power failed; in Unit 2, in contrast, the valves for the reactor core isolation cooling system remained open. These different outcomes were apparently determined by small differences in the timing and sequence of power failures resulting from the flooding of multiple power sources and distribution systems.

The situation where two signals compete to perform actions is known as a *race condition*. This condition can occur whenever electronic logic circuits and computers are used to control safety systems. Such systems can be found in technologies ranging from nuclear power plants to your automobile. When a race condition is not anticipated or correctly resolved, the consequences can range from merely annoying (e.g., causing your personal computer to "blue screen") to catastrophic (e.g., disabling the isolation condenser in Unit 1 at the Fukushima Daiichi plant).

Understanding race conditions is of increasing importance in both system design and safety analysis (Levenson, 1995). The control system must not only handle all permutations of input states under normal operating conditions but also the failure of power supplies for the logic controller and all controlled systems. It is essential that the logic controller and controlled systems wind up in predictable and safe states following a power loss or transient. This did not happen at the Fukushima Daiichi plant: following the complete loss of AC and DC power, operators had no idea of the status of almost all critical systems.

The inclusion of race conditions in risk analysis is complicated by several factors. First, it requires a more detailed analysis of the logic controller software and hardware, power circuits, and structures, systems, and components than are usually considered in a plant-level risk assessment. Second, race conditions often happen when multiple abnormal conditions and seemingly unlikely combinations of events take place. These combinations are frequently found at extreme values of parameters, sometimes referred to as *corner conditions*, within the event space and fault sequences being considered as part of a risk analysis. Third, many

- Economic: loss of economic activity as well as support for evacuated populations, cleanup of contaminated areas, and relocation and/or resettlement of affected populations.
- Social: disruptions to families and communities; loss of trust.

power plant systems are large and respond slowly because of the inertia in the plant's systems and components—except for the logic controller and electrical circuits. This creates a mismatch that has to be analyzed carefully; specialized engineering analysis may be required to examine high-consequence, low-probability corner conditions (e.g., multiple, nearly simultaneous power failures on buses that are expected to be independent).

The simultaneous loss of all AC and DC power at Fukushima Daiichi appears to be a corner condition that the plant's engineering systems were not designed to handle. The unknown state of multiple safety-related components and the inability to actuate those components greatly complicated management of the accident and may have contributed to its severity. This condition was manifested in at least three plant systems (see Chapter 4):

- The closure of isolation valves for Unit 1 and 2 cooling systems, as discussed previously.
- The interaction of containment venting with the standby gas treatment systems. Because the AC-powered dampers used to close the standby gas treatment systems were in an unknown position and could not be operated, the venting of the containments may have allowed hydrogen gas to enter the plant's reactor buildings. Gravity-operated dampers in Units 1, 2, and 3 appeared to be effective in preventing hydrogen backflow into those units. Hydrogen backflow into Unit 4 apparently did result because dampers were never installed because they were considered unnecessary (TEPCO, 2012b, p. 351).
- The interaction of water injection by fire truck pumps with the condensate makeup water system. A sequence of valves was used to connect the fire protection plumbing to the reactor pressure vessel using components of the condensate makeup water system. Unfortunately, the valves leading to the condensate storage tank were open, diverting water flow from the reactor and reducing the effectiveness of core cooling.

The increased use of embedded controllers in process control, including the ongoing upgrades of nuclear power plant control rooms, and the unanticipated corner and race conditions at Fukushima indicate that increased attention to race and corner conditions is needed in future risk assessment for nuclear power plants as well as future design and verification and validation activities for next-generation nuclear power plants.

7. Include quantitative uncertainty estimates for event probabilities. Extreme events are understood to have low probabilities of occurrence, but those probabilities frequently have high associated uncertainties. Such events must not be prematurely screened out of PRAs without good justifi-

SIDEBAR 5.4
East Coast Tsunamis

Although tsunamis in the Atlantic Ocean Basin do not occur with the frequency of those in the Pacific and Indian Ocean Basins, the potential for tsunami generation is high in some locations. One such location is the eastern margin of the United States.

This margin has a broad and gently sloping continental shelf comprising sediments derived from erosion of the North American continent. Sediment slumps and slides along the outer edges of this shelf have the potential to create large tsunamis. Slumps and slides could be initiated by earthquake shaking or by the release of methane hydrates, which are plentiful along the continental margin. (Hydrate release could be caused by ocean warming or by uncovering by a previous slide.)

Driscoll et al. (2000) propose that the sediment slides along the shelf margin can be characterized by power-law distributions—that is, by a large number of small-scale slides and a small number of large-scale slides. An example of such a large-scale slide is the Albemarle-Currituck slide shown in Figure S5.2. This slide displaced approximately 150 km^3 of sediment, similar to the Grand Banks slide described in Appendix K. Such slides are both infrequent and unpredictable. Within about half an hour of such a slide, ocean surface levels above the slide will decrease rapidly by a few meters. This would be followed by a rapid increase in ocean surface levels minutes to about an hour later. A large, rapidly moving coherent slide has the potential to produce a tsunami of considerable size. Its effect on coastal regions, however, will depend on factors such as the tidal cycle, ocean floor topography, and coastal geometry.

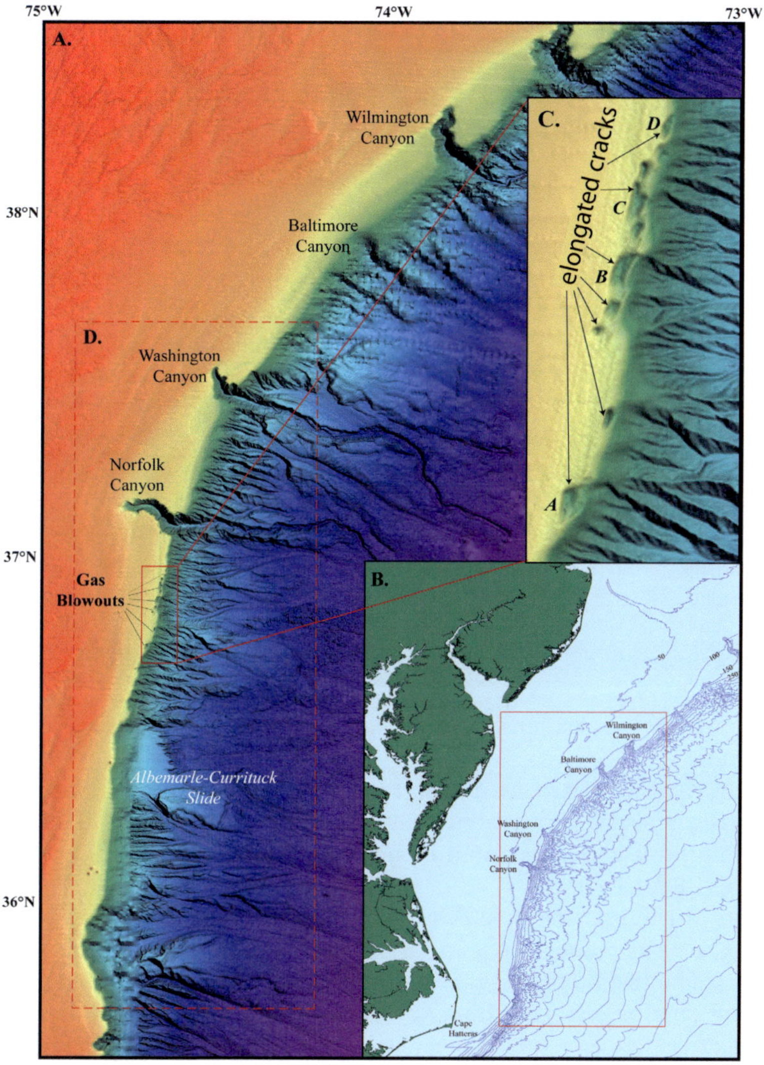

FIGURE S5.2 (A) High-resolution image of the continental shelf and slope offshore of Virginia and North Carolina showing the Albemarle-Currituck slide and several large canyons. (B) Inset map showing location of image in (A). (C) Close-up image of continental shelf edge showing gas blowouts. SOURCE: Driscoll et al. (2000).

cation. A good example of a high-uncertainty external event is a large tsunami on the East Coast of the United States (Sidebar 5.4 and Appendix K).[23]

There are advantages and disadvantages to using PRA to evaluate risks from beyond-design-basis external events. The primary advantages are the following:

- PRAs are based on well-established risk assessment principles.
- PRAs are already being used to assess and mitigate internal hazards at nuclear plants and to establish maintenance and test protocols. Consequently, plant licensees are familiar with their use.
- PRAs can be used to identify non-rare-event scenarios that result from plant design or operational flaws that are not uncovered in the design-basis regulatory review.
- PRAs can provide an integrated examination of plant design and operations.
- If executed properly, PRAs can provide a systematic examination of external hazards and their potential consequences. They can be useful for examining hazard mitigation strategies, for making backfit-rule decisions (Sidebar 5.5), and for emergency planning (see Chapter 6).

The primary disadvantages are the following:

- PRAs are expensive and can be time-consuming to produce and maintain.
- Extending the scope of PRAs will require additional technical expertise, especially in containment response analysis and offsite impacts. Obtaining this expertise could be difficult for industry and the USNRC.
- PRAs that have been performed generally do not adequately account for human error in design, construction, maintenance, and operation of nuclear plants (Appendix I) or for intentional sabotage.
- The results of PRAs are limited by experts' ability to recognize all relevant phenomena, including potentially important external hazards, and by uncertainties and incompleteness of estimates of accident probabilities and consequences.
- The results of full-scope (i.e., Level 3) PRAs (see Appendix I) are also limited by the ability to validate phenomenological modeling of core damage and radioactive release as well as consequence modeling.

---

[23] Progress Energy (2008) has developed a PRA for tsunamis for a new plant that it proposes to build in Levy County, Florida. Pacific Gas and Electric Company (2010) has produced a trial probabilistic tsunami hazard analysis for its Diablo Canyon nuclear plant.

SIDEBAR 5.5
Backfit Analysis

"Backfitting" is any mandated modification to the design or operations of an already-licensed nuclear plant under 10 CFR § 50.109. Except in some narrowly designed circumstances,[a] the USNRC requires that its staff estimate all the costs to the licensee and the USNRC of the proposed backfit and balance these costs against the potential benefits in reduced risks to the facility, its employees, and the public. If the benefits exceed the costs then the proposed backfit is determined to be cost-effective.

The USNRC's Regulatory Analysis Technical Evaluation Handbook (USNRC, 1997b) provides guidance on how to carry out the required analysis. PRA plays a central role in estimating the risk reduction for the proposed backfit.

A backfit analysis was carried out recently by the USNRC for adding filtered vents to the containments of Mark I and Mark II boiling water reactors (see Chapter 2 for a description of these reactors) to reduce the release of radioactive materials to the environment following a core meltdown. The analysis used a simple PRA for containment failure modes and the MELCOR code for estimating how much radioactivity would escape from containment for each failure mode. Population radiation doses, population evacuations, and land contamination areas were calculated for a reference plant (the Peach Bottom plant in Pennsylvania) and averaged over weather and wind conditions at the plant location. The quantitative analysis (see Appendix L) concluded that the cost of installing filtered vents on reactors with Mark I and Mark II containments would exceed the benefits. Installation of filtered vents therefore failed the backfit cost-benefit test based on this quantitative analysis.

Appendix L describes the hypothetical costs for the accident at the Peach Bottom plant used in the USNRC's backfit analysis and compares them with the projected costs for the accident at the Fukushima Daiichi plant. This comparison illustrates the sensitivity of cost estimates to assumptions made about the accident scenario, the plant, and its location. As shown in Appendix L, the likely costs for the Fukushima Daiichi nuclear accident exceed the estimated costs for the hypothetical accident at the Peach Bottom plant by a factor of about 33.

---

[a] Section 50.109 states that "The Commission shall always require the backfitting of a facility if it determines that such regulatory action is necessary to ensure that the facility provides adequate protection to the health and safety of the public and is in accord with the common defense and security."

Dr. Kiyoshi Kurokawa, chair of the Japanese Diet report on the Fukushima Daiichi accident (NAIIC, 2012), commented to the committee at its Tokyo meeting that the problem is not how to estimate rare events, but rather how to identify events that are not rare but go unrecognized. The insight captured by this remark is that unrecognized events need not be low probability. There is a need to guard against missing events, even higher probability events, that result from limitations in identification processes.

If used mechanically without recognizing and acknowledging these limitations, PRAs can supplant judgment and undermine the regulatory policymaking process.[24] For example, PRAs that underestimate the uncertainties in event probabilities or that contain incomplete consequence estimates can result in misleading cost-benefit evaluations for regulatory decisions under the backfit rule (Sidebar 5.5). Appendix L compares the estimated costs of the Fukushima Daiichi accident to the hypothetical costs for a core-melt accident at a U.S. nuclear plant to illustrate the sensitivity of cost estimates to PRA assumptions. Appendix L suggests that USNRC cost estimates for backfit analyses do not include a full accounting of costs and consequences arising from severe nuclear accidents. It is essential that the USNRC fully account for the costs of severe nuclear accidents when making backfit decisions.

An opportunity exists to use the accident progression at the Fukushima Daiichi nuclear plant to validate and improve severe accident system models (e.g., MAAP and MELCOR; see Chapter 4) and thereby enable higher-fidelity consequence modeling, for both onsite events and for offsite releases of radioactive materials during accidents, including the types of long-term releases to groundwater that are occurring at the Fukushima Daiichi plant. Efforts to improve these models have already been initiated (Gauntt et al., 2012b; EPRI, 2013). An extensive post-Fukushima code validation effort is being carried out in Japan (Yamanaka, 2012) and the Nuclear Energy Agency has initiated a code benchmark effort involving eight member states including the United States.[25] It is important that these efforts be taken to completion for the reasons noted above.

In the course of this study, the committee identified three types of external events that merit attention in the recommended risk assessments: large earthquakes; large floods, including those caused by tsunamis (see Sidebar 5.4 and Appendix K); and geomagnetic disturbances produced when coronal mass ejections encounter Earth's magnetosphere (see Sidebar 5.2). The latter two types of events have the potential to affect large geographic regions and offsite electrical power supplies to multiple nuclear plants. Adequate preparation requires the identification of these events and, to

[24] Although risk assessment is an integral part of evaluating nuclear plant safety, the principal strategy for designing and regulating nuclear plants remains the long-standing defense-in-depth philosophy. This strategy involves the use of multiple redundant components and systems to compensate for potential mechanical and human failures as well as providing a buffer against the uncertainties inherent in risk assessment. See Keller and Modarres (2005) for further discussion of the relationship between defense in depth and PRA.

[25] See http://www.oecd-nea.org/jointproj/bsaf.html.

the extent possible, quantification of their expected frequencies, including uncertainties, and consequences.

FINDING 5.3: Four decades of analysis and operating experience have demonstrated that nuclear plant core-damage risks are dominated by beyond-design-basis accidents. Such accidents can arise, for example, from multiple human and equipment failures, violations of operational protocols, and extreme external events. Current approaches for regulating nuclear plant safety, which traditionally have been based on deterministic concepts such as the design-basis accident, are clearly inadequate for preventing core-melt accidents and mitigating their consequences. Modern risk assessment principles are beginning to be applied in nuclear reactor licensing and regulation. The more complete application of these principles in licensing and regulation could help to further reduce core-melt risks and their consequences and enhance the overall safety of all nuclear plants, especially currently operating plants.

RECOMMENDATION 5.3: The U.S. Nuclear Regulatory Commission should further incorporate modern risk concepts into its nuclear reactor safety regulations. This effort should utilize the strengthened capabilities for identifying and evaluating risks that are described in Recommendation 5.2A.

The committee uses the term *modern risk concepts* to mean risk that is defined in terms of the risk triplet (What can go wrong? How likely is that to happen? What are the consequences if it does happen? [see Appendix I]) and subject to the limitations for quantitative analyses described in Section 5.2, especially with respect to uncertainties and incompleteness of estimates of accident probabilities and consequences. Implementing this recommendation fully would likely require changes to some current USNRC regulatory procedures, for example, those used for backfit analyses (see Sidebar 5.5 and Appendix L).

It has been recognized since the 1950s that risks to the public from the operation of nuclear power plants are dominated by accidents involving core damage and radioactive material releases.[26] Nuclear plants were

[26] Radioactive material releases from spent fuel pools that lose their water inventories have also been suggested as a source of risk to the public (see, e.g., Alvarez et al., 2003; NRC, 2004b). Spent fuel safety and security will be examined in next phase of this study (see Chapter 1).

initially sited at large distances from population centers[27] to reduce these risks—in the event of a severe accident any released radioactive materials would be dispersed (i.e., diluted) in the atmosphere before reaching the public (Okrent, 1981). The remote siting of plants also minimized the need for detailed investigations of accident sequences, simplified the modeling of health consequences, and resulted in the development of conservative hazard analyses (e.g., AEC, 1957). The nuclear reactor siting criteria produced as a result of these analyses (i.e., 10 CFR Part 100[28] and TID-14844[29]) were adopted in some fashion by other national regulatory authorities.

The siting criteria recognized the *maximum credible accident* as a core-melt accident with a specified radioactive fission-product source term. The source term was used to calculate doses at a nuclear plant's site boundary and for assessing the plant's containment performance.

In contrast, nuclear reactor systems were designed using a different concept, namely the design-basis accident (DBA). A DBA is a stylized accident, for example, a loss-of-coolant accident or transient overpower accident,[30] that is required (by regulation) to be considered in a reactor system's design. The current generation of U.S. nuclear plants were designed, licensed, and built under these different siting and design criteria.

It was recognized in the 1960s that accident likelihoods (i.e., the probability that a postulated accident would occur) needed to be considered in nuclear plant safety analyses. Farmer (1967) suggested a probability-based technique for analyzing nuclear plant safety. This technique, referred to then as "probabilistic reliability analysis" but known today as "probabilistic risk assessment" (i.e., PRA), was beginning to be used in the British aeronautical industry. In the United States, work on PRA was advanced by reports from a U.S. engineering firm (Holmes and Narver, Inc.) that advocated for the use of advanced systems-engineering methods for modeling the reliability of safety systems; by Garrick (1968), who developed a probabilistic-based methodology for analyzing the safety of nuclear power plants; and finally by the first Reactor Safety Study (USNRC, 1975). The latter study inspired many first-of-a-kind nuclear plant risk assessments (Garrick, 2008).

Although these early nuclear plant PRAs were recognized for their

[27] Of course, population growth has occurred around many nuclear plants since they were constructed, and so some plants may no longer be located far from population centers. The risk implications of such growth may not have been anticipated during the original licensing proceedings for some plants.

[28] Reactor site criteria; available at http://www.nrc.gov/reading-rm/doc-collections/cfr/part100/.

[29] AEC (1962).

[30] A transient is a "change in the reactor coolant system temperature, pressure, or both, attributed to a change in the reactor's power output." See http://www.nrc.gov/reading-rm/basic-ref/glossary/transient.html.

innovative methodologies, they also were found to be lacking in some important respects. For example, the General Accounting Office (now the Government Accountability Office), in commenting on the first PRA for the Indian Point nuclear plant in New York, noted that:

> While many analysts consider the Indian Point PRA to be the state of the art in risk assessment, it suffers from the same fundamental problems as all PRAs: uncertainty and incomparability of results. Also, although the study identified the dominant contributors to risk, it did not identify the precise level of risk from operating the Indian Point nuclear powerplants. (USGAO, 1983, p. 2)

PRA is fundamentally different than DBA analysis: in DBA analysis a particular accident is postulated and deterministically analyzed. In contrast, a PRA considers a myriad of possible accident sequences, each having greater-than-zero probability values, even though they may be small.

In 1975 the USNRC applied PRA to a pressurized water reactor at the Surry Power Station in Virginia and a boiling water reactor at the Peach Bottom Atomic Power Station in Pennsylvania to estimate accident risks and their sources. This analysis (USNRC, 1975) provided a standard methodology as well as a benchmark for future studies; it also reaffirmed the conclusion that severe accidents involving core melts and radioactive material releases dominated risks to the public from nuclear plants. It also spurred research in Britain, Germany, and the United States to better understand physical core-melt processes.

The USNRC (1975) analysis had a minimal impact on nuclear plant regulation and operation in the years immediately following its publication. However, its influence has grown steadily. Deterministic analyses continued to be used to analyze DBAs and assess containment systems. Deterministic approaches were thought to be sufficiently conservative to be protective of public health.

The 1979 Three Mile Island accident (Walker, 2004) altered this perception. This accident involved multiple equipment malfunctions and operator actions that allowed an operational transient to evolve into a core-melt accident over a period of a few hours. The accident resulted in a significant amount of fuel melting in the reactor core and fission product release from the core into containment. The containment successfully prevented any major releases to the environment. However, the accident raised doubts about the comprehensiveness of deterministic approaches for nuclear plant safety analyses and assumptions about operator performance. It also highlights a lack of understanding of the physical processes that can threaten containment integrity. Deterministic approaches for nuclear safety analyses reflect a major effort in understanding complex processes but they do not cover all scenarios.

The 1986 Chernobyl accident in Ukraine reinforced these doubts. This accident occurred in a Soviet-era reactor having an unstable design[31] and was initiated by a series of inappropriate operator actions. The accident resulted in major offsite radioactive material releases with acute fatalities and long-term health effects (see Chapter 6).

Following these accidents, the USNRC established a policy (USNRC, 1995) on using PRA to complement regulations. The policy states that

> 1. The use of PRA technology should be increased in all regulatory matters to the extent supported by the state-of-the-art in PRA methods and data and in a manner that complements the [US]NRC's deterministic approach and supports the [US]NRC's traditional defense-in-depth philosophy.
> 2. PRA and associated analyses (e.g., sensitivity studies, uncertainty analyses, and importance measures) should be used in regulatory matters, where practical within the bounds of the state-of-the-art, to reduce unnecessary conservatism associated with current regulatory requirements, regulatory guides, license commitments, and staff practices. Where appropriate, PRA should be used to support the proposal of additional regulatory requirements in accordance with 10 CFR 50.109 (Backfit Rule). Appropriate procedures for including PRA in the process for changing regulatory requirements should be developed and followed. It is, of course, understood that the intent of this policy is that existing rules and regulations shall be complied with unless these rules and regulations are revised.
> 3. PRA evaluations in support of regulatory decisions should be as realistic as practicable and appropriate supporting data should be publicly available for review.
> 4. The Commission's safety goals for nuclear power plants and subsidiary numerical objectives are to be used with appropriate consideration of uncertainties in making regulatory judgments on the need for proposing and backfitting new generic requirements on nuclear power plant licensees.

This policy, coupled with additional Commission guidance issued in 1999, has resulted in a variety of risk-informed program-specific improvements: for example, the maintenance rule for operating reactors,[32] the

---

[31] This Soviet-designed RBMK reactor has a positive void coefficient, so it becomes more reactive with increasing steam content in the core. RBMK reactors also do not have containments.

[32] 10 CFR § 50.65, Requirements for Monitoring the Effectiveness of Maintenance at Nuclear Power Plants. Available at http://www.nrc.gov/reading-rm/doc-collections/cfr/part050/part050-0065.html.

pressurized thermal shock rule,[33] and the backfit rule (Sidebar 5.5 and Appendix L). Nevertheless, slow progress has been made in risk-informing the USNRC's regulations.

The Fukushima accident, which was initiated by an extreme external event, further confirms the need for more expeditious consideration of risk-informed approaches to safety, particularly for beyond-design-basis events. The USNRC's Near-Term Task Force (USNRC NTTF, 2011; see Appendix F) recommended that the agency establish "a logical, systematic, and coherent regulatory framework for adequate protection that appropriately balances defense-in-depth and risk considerations" (USNRC NTTF, 2011, p. ix). Another USNRC task force (USNRC, 2012a) has recommended that a risk management regulatory framework be adopted by the Commission.

The Nuclear Energy Institute has commented on the lack of progress in implementing risk-informed regulations (RIRs):

> Over the past five years, progress in RIR has been stunted. A variety of factors have contributed to this, but the result has been a growing distrust of risk-informed processes. Ironically in the post-Fukushima era, where nuclear power faces many decisions that could be better informed by a risk perspective, the reluctance to use PRA in new regulatory activities has removed a valuable tool from the process. (NEI, 2013, p. 1)

The difficulty in expanding risk-informed regulations has been greater than some had anticipated. On the other hand, expansion has continued steadily in spite of resistance in some quarters of the USNRC and industry.

The committee judges that the broader use and expanded scope of modern risk concepts in nuclear reactor safety regulations could improve safety and lead to better policy decisions.

---

[33] 10 CFR § 50.61, Fracture Toughness Requirements for Protection Against Pressurized Thermal Shock Events. Available at http://www.nrc.gov/reading-rm/doc-collections/cfr/part050/part050-0061.html.

6

Lessons Learned: Offsite Emergency Management

The focus of this chapter is on offsite emergency responses to the Fukushima Daiichi accident and lessons learned for emergency preparedness in the United States. The chapter focuses primarily on offsite responses during the first few critical days of the accident (early phase, see Sidebar 6.1). However, information about the early-phase response was useful to the committee for identifying lessons learned for the intermediate and late (or recovery) phases.

The information used in this chapter was obtained from several sources: independent examinations of the Fukushima Daiichi accident carried out in Japan, the United States, and other countries (see Table 1.1 in Chapter 1); Japanese regulations related to offsite emergency management in Japan at the time of the accident; and a number of scientific publications. The committee's review of Japanese documents was limited to those translated to English. At the committee's request, the National Academy of Sciences arranged for English translations of selected sections of the Japanese government's 2007 version of the "Basic Plan for Emergency Preparedness" (NSC, 2013).

At the time of the March 2011 Fukushima Daiichi accident, Japanese[1] and U.S. approaches to offsite emergency response had many common features. These included specified incident notification levels; guidance on conditions for each notification level; designation of specific emergency planning zones; protective action guidelines (PAGs) for decisions relating to

[1] Described in NSC (2013).

SIDEBAR 6.1
Phases of a Nuclear Power Plant Accident

During a radiological emergency in the United States, there is a generic framework for structuring responses following a disaster based on three phases: early, intermediate, and late.

According to the USEPA (2013), the early phase (also referred to as the emergency phase) lasts from several hours to several days. During this phase, conditions at the location of the incident are evaluated, responsible authorities are notified, and the potential consequences of the incident to members of the public are predicted or evaluated. Decisions on protective actions such as evacuation, sheltering in place, and taking KI for thyroid protection are made primarily on the basis of the status of the nuclear power plant and the prognosis of changes in the conditions.

The intermediate phase lasts from weeks to months. During this phase, the source and releases from the plant have been brought under control. Also, environmental measurements of radioactivity and dose models are available to project doses to members of the public and base decisions on additional protective actions such as food and water interdictions.

The late phase (also referred to as the recovery phase) can last from months to years. It begins sometime after the initiation of the intermediate phase and proceeds independently of the protective actions implemented during that phase. During the late phase, recovery actions designed to reduce radiation levels in the environment are commenced and end when all recovery actions have been completed.

Because of the possible overlap, phases of the emergency response are not viewed in terms of time but instead in terms of activities performed.

shelter, evacuation, and distribution and administration of potassium iodide (KI)[2]; and guidelines for food and water intake.

However, approaches to managing offsite responses in Japan and the United States were different in some notable ways: the United States uses a "bottom-up" approach to managing offsite emergency response. That is, the responsibility for responding to a disaster begins at the local level, extends to state and tribal governments, and can include the federal government as supplemental resources are requested (Sidebar 6.2).[3] The Japanese

[2] KI is a prophylactic agent that prevents the uptake of radioactive iodine (i.e., radioiodine) into the thyroid gland and thus reduces the risk of thyroid cancer.

[3] There are limited exceptions to this approach. The President of the United States is authorized to support precautionary evacuation measures, accelerate federal emergency response and recovery aid, and provide expedited federal assistance (coordinated with the state to the extent possible) in the absence of a specific request from state officials (Robert T. Stafford

SIDEBAR 6.2
National Response Framework

The roles of federal agencies in U.S. nuclear emergencies are laid out in the National Response Framework (NRF) (USDHS, 2013). The NRF creates a broad-based "all-hazards response" emergency planning process to address a wide variety of emergencies including natural disasters, terrorism, and other human-initiated accidents and events, encompassing nuclear and radiological events. Nuclear power plant accidents involving radioactive material releases are just one of the many potential emergencies to which this all-hazards approach applies.

The all-hazards approach is based on the notion that there are common features among disasters irrespective of their initiating events; therefore, many of the same planning strategies can apply to all emergencies. These features include the need for robust communication channels; collection of adequate data; information exchange and interpretation; requisitioning of resources and expertise; assessment and management of offsite impacts; and community involvement. Many elements necessary to an effective response to a nuclear incident are common to other types of emergencies, such as sheltering or evacuating a specific population, establishing an emergency communication network, or implementing mutual-aid agreements with nearby (but unaffected) jurisdictions.

Thus, embedding planning for a nuclear-related event in an overall emergency response plan for all types of natural and man-made emergencies provides the framework for a scalable, flexible, and adaptable plan that is expected to be responsive to small, common, and well-defined events as well as large, rare, and complex events. An additional advantage of the all-hazards approach is that it maintains a higher state of readiness, because the plan is implemented more often and because all response agencies, nongovernmental organizations, and private entities are working within the same response framework with a common command structure.

approach at the time of the Fukushima Daiichi accident was "top down," with the central government providing direction and national resources to local communities (NSC, 2013). Despite these differences in approaches, the Japanese response to the Fukushima Daiichi accident provides valuable lessons for the United States.

The committee did not have the time or resources to perform an in-depth examination of U.S. preparedness for severe nuclear accidents.[4] Also, many U.S. agencies are still in the process of developing lessons learned from the Fukushima Daiichi accident. The committee engaged in discus-

Disaster Relief and Emergency Assistance Act, Public Law 93-288, as amended, 42 U.S.C. §§ 5121 et seq.).

[4] See Chapter 1, Sidebar 1.2 for a definition of "severe accident."

sions with several U.S. agencies that have emergency management responsibilities (see Sidebar 6.3) to become better informed about these ongoing efforts: the U.S. Nuclear Regulatory Commission (USNRC), the Federal Emergency Management Agency (FEMA), and the U.S. Environmental Protection Agency (USEPA) (see meeting agendas in Appendix B). In addition, the committee requested information from the U.S. Centers for Disease Control and Prevention (CDC).

This chapter is organized into five sections. The first and second sections aim to put the radiological consequences of the Fukushima Daiichi accident and the difficulties in responding to the accident due to the competing natural disasters—the earthquake and tsunami—into perspective. The third section provides a brief description of the offsite emergency response during the first few days of the Fukushima Daiichi accident. The fourth section discusses some key issues that arose from the committee's analysis of emergency management in Japan. The fifth and final section provides the committee's lessons learned for nuclear emergency preparedness in the United States. These lessons learned are presented as findings and recommendations and are directed to the U.S. nuclear industry, state and local governments, and federal agencies with emergency preparedness responsibilities.

6.1 RADIOLOGICAL CONSEQUENCES OF THE FUKUSHIMA DAIICHI ACCIDENT

The Fukushima Daiichi accident is one of the major accidents in the history of commercial nuclear power. The accident resulted in the most extensive release of radioactive materials into the environment since the 1986 Chernobyl accident in Ukraine. Radioactive releases into the environment started on March 12, 2011, and the significant discharge phase ended at midnight on March 25 (IRSN, 2011). However, minimal releases of radioactive material to the atmosphere continued until December 2011 when cold shutdown of the last impacted reactor at the Fukushima Daiichi plant was achieved (Brumfiel, 2011). Releases to the ocean have continued to the present.

Both the Fukushima Daiichi and Chernobyl nuclear accidents were designated as Category 7 on the International Atomic Energy Agency's (IAEA's) International Nuclear and Radiological Event Scale.[5] However, the physical health-related radiological consequences of the Fukushima Daiichi accident are less severe than those for the Chernobyl accident for four main reasons:

[5] Category 7 is the highest level of the scale.

SIDEBAR 6.3
Nuclear Power Plant Accident Preparedness in the United States

In the United States, state and local governments have the primary responsibility for making protective action decisions and communicating health and safety instructions to affected populations during a nuclear power plant accident. As laid out in the National Response Framework (NRF; see Sidebar 6.2), a number of federal agencies also play an important role in responding to the accident (USDHS, 2013).

USNRC and FEMA

The USNRC and FEMA are the primary federal agencies responsible for radiological emergency preparedness in the United States. The USNRC is responsible for ensuring that nuclear plants are prepared for radiological emergencies. The USNRC coordinates with FEMA, which oversees state and local agencies' preparedness for offsite actions. FEMA also provides guidance and support to local and state authorities through its Radiological Emergency Preparedness (REP) program (FEMA, 2013b).

It is not practical for emergency plans to address every possible combination of events (no matter how unlikely) or to present every possible action that can or should be taken in response to an evolving event. Instead, a "planning basis" is available for nuclear power plant events in the United States and is described in a USNRC and USEPA Task Force report (USNRC and USEPA, 1978). The planning basis is utilized in the joint USNRC and FEMA document "Criteria for Preparation and Evaluation of Radiological Emergency Response Plans and Preparedness in Support of Nuclear Power Plants" (USNRC and FEMA, 1980). This document is currently undergoing review; a revised draft is expected to be available for public comment in November 2014.

USEPA

One of the USEPA's roles in radiological emergency preparedness is to establish protective action guidelines (PAGs) and provide guidance on implementing them, including recommendations on protective actions. USEPA's PAGs are expressed in terms of projected doses at which protective actions should be taken to reduce or eliminate exposures (USEPA, 2013). In setting the range of values for its PAGs, USEPA considered the following four principles (Conklin and Edwards, 2000):

1. Avoid acute radiation health effects.
2. Minimize the risk of delayed health effects.
3. Dose values should not be higher than justified by a cost-benefit analysis.
4. Risks to health from implementing the protective action should not be greater than the risk from the dose avoided.

Emergency responders can use the PAGs for any radiation incident involving relatively significant releases of radioactive materials, including nuclear power plant accidents for the early and intermediary phases.

CDC

The CDC's roles in radiological emergency preparedness include:

1. Providing guidance to state and local governments on the health effects from exposure to radiation and guidance on how to minimize adverse health effects, including psychological health effects from exposure to radiation.
2. Providing medical treatment of exposed individuals and epidemiological surveillance of exposed populations.
3. Participating in the Advisory Team for Environment, Food and Health, a radiological emergency response group tasked with issuing protective action recommendations to prevent or minimize radiation exposure through ingestion by preventing or minimizing contamination of milk, food, and water.

USDOE

USDOE's role in a radiological emergency is to coordinate federal environmental radiological monitoring and produce predictive plume models and dose assessments. USDOE makes use of a variety of emergency response assets to estimate the probable or actual spread of radioactivity in the environment. The assets include the National Atmospheric Release Advisory Center (NARAC) for plume and deposition modeling and the Aerial Measuring System (AMS) for measurements of ground deposition with aircraft-mounted detectors. USDOE can create a federal radiological monitoring and assessment center (FRMAC) to help integrate consequence management resources and coordinate the development of a common operating framework.

1. Radioactive releases from the Fukushima Daiichi accident (approximately 100-500 Petabecquerel [PBq][6] of iodine-131 and 6-20 PBq of cesium-137[7]; UNSCEAR, 2013b) are estimated to be less than 15 percent of those from Chernobyl (approximately 1,760 PBq iodine-131 and 85 PBq cesium-137; UNSCEAR, 2011; Povinec et al., 2013).

2. Prevailing winds at the time of the accident appear to have blown about 80 percent of the radioactive material released from the Fukushima Daiichi plant out to the Pacific Ocean (Kawamura et al., 2011; Morino et al., 2011). The majority of the radioactive material deposited over land was dispersed along a track stretching about 50 km to the northwest of the plant. In contrast, radioactive material from Chernobyl was largely deposited over land (UNSCEAR, 2011).

3. Evacuation of those living in proximity (within 3 km) to the Fukushima Daiichi plant was ordered a few hours after the accident began and at least 12 hours before major releases of radioactive materials from the reactors started (Investigation Committee, 2011). At Chernobyl, evacuations started almost a day after the accident began at which point releases had already started (NEA, 2002; UNSCEAR, 2011).

4. Government restrictions put into place after the Fukushima Daiichi accident kept most contaminated foodstuffs off the market (IAEA, 2011). After the Chernobyl accident there were long delays in implementing appropriate food restrictions in some local areas (UNSCEAR, 2011).

The grave consequences of the Chernobyl accident included the immediate deaths of 28 first responders and firefighters from acute radiation sickness and an epidemic of thyroid cancer in children in Ukraine and neighboring countries.[8] With respect to the Fukushima Daiichi accident, there is general agreement in the scientific community that no worker received a dose that resulted in acute radiation death or sickness. Also, doses received by members of the public are estimated to be generally low; therefore, any increase in an individual's risk of developing cancer in the

---

[6] Becquerel (Bq) is the international (SI) name for the unit of activity; one Bq is equal to one disintegration per second, or 2.7×10^{-11} curies (Ci). 1 PBq = 1.0×10^{15} Bq.

[7] The entire inventory of the Fukushima Daiichi Units 1-3 was estimated to be 6,000 PBq iodine-131 and 700 PBq cesium-137 (UNSCEAR, 2013b).

[8] About 6,000 excess thyroid cancers were reported up to the year 2005 and many more were projected in the future resulting from exposure to radioactive iodine releases during the Chernobyl accident, mostly through ingestion of contaminated cow's milk (Cardis et al., 2006; UNSCEAR, 2011).

future is also low (UNSCEAR, 2013b; WHO, 2013; Steinhauser et al., 2014).[9]

According to reports by the World Health Organization (WHO, 2013) and the United Nations Scientific Committee on the Effects of Atomic Radiation (UNSCEAR, 2013b), most people in Fukushima Prefecture received an effective dose[10] between 1 and 10 mSv in the first year following the accident. People from Namie, which is located inside the evacuation zone, and Iitate, which is located 40 km (25 miles) northwest of the Fukushima Daiichi plant, may have received the highest effective doses; those doses are estimated to be between 10 to 50 mSv, all delivered in the first year.[11] However, infants in Namie were thought to have received higher thyroid radiation doses—between 100 and 200 mSv. The authors of the WHO (2013) report conclude that

> The present results suggest that the increases in the incidence of human disease attributable to the additional radiation exposure from the Fukushima Daiichi nuclear power plant accident are likely to remain below detectable levels.

Nevertheless, the government of Japan has launched a 30-year-long health survey of the 2 million residents of Fukushima Prefecture. The survey includes pediatric thyroid monitoring (Yasumura et al., 2012).

This discussion of the physical health-related radiologic consequences

---

[9] This conclusion is based on the linear no-threshold (LNT) model of risk assessment. According to this model, the risk of developing cancer is proportional to dose received, and even a small dose can result in a small increase in lifetime risk of developing cancer.

Using risk-projection models, estimates of the number of cancer cases and deaths possibly attributable to the Fukushima Daiichi accident globally or locally have been published in peer-reviewed journals (Ten Hoeve and Jacobson, 2012; Beyea et al., 2013; Evangeliou et al., 2014). These estimates, which should be considered preliminary, are based on LNT risk models developed by the U.S. National Academy of Sciences' Committee on the Biological Effects of Ionizing Radiation (BEIR VII) (NRC, 2006a). The central estimates range from a few hundred to 1,700 cases depending on the specific LNT model used and do not fully account for uncertainties in the model at low doses. Future revisions to the estimates are likely as doses from the Fukushima Daiichi accident are better assessed (similar to the Chernobyl dose assessments; see Cardis et al., 2006).

[10] Effective dose, expressed in millisieverts (mSv), is a dose parameter used to normalize partial-body radiation exposures relative to whole-body exposures to facilitate radiation protection activities (ICRP, 1991). For nuclear power plant accidents where populations are exposed primarily to gamma radiation, such as occurred as a result of the Fukushima Daiichi accident, whole-body dose expressed as effective dose and reported in millisieverts and organ absorbed dose reported in milligrays are numerically equivalent (NRC, 2006a). For consistency throughout this chapter, discussions of dose are in terms of effective dose and reported in millisieverts.

[11] For comparison, the average radiation background in Japan is 2.4 mSv/yr. See http://www.jaea.go.jp/04/ztokai/kankyo_e/kaisetsu/expln_1.html. Last accessed on June 12, 2014.

TABLE 6.1 Chronologies of Evacuation and Shelter-in-Place Orders Following the Fukushima Daiichi Accident

Date in 2011 (time)	Distance from Plant	Orders[a]	Area Designation
March 11 (20:50)	2 km	Compulsory evacuation issued by the Fukushima prefectural government	Restricted Zone
(21:23)	3 km	Compulsory evacuation	Restricted Zone
March 12 (05:44)	10 km	Compulsory evacuation	Restricted Zone
(18:25)	20 km	Compulsory evacuation	Restricted Zone
March 15	20-30 km	Shelter in home	Evacuation Prepared Area
March 25	20-30 km	Self-evacuation	Evacuation Prepared Area
April 22	Areas with dose >20 mSv/yr	Evacuation within 1 month	Deliberate Evacuation Area
June 16	Hotspots with dose >20 mSv/yr	Recommended for evacuation	Specific spots recommended for evacuation
September 30	20-30 km	Lifted order to shelter indoors or self-evacuate	Lifting of Evacuation Prepared Area

[a] Issued by the central government unless otherwise stated; order unless otherwise stated.
SOURCE: Adapted from Hasegawa (2013).

of the Fukushima Daiichi accident is not intended to downplay other severe long-term health impacts. Of the approximately 150,000 people who were evacuated as a result of the accident[12] (UNSCEAR, 2013b), over 80,000 (WNA, 2014) still lived in shelters or temporary locations 3 years after the accident with continuing uncertainties about the future. The difficulties in

[12] About 78,000 people living within a 20-km radius of the Fukushima Daiichi plant and 62,000 people living between 20 and 30 km from the plant were evacuated during the first few days of the accident. In April 2011, the government of Japan recommended the evacuation of about 10,000 more people living farther to the northwest of the plant (UNSCEAR, 2013b). See Table 6.1 for the evacuation time line.

evacuees' daily lives, possible separation from family members, and loss of property and business or employment are further complicated by the fear of developing cancer from accident-related radiation exposures and the societal stigma resulting from those exposures (NRA, 2013b). As with the Chernobyl accident, mental health effects, which include depression, anxiety, and post-traumatic symptoms, are considered to be the largest public health problem from the accident (González et al., 2013; Bromet, 2014).

The environmental and economic consequences of the accident are also severe. About 13,000 km^2 of land (about the size of the U.S. state of Connecticut) are contaminated such that the average annual dose to occupants would exceed the 1-mSv/yr long-term cleanup goal (Chen and Tenforde, 2012).[13] Cleanup of such a large area is proving to be challenging due to the limited effectiveness of decontamination techniques (Yasutaka et al., 2013), lack of short- or long-term plans for disposal of the radioactive waste created by cleanup, and ongoing negotiations among stakeholders about acceptable radiation dose criteria for resettlement. The final determination of how much residential land will be off limits indefinitely has still not been made.[14] Return of evacuated persons—although a high priority of the Japanese government—remains an unresolved issue 3 years after the accident.

6.2 CHALLENGES OF RESPONDING TO THE ACCIDENT

Emergency response to the Fukushima Daiichi accident was greatly inhibited by the widespread and severe destruction caused by the Great East Japan Earthquake and tsunami: local electrical power and regional communication infrastructure were knocked out and the transportation infrastructure (roads, bridges, ports, and railroads) was damaged. Japan is known to be well prepared for natural hazards; however, the earthquake and tsunami caused devastation on a scale beyond what was expected and prepared for. More than 20 prefectures were affected by the natural disaster. The National Police Agency of Japan reports 15,883 confirmed deaths and 2,652 people missing due to the earthquake and tsunami. Damage to buildings was extensive: over 126,000 buildings totally collapsed and about 1 million buildings were partially damaged (National Police Agency of Japan, 2014).

Responses to the earthquake and tsunami diverted emergency response

[13] The IAEA has recommended a short-term goal of achieving effective doses of 1-20 mSv per year with the ultimate goal of achieving residual effective doses at or below 1 mSv per year (IAEA, 2013b).

[14] There are areas where the estimated annual dose level is over 50 mSv per year due to cesium-137 (30-year half-life) and cesium-134 (2-year half-life) deposition. According to IRSN (2012b), the population's return "seems barely feasible in the long term."

teams that could have otherwise focused on responding to the Fukushima Daiichi accident. Immediately after the earthquake and tsunami, the government established an emergency response team headed by the prime minister. (The prime minister also acted as the director-general for the offsite response to the nuclear accident.) Within a day of the disaster, the Ministry of Defense ordered the dispatch of the country's military, the Japanese Self-Defense Forces (SDF), which included 110,000 active and reserve troops, along with 28,000 members of the National Police Force as well as the Fire and Disaster Management Agency (Carafano, 2011). These three forces were also called on during the period March 14-17 to help inject water into the Fukushima Daiichi plant's cooling systems and spent fuel pools (NERHQ-TEPCO, 2011). In addition, SDF provided air transport within the 20-km evacuation zone to people who needed help to evacuate (Mizushima, 2012). Similarly, the national police assisted with environmental radiation monitoring (NERHQ-TEPCO, 2011), and the Japanese Red Cross Society provided medical and psychological support to earthquake and tsunami victims as well as those affected by the nuclear accident.

In addition to the overwhelming relief demands on the emergency response teams, which had to deal with three simultaneous disasters of unexpected scale, emergency response to the Fukushima Daiichi accident was conducted with limited information on the status of the nuclear plant itself. As described in Chapter 4 of this report, many monitoring and control systems at the plant were not functional because of tsunami-related flooding. Additionally, some offsite instrumentation also was not functional. Consequently, decisions on protective actions for affected offsite populations (e.g., evacuations, sheltering in place, and KI distribution) were made under great stress and great uncertainty about the status of the plant, accident progression prospects, and projected doses to nearby populations.

6.3 OFFSITE EMERGENCY RESPONSE

The following sections describe the offsite emergency response to the accident at the Fukushima Daiichi plant.

6.3.1 Declaration of Emergency

Immediately after the arrival of the second (main) tsunami wave at 15:36-15:37 on March 11 (see Sidebar 3.1 in Chapter 3), TEPCO, in accordance with Article 10, Paragraph 1, of the Act on Special Measures Concerning Nuclear Emergency Preparedness (Cabinet Secretariat, Government of Japan, 1999), informed the Nuclear Industry Safety Agency (NISA) of the plant's total loss of alternating current (AC) power. This notification was made at 15:42 on March 11. There are two different accounts of the step that followed:

- By one account (Investigation Committee, 2011), NISA, in consultation with the Ministry of Economy, Trade and Industry (METI), determined at 16:36 on March 11 that the incident rose to the level of a nuclear emergency situation as defined in Article 15, Paragraph 1, of the Act. This Act calls for the Japanese prime minister to immediately give public notice of the occurrence of a nuclear emergency situation.
- By another account, TEPCO informed NISA at 16:45 on March 11 that the situation required the Article 15 public notice.

In either case, at around 17:42 on March 11, NISA and METI reported the situation to the prime minister and provided him with a draft public notice. The prime minister gave the required public notice at 19:03 on March 11.

Authorities in Japan acted immediately to reduce the consequences of potential releases of radioactive materials from the Fukushima Daiichi plant. Their actions were to be coordinated through the Nuclear Emergency Response Headquarters (NERHQ), which was established near the prime minister's office in Tokyo and was led by the prime minister. In addition, the local NERHQ was established in Fukushima Prefecture about 5 km west of the Fukushima Daiichi plant and was led by METI's senior vice-minister. However, full operation of the local NERHQ was delayed until about March 15 (JNES, 2013). This delay was due to the lack of electrical power and damage to highways and roads, which made local travel difficult.[15] Because of this delay, coordination between the national and local governments for ordering, implementing, and confirming evacuations and other protective actions was difficult.

6.3.2 Issuance of Protective Actions

Instrumentation that would normally have been used to inform protective-action decisions following the accident were unavailable due to the loss of electrical power and damage from the earthquake and tsunami. This instrumentation included

- Twenty-four radiation monitoring stations on the Fukushima

---

[15] The alternative location for the offsite center (in the Minamisoma City Hall) was already being used as an emergency response center for the earthquake and tsunami. The local NERHQ was therefore established in the Fukushima Prefectural Building.

Daiichi plant site; 23 of these stations were rendered nonfunctional by the tsunami (WNA, 2014).

- The Emergency Response Support System (ERSS), which provides data on plant status to multiple offsite centers. This system malfunctioned immediately after the accident (NERHQ-TEPCO, 2011). Consequently, critical information about the status of the Fukushima Daiichi plant could not be obtained.
- The System for Prediction of Environmental Emergency Dose Information (SPEEDI) is used during emergencies to predict atmospheric concentrations of radioactive materials, dose rates, and environmental exposures. These predictions are used to inform decisions by authorities on protective actions.[16] The ERSS feeds information on radioactive release sources to SPEEDI; but, as noted previously, ERSS was not functional.

Reliable real-time estimates of sources and magnitudes of radioactive material releases from the Fukushima Daiichi plant were therefore unavailable.

As discussed in Chapter 4, some releases of radioactivity to the atmosphere from the plant occurred through uncontrolled pathways (see also Narabayashi et al., 2012). An instrument at the main gate of the plant produced a continuous record of gamma dose rate from these releases (NERHQ-TEPCO, 2011) and cars with measuring instruments produced some scattered measurements elsewhere on the site. However, these data could not be analyzed in real time.

The Ministry of Education, Culture, Sports, Science and Technology (MEXT) obtained some measurements on March 15 from a car located 20 km to the northwest of the plant (NERHQ-TEPCO, 2011). A number of monitoring instruments were also set up beyond the 20-km evacuation radius starting on March 16 (NERHQ-TEPCO, 2011). Gross gamma dose-rate measurements from these instruments were adequate to determine whether populations should be moved from already-contaminated areas.

Airborne measurements of ground contamination levels were made by the U.S. Department of Energy (USDOE) starting March 17 (Lyons and

[16] Weather forecasting is uncertain, so any projection of plume transport using SPEEDI becomes increasingly uncertain as the forecast time for the projection increases. Also, the timing of multiple, prolonged releases with respect to wind patterns complicates predictions. As a result, projections with SPEEDI or other similar systems can only be probabilistic. The uncertainties increase in situations with multiple releases occurring at apparently random times.

Colton, 2012).[17] USDOE focused on measuring radionuclides that had been deposited on the ground after passage of the plume.[18]

Given the sparse information on the status of the plant and uncertainties about projected doses, decision makers who issued protective actions showed a preference for evacuation of populations located near the plant rather than sheltering in place.

6.3.2.1 *Evacuation Orders*

Several evacuation orders were issued following the prime minister's declaration of a nuclear emergency (see Table 6.1). The evacuation zones were gradually expanded over time, and residents were ordered to evacuate repeatedly from one place to another. Prior to instructions from the NERHQ (see Section 6.3.1), the governor of Fukushima Prefecture instructed Okuma Town and Futaba Town—the two towns nearest to the Fukushima Daiichi plant (see Figure 6.1)—to evacuate residents living within a 2-km radius of the Fukushima Daiichi plant. Approximately 30 minutes later, the NERHQ instructed the Fukushima Prefectural governor and all relevant local governments to issue an evacuation order to citizens within a 3-km radius of the plant and to issue a shelter-in-place order to citizens between 3 and 10 km of the plant. These evacuation orders were preemptive; there were no data at the time indicating that there had been a release of radioactive material from the plant or that such a release was imminent.

Following instructions by the prime minister to the heads of relevant municipalities, the evacuation area was increased to a 10-km radius the morning of March 12 because of fears that potentially large quantities of radioactive materials would be released. The evacuation zone was further increased to 20 km that afternoon following the hydrogen explosion at Unit 1 (see Sidebar 3.1 in Chapter 3). Fukushima Prefecture, Okuma town, Futaba town, Tomioka town, Namie town, Kawauchi town, Naraha town, Minamisoma city, Tamura city, and Katsurao village were among the municipalities evacuated (NERHQ-TEPCO, 2011). An estimated 78,000 people evacuated from the 20-km-radius zone around the plant

---

[17] Measurements were made from altitude bands of 152-305 m (helicopter) and 550-700 m (fixed-wing aircraft). Each aircraft used detectors equipped with a total of 12 large-volume (5 cm × 10 cm × 40 cm) sodium iodide scintillator crystals.

[18] Such measurements are made to identify areas that should be subject to long-term evacuations because of contamination by the long-half-life isotopes cesium-134 (2-year half-life) and cesium-137 (30-year half-life). These measurements are not intended to inform decisions on short-term precautionary evacuations (i.e., evacuations to protect populations from exposures to high-dose-rate, short-lived fission products, or decisions to advise populations to take KI to prevent thyroid uptake of inhaled radioactive iodine when evacuations cannot occur in time to avoid such inhalation).

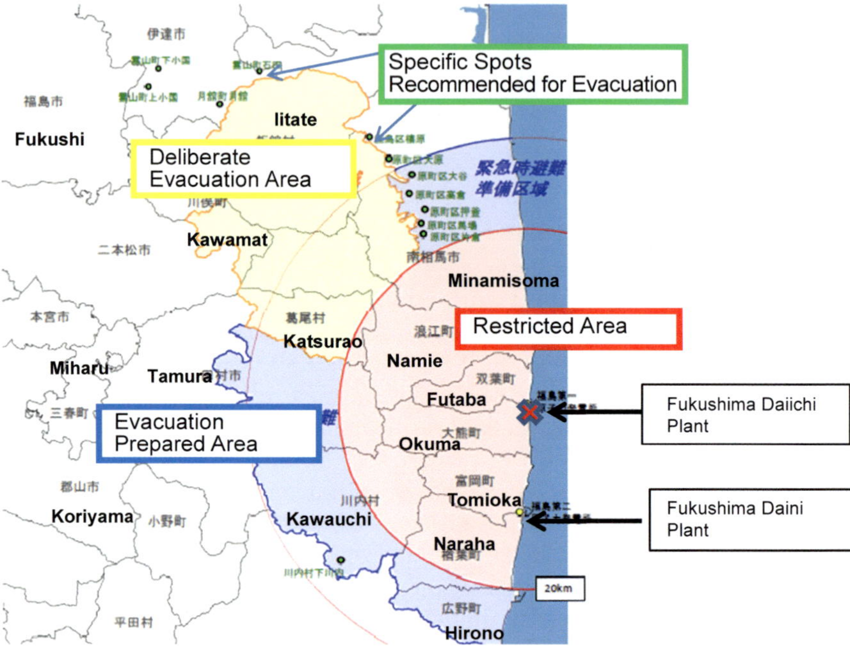

FIGURE 6.1 Evacuation zones established by the Japanese government following the Fukushima Daiichi accident. Fukushima Nuclear Power Plant (No. 1) is the Fukushima Daiichi plant; Fukushima Nuclear Power Plant (No. 2) is the Fukushima Daini plant. SOURCE: Adapted from Hasegawa (2013).

(UNSCEAR, 2013b). This area was designated as a "Restricted Zone" with entry initially prohibited.[19]

The hydrogen explosions in Unit 1 (15:36 on March 12), Unit 3 (11:01 on March 14), and Unit 4 (06:14 on March 15) (see Chapter 3, Sidebar 3.1) led the prime minister to issue new instructions to the heads of relevant local governments on March 15 including Fukushima Prefecture, Okuma town, Futaba town, Tomioka town, Namie town, Kawauchi town, Minamisoma city, Katsurao village, Hirono town, and Iitate village (see Figure 6.1), to order residents within the 20- to 30-km radius from the plant to shelter in place in what was designated as an "Evacuation Prepared Area."

[19] Some progress has been made with respect to the resettlement of parts of this area (METI, 2013).

Approximately 60,000 people lived within the 20- to 30-km shelter-in-place zone (UNSCEAR, 2013b). On March 25, these residents were advised by the government to begin voluntary evacuations.

On April 22, 2011, the central government issued a new evacuation order to residents of Iitate, located outside the 20-km radius evacuation zone, where high radiation levels had been detected. Residents of that village were given 1 month to evacuate. The area was designated as a "Deliberate Evacuation Area."

At this point onward the government switched from communicating evacuation orders on the basis of distance from the plant to using a threshold radiation dose of 20 mSv/yr as a basis for evacuation (Hasegawa, 2013). In June 2011, the government began to identify hotspots[20] where radiation levels exceeded this 20-mSv/yr threshold. These hotspots were named "Specific Spots *Recommended*[21] for Evacuation." These hotspots were more than 20 km away from the Fukushima Daiichi plant and outside the Deliberate Evacuation Area (UNSCEAR, 2013b).

6.3.2.2 *Potassium Iodide Distribution*

In addition to evacuation and shelter-in-place orders, residents leaving the 20-km Restricted Zone were instructed to take potassium iodide (KI). This instruction was issued on March 16, four days after major releases of radioactive iodine (iodine-131) had begun and after about half of the iodine release had occurred (TEPCO, 2012b, Fig. 27). This was also 4 days after residents within the Restricted Zone were instructed to evacuate and a day after residents in the 20- to 30-km Evacuation Prepared Area were instructed to shelter in place. Upon issuing this instruction, KI was made available for distribution. The KI consisted of 1.51 million pills for 750,000 people and 6.1 kg powder for 120,000–180,000 people. However, the KI was likely not distributed because the evacuation had already been completed (Hamada et al., 2012).

On March 15, four towns close to the plant, Futaba, Tomioka, Iwaki,[22] and Miharu, distributed in-stock KI pills to local residents without awaiting distribution instructions from the government. Futaba and Tomioka also instructed their residents to take the pills (Hayashi, 2011).

---

[20] Hot spots are defined based on radioactive contamination levels. They are regions where contamination levels significantly exceed those in surrounding areas.

[21] In other words, evacuations in these areas were not ordered.

[22] Iwaki is located south of the area shown in Figure 6.1.

6.3.2.3 *Food Interdictions*

On March 15, 2011, high levels of radioactive iodine (iodine-131) and radioactive cesium (cesium-134, -137) were detected in topsoil and vegetation near the Fukushima Daiichi plant (Hamada et al., 2012).[23] The Nuclear Safety Commission (NSC) advised that monitoring surveys of food and water should begin immediately. Food and water samples were collected beginning on March 16, 2011. On March 17, the Ministry of Health, Labor and Welfare (MHLW) set regulatory limits for contaminated food and water; these limits were stipulated as *provisional regulatory values* (PRVs).

PRVs were adopted from the index values preset by NSC except for radioactive iodine (iodine-131) in water and milk ingested by infants and in seafood[24] (Hamada and Ogino, 2012). PRVs for foodstuffs and drinkable liquids contaminated with radioactive cesium (cesium-134, -137), uranium, plutonium, and some other transuranic isotopes were based on an effective dose limit not to exceed 5 mSv/yr (Hamada and Ogino, 2012). The Food Safety Commission of Japan decided on March 20, 2011, that these PRVs were effective enough to ensure public safety. These PRVs were applied in various districts of Fukushima, Ibaraki, Chiba, Miyagi, Tochigi, Iwate, Gunma, and Kanagawa Prefectures starting on March 21 (FSC, 2011; Hamada and Ogino, 2012; IRSN, 2012b).

These initial PRVs were in place until March 31, 2012. New regulatory values went into effect on April 1, 2012. These new values were expressed as radioactive concentrations of cesium-134 and -137, but also considering the contributions of strontium-90, plutonium-238, -239, -241, and ruthenium-106, not to exceed a committed effective dose of 1 mSv/yr (Hamada and Ogino, 2012).

6.3.3 Accident Recovery

From March 11, 2011, to August 2011, implementation of an integrated recovery plan was hampered by administrative delays. In particular, time was needed to establish the required administrative structures, regulations, and a budget framework for those recovery actions that were not covered in existing disaster management plans (Hardie and McKinley, 2013). Decontamination activities during this period were conducted outside of the evacuation zones with a focus on high-sensitivity areas, such as schools

[23] Hamada et al. (2012) do not specify the location where high levels of cesium were found.

[24] Contamination of foodstuffs and liquids with iodine-131 became less of a public health concern with time owing to that isotope's short half-life (approximately 8 days). This was not the case for foodstuffs and liquids contaminated with cesium-134 and cesium-137 which have much longer half-lives.

and playgrounds, associated with radioactive hotspots (see Footnote 23). These decontamination activities were carried out primarily by local groups coordinated at a community or municipality level; technical support for these activities was provided by organizations such as the Japan Atomic Energy Agency (JAEA) (Hardie and McKinley, 2013).

In August 2011, the Japanese government passed the Act on Special Measures Concerning the Handling of Radioactive Pollution.[25] Pursuant to the Act, Japanese agencies developed a framework and guidance for remediating contaminated areas. These guidelines cover methods for surveying and measuring contamination levels as well as strategies for decontamination and storage of contaminated materials (Yasutaka et al., 2013). The Act took full effect in January 2012; it established JAEA as the responsible organization for coordinating the development of a technical basis for the regional remediation plan to be developed under the Act (Hardie and McKinley, 2013).

According to the Act, contaminated areas were to be grouped into two categories:

- *Special decontamination areas*. These areas comprise the Restricted Zone (i.e., areas within 20 km of the plant) as well as the Deliberate Evacuation Area (i.e., area beyond 20 km where the annual effective dose for individuals was anticipated to exceed 20 mSv[26]). The national government is responsible for the decontamination of these areas with a goal to reduce annual cumulative doses to less than 20 mSv. The long-term goal is to reduce annual cumulative dose to less than 1 mSv.
- *Intensive contamination survey areas*. These include all other contaminated areas in which the cumulative radiation dose for individuals was anticipated to range between 1 mSv and 20 mSv annually. Decontamination is to be overseen primarily by local municipalities with the goal to reduce the "air dose rate" to less than 1 mSv/yr.

The special decontamination area has been further subdivided into three areas (see Figure 6.2 for the most current (April 2014) map of these areas)[27]:

[25] http://josen.env.go.jp/en/. Accessed July 2014.

[26] According to ICRP (2011), 1-20 mSv/yr is the reference dose recommendation for exposure situations involving, for example, people living in long-term contaminated areas after a nuclear accident or a nuclear emergency.

[27] http://www.meti.go.jp/english/earthquake/nuclear/roadmap/pdf/20120330_01b.pdf.

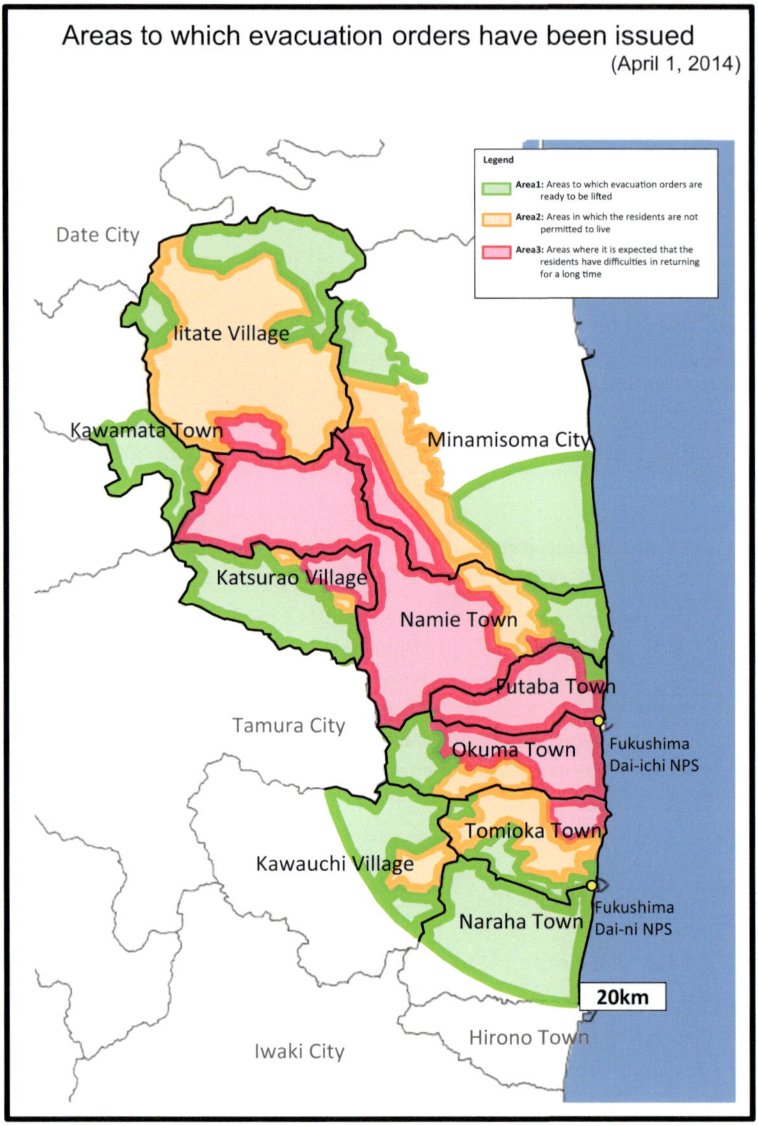

FIGURE 6.2 METI projections for land decontamination end states in regions affected by the Fukushima Daiichi accident. Area 1 (green): estimated annual dose level is below 20 mSv; Area 2 (orange): estimated annual dose level is 20-50 mSv; Area 3 (pink): estimated annual dose level is over 50 mSv and residents are not allowed entry. SOURCE: METI. Available at http://www.meti.go.jp/english/earthquake/nuclear/roadmap/pdf/140401MapOfAreas.pdf.

- *Area 1:* Estimated annual dose level is below 20 mSv and residents can return home temporarily. Evacuation orders within this area are "ready to be lifted."[28]
- *Area 2:* Estimated annual dose level is 20-50 mSv; residents are allowed entry for specific purposes but are ordered to remain evacuated.
- *Area 3:* Estimated annual dose level is over 50 mSv and residents are legally required to remain outside these areas. Levels are not expected to drop below 20 mSv/yr before about March 2016, five years after the Fukushima Daiichi accident.

Decontamination of these areas involves the cleaning of structures and removal of contaminated soil. The removed soil and other contaminated wastes are being stored at remediation locations or at temporary sites.[29] Incineration is being used for volume reduction of some contaminated materials (while meeting applicable emission standards for limiting public exposures) (IAEA, 2014b). Contaminated soil and waste are to be gathered and placed into interim storage facilities until transferred to a long-term disposal site outside of the Fukushima area. The national government aims to have these interim storage facilities in operation by early 2015.[30]

6.4 VULNERABILITIES IN EMERGENCY RESPONSE IN JAPAN

FINDING 6.1: The Fukushima Daiichi accident revealed vulnerabilities in Japan's offsite emergency management. The competing demands of the earthquake and tsunami diminished the available response capacity for the accident. Implementation of existing nuclear emergency plans was overwhelmed by the extreme natural events that affected large regions, producing widespread disruption of communications, electrical power, and other critical infrastructure over an extended period of time. Additionally:

[28] METI's April 2014 map of Area 1 has remained unchanged, with few exceptions, since the previous update provided by the agency in August 2013. Therefore METI's designation of Area 1 as "evacuation orders are ready to be lifted" may be misleading. (See METI maps http://www.meti.go.jp/english/earthquake/nuclear/roadmap/pdf/20130807_01.pdf and http://www.meti.go.jp/english/earthquake/nuclear/roadmap/pdf/140401MapOfAreas.pdf for a direct comparison of the areas.)

[29] Available at http://josen.env.go.jp/en/ from the Ministry of the Environment, Government of Japan. Accessed on June 4, 2014.

[30] See https://www.reconstruction.go.jp/english/topics/2013/03/decontamination-process.html. Accessed on July 17, 2014.

- **Emergency management plans in Japan at the time of the Fukushima Daiichi accident were inadequate to deal with the magnitude of the accident, requiring emergency responders to improvise.**
- **Decision-making processes by government and industry officials were challenged by the lack of reliable, real-time information on the status of the plant, offsite releases, accident progression, and projected doses to nearby populations.**
- **Coordination among the central and local governments was hampered by limited and poor communications.**
- **Protective actions were improvised and uncoordinated, particularly when evacuating vulnerable populations (e.g., the elderly and sick) and providing potassium iodide.**
- **Different and revised radiation standards and changes in decontamination criteria and policies added to the public's confusion and distrust of the Japanese government.**
- **Cleanup of contaminated areas and possible resettlement of populations are ongoing efforts 3 years after the accident with uncertain completion time lines and outcomes.**
- **Failure to prepare and implement an effective strategy for communication during the emergency contributed to the erosion of trust among the public for Japan's government, regulatory agencies, and the nuclear industry.**

6.4.1 Lack of Planning for a Severe Nuclear Accident

According to an independent Diet investigation of the Fukushima Daiichi accident (NAIIC, 2012), Japan was not prepared for the severe demands of the triple disaster that occurred on March 11, 2011. Moreover, Japan's preparedness for a nuclear disaster would have been deficient even if it had occurred in isolation of the natural disasters. The possibility of a reactor-core-damaging event at a nuclear plant in Japan was considered implausible (see Chapter 4). Consequently, planning for such an event was not treated seriously, leaving Japan unprepared for the scope and extent of the required emergency response. For example, the 2007 NSC guide for emergency preparedness describes the basis for establishing an Emergency Planning Zone[31] as being a result of a *hypothesized* situation of releases that could not possibly occur (NSC, 2013). The guide further states that actions such as sheltering in place or evacuation would not be needed outside of the 8- to 10-km radius (NSC, 2013).

The belief that design, engineering, and administrative controls related to nuclear plant operation excluded the possibility of a severe accident may

[31] This zone extends out to approximately 10 km from a nuclear plant.

have contributed to the many difficulties faced during implementation of the Japanese government's Basic Plan for Emergency Preparedness. In addition, the plan did not address contingencies, such as the loss of electrical power and communications, or diversion of response staff (e.g., local police and fire response resources) by competing events such as an earthquake and tsunami (NSC, 2013).

In spite of these limitations, the Japanese government was able to substantially decrease radiation exposure risks to the public using standard protective actions: evacuation, sheltering in place, and food and water interdictions (NSC, 2013). While some KI was distributed by Japanese prefectures and towns near the Fukushima Daiichi plant (see Section 6.3.2.2), it is not clear whether this KI was taken, and, if it was, whether its administration resulted in dose savings.

6.4.2 Lack of Reliable Information to Make Informed Decisions

As noted in Section 6.3.2, SPEEDI was not functioning at its full capacity after the accident.[32] Moreover, much of the radioactive material releases from the Fukushima Daiichi plant were through uncontrolled pathways. The loss of onsite power made measurements of radioactive material releases through controlled-release pathways (e.g., through the plant stack) impossible. SPEEDI could still be used to estimate the offsite atmospheric dispersion of radioactive noble gases and iodine using reference release rates (Investigation Committee, 2011). However, these estimates were not always communicated to relevant organizations (e.g., MEXT, NSC, and Fukushima Prefecture) (Investigation Committee, 2011). Also, SPEEDI results were not initially made public; therefore, local governments could not use those results to plan evacuations.[33]

6.4.3 Uncoordinated Issuance of Protective Actions

As noted in Section 6.3.2.1, a series of evacuation decisions were made as the accident at the Fukushima Daiichi plant unfolded. However, because of the lack of reliable and timely information about radioactive material releases from the plant, coupled with the loss of electrical power and gen-

---

[32] Normally, if the releases were via the reactor stacks and power was available, the rates of release of radioactive materials could be monitored during an accident. See USNRC (2006).

[33] SPEEDI estimates were not made public for several reasons. First, officials did not trust the SPEEDI estimates because they were made using assumed scenarios. Second, until March 16, it was not clear whether MEXT or NSC had the responsibility for operating SPEEDI. On March 16, NSC became the responsible organization for operating and maintaining SPEEDI and making its estimates public. The first SPEEDI estimates were made public on March 23 (Investigation Committee, 2011).

eral disruption of infrastructure, these decisions had to be made on an ad hoc basis. The evacuations were considered a precautionary response in light of the uncertainties about the status of the Fukushima Daiichi reactors and the potential offsite doses to surrounding populations if the reactors could not be stabilized.

The protective actions issued during the accident were generally successful due to the combination of good execution by the organizations involved, improvisation, and good luck. With respect to "good luck," as noted previously, about 80 percent of the radioactivity released from the Fukushima Daiichi plant was transported to the Pacific Ocean (Kawamura et al., 2011; Morino et al., 2011). Nevertheless, according to UNSCEAR (2013b), evacuations reduced by up to a factor of 10 the doses that would have been received by those living in the evacuated areas.

6.4.3.1 Evacuations

Independent of how successful the evacuation orders were in reducing radiation doses, evacuation instructions issued by the central and local governments lacked coordination, primarily because of disruptions to the communication infrastructure. As a result, many (and perhaps most) residents ordered to evacuate had to do so repeatedly, moving from one place to another (Hasegawa, 2013; Kurihara, 2013).

Additional confusion was caused by a March 16 recommendation issued by the U.S. Department of State that Americans located within about 80 km (50 miles) of the Fukushima Daiichi plant evacuate because of concerns that the situation could worsen (USNRC, 2011d; U.S. Embassy, 2011). This order sowed confusion and anxiety among the Japanese people who were living in this zone, because they had not been told by the Japanese government to evacuate. It also prompts the question "What is the appropriate role of foreign authorities in providing recommendations to its traveling or relocated citizens in a nuclear emergency?" This question is particularly relevant when recommendations are contradictory to those made by the host country government (González et al., 2013). This complex issue needs to be studied and resolved, not only for potential future nuclear accidents or events involving different national governments,[34] but also for those involving national and local authorities of the same country.

A major issue with the evacuations during the Fukushima Daiichi acci-

[34] Understanding this complex issue is particularly important for future national or international U.S. responses that involve a neighboring country, for example, Canada. The committee was told that the United States and Canada recognize the need for cooperation in developing responses to nuclear plant accidents and other radiological emergencies (FEMA, oral communication with the committee, May 17, 2013).

dent was the lack of detailed planning for vulnerable populations such as the elderly and the hospitalized (Nomura et al., 2013). Tanigawa et al. (2012) describe the chaotic evacuation of bedridden patients, some of whom died before reaching admitting facilities because of evacuation-related trauma or their own medical conditions. Patients could not take personal belongings because of space restrictions, and many patients were transferred several times to different locations over the period of a few months. Some institutions denied entrance to evacuees due to fears that they could contaminate others with radioactivity (Tanigawa et al., 2012; Tominaga et al., 2014). According to a recent article (Tominaga et al., 2014), there was insufficient education and training of emergency responders and physicians on radiation and its health effects.

Healthy elderly were also at risk. Yasumura et al. (2013) analyzed monthly mortality data among the 1,770 institutionalized elderly who were relocated from nursing homes, geriatric health service facilities, and other facilities within the 20-km evacuation zone. They estimated that there were 109 excess deaths among persons in this group.[35] The most common cause of death was pneumonia (41 percent), possibly due to low temperatures and poor nutrition (Yasumura et al., 2013). This high number of excess deaths emphasizes the vulnerability of geriatric populations to relocation.

6.4.3.2 Potassium Iodide

As noted in Section 6.3.2.2, little KI was administered to populations living near the Fukushima Daiichi plant because they had already evacuated. Reports indicate that radiation thresholds for KI administration were not exceeded as a result of timely evacuations (UNSCEAR, 2013b). Consequently, discussions of KI's medical effectiveness during the Fukushima Daiichi accident are irrelevant. However, the efforts to distribute KI during the accident highlight problems with planning for KI distribution and communication and coordination between the national and local governments. These problems were due in part to the loss of the communication infrastructure because of the earthquake and tsunami.

For example, on the night of March 14, the NERHQ and NSC were informed that not all hospitalized patients within the 20-km-radius Deliberate Evacuation Area had actually evacuated. The next day, NERHQ sent a fax to its local NERHQ office advising that these patients should take KI as they evacuate. However, the fax arrived as the local NERHQ office was relocating to the Fukushima Prefectural Office building; it was not discovered until later that evening. The local NERHQ considered it highly

[35] The focus here was on those affected by the nuclear emergency, not the earthquake and tsunami.

likely that many other elderly citizens and hospital staff still remained in this zone, so it created an instruction draft advising everyone that remained to take KI. The Fukushima prefectural government had confirmed, however, that there were no remaining elderly citizens or hospital staff within this zone (Investigation Committee, 2011).

6.4.4 Revisions to Radiation Standards

Confusion among residents living in the evacuation zones was exacerbated by the changing radiation safety standards established by the central government in the days and weeks following the accident. As noted in Section 6.3.2.1, from April 22, 2011, onward, the government started communicating relocation and sheltering orders to the public based on a threshold radiation dose of 20 mSv/yr (Hasegawa, 2013). This dose is 20 times higher than the allowable dose limit for public exposures resulting from normal nuclear power plant operations (1 mSv/yr in Japan). Although within the range set by the International Commission on Radiological Protection (ICRP) of 1-20 mSv/yr for "emergency or existing radiation exposure conditions" including nuclear accidents (ICRP, 2007), it was difficult for the public to understand "why the dose limit of 1 mSv/yr, which was valid before the accident, could be exceeded after the accident—at a time when people expect[ed] to be better protected" (González, 2012).

The public's trust in the government was also affected by the resignation of an academic advisor to the government over the 20-mSv/yr threshold. The advisor judged that this threshold was not acceptable for children (Kai, 2012).[36] This example suggests an important lesson: individuals responsible for informing the establishment of radiation protection guidelines during an accident need to agree on the technical criteria for establishing such guidelines in advance, and they also need to support the guidelines subsequently. However, agreeing on criteria that will cover all conceivable situations ahead of time is challenging because of the accepted standard practice of reducing doses to a level as low as deemed reasonably achievable (ALARA). It is not possible to know ahead of time what "reasonably achievable" may mean in a particular situation. Moreover, decisions about "what is reasonable" are themselves somewhat subjective, which can breed suspicion in situations where there is a lack of trust. Nevertheless, the solution is not to defer the establishment of standards or develop comprehensible explanations until a crisis forces action.

Additional public confusion and reported loss of trust in the Japanese government in its management of the response to the accident (Hasegawa,

[36] It is understood that for a given radiation dose, children are generally at a higher risk of developing cancer compared with adults (UNSCEAR, 2013a.)

2013) relates to the change in decontamination criteria for evacuees before and during the accident. Prior to March 11, 2011, a decontamination criterion for evacuees was established at 40 Bq/cm^2. Using various assumptions, and for practical purposes, this was translated to correspond to a reading of 13,000 counts per minute on a widely used Geiger-Muller radiation survey meter (Ogino and Hattori, 2013). On March 14, 2011, the Fukushima prefectural government decided that only partial decontamination would be undertaken for evacuees with readings between 13,000 and 100,000 counts per minute, and also that full-body decontamination would be undertaken only for evacuees that exceeded 100,000 counts per minute. This revision was made out of two concerns: (1) that too many persons would be considered "contaminated" at low levels (Investigation Committee, 2012)[37]; and (2) that there were not enough decontamination tools such as tents and water to process the large numbers of persons who would need to be decontaminated (Ogino and Hattori, 2013).[38]

Another example of changing radiation standards following the accident relates to food interdictions (see Section 6.3.2.3). As the accident evolved, the Japanese government issued PRVs to limit the intake of contaminated food and water. These "provisional" limits aimed to maintain doses to below 5 mSv/yr. These were later revised downward to a maximum permissible dose of 1 mSv/yr (MHLW, 2012, Slide 3).

The bases and reasons for revising these radiation standards continue to be an issue during the recovery phase of the accident. Cleanup of the large areas contaminated by releases of radioactive materials from the Fukushima Daiichi plant is proving to be challenging due to the limited effectiveness of decontamination techniques (Yasutaka et al., 2013) and lack of short- and long-term plans for disposal of the radioactive waste created during the cleanup.

6.4.5 Lack of an Effective Communication Strategy

A wide number of sources indicate that public trust was challenged due to the perception of partial, incorrect, delayed, and ambiguous information about the accident (Nakamura and Kikuchi, 2011; NERHQ-TEPCO, 2011;

[37] The written materials reviewed by the committee did not specify the area over which the 13,000 (or 100,000) counts per minute should be measured. However, it may be reasonably presumed that the counts correspond to measurement of an area equal to the active area of the probe used.

[38] According to the IAEA, it is unlikely that skin or clothing contamination from radioactive materials released during a severe nuclear accident would pose a significant health concern to offsite populations. Members of the general public can protect themselves from radioactive material on the skin and clothing by taking precautions such as showering and changing their clothes at the first opportunity (IAEA, 2003).

Fitzgerald et al., 2012; Kai, 2012; Ng and Lean, 2012; Figueroa, 2013; Hosono et al., 2013; Tateno and Yokoyama, 2013). According to these sources, the government

- Failed to characterize accurately the conditions at the plant in terms of safe shutdown of reactors and radioactive material releases (Imtihani and Mariko, 2013);
- Rejected as premature a report by a press secretary, later admitted to be correct, that there had been a reactor meltdown early after the tsunami hit the plant (Nakamura and Kikuchi, 2011);
- Withheld for considerable time rough (albeit questionable) predictions of doses based on hypothetical release magnitudes (Tateno and Yokoyama, 2013);
- Provided vague explanations about associated risks (RJIF, 2014); and
- Did not provide adequate information on food contamination and internal exposure (Tateno and Yokoyama, 2013).

Not surprisingly, these perceptions of mistrust increased the tendency of the media to seek out alternative, nongovernmental information sources. The messages from these alternative sources sometimes conflicted with, or appeared to be in conflict with, government statements (Sasakawa, 2012). This was the first major nuclear power plant accident in the Internet age. According to one view, Internet information about the risks of radiation exposure increased public concerns about the risks to children (Kai, 2012). However, a survey of residents living within 300 km of the Fukushima Daiichi plant indicates that the public was skeptical about Internet information and placed NHK[39] at the top of the list of credible sources (Tateno and Yokoyama, 2013).

6.5 LESSONS LEARNED FROM THE FUKUSHIMA DAIICHI ACCIDENT FOR THE UNITED STATES

FINDING 6.2: The committee did not have the time or resources to perform an in-depth examination of U.S. preparedness for severe nuclear accidents. Nevertheless, the accident raises the question of whether a severe nuclear accident such as occurred at the Fukushima Daiichi plant would challenge U.S. emergency response capabilities because of its severity, duration, and association with a regional-scale

[39] NHK is Japan's national public broadcasting organization known in English as Japan Broadcasting Corporation.

natural disaster. The natural disaster damaged critical infrastructure and diverted emergency response resources.

RECOMMENDATION 6.2A: The nuclear industry and organizations with emergency management responsibilities in the United States should assess their preparedness for severe nuclear accidents associated with offsite regional-scale disasters. Emergency response plans, including plans for communicating with affected populations, should be revised or supplemented as necessary to ensure that there are scalable and effective strategies, well-trained personnel, and adequate resources for responding to long-duration accident and/or disaster scenarios involving

- **Widespread loss of offsite electrical power and severe damage to other critical offsite infrastructure, for example, communications, transportation, and emergency response infrastructure;**
- **Lack of real-time information about conditions at nuclear plants, particularly with respect to releases of radioactive material from reactors and/or spent fuel pools; and**
- **Dispersion of radioactive materials beyond the 10-mile emergency planning zones for nuclear plants that could result in doses exceeding one or more of the protective action guidelines.**

RECOMMENDATION 6.2B: The nuclear industry and organizations with emergency management responsibilities in the United States should assess the balance of protective actions (e.g., sheltering in place, evacuation, relocation, and distribution of potassium iodide) for offsite populations affected by severe nuclear accidents and revise the guidelines as appropriate. Particular attention should be given to the following issues:

- **Protective actions for special populations (children, ill, elderly) and their caregivers;**
- **Long-term impacts of sheltering in place, evacuation, and/or relocation, including social, psychological, and economic impacts; and**
- **Decision making for resettlement of evacuated populations in areas contaminated by radioactive material releases from nuclear plant accidents.**

The Fukushima Daiichi accident revealed that existing Japanese emergency response plans for dealing with nuclear accidents were inadequate, and it brought to the surface problems with the coordination and decision-making processes used by government and industry officials. The difficulties in responding to the accident were exacerbated by the lack of reliable, real-time information on the status of the plant, accident progression, and projected doses to nearby populations.

The committee did not have the time or resources to perform an in-depth examination of U.S. preparedness for severe nuclear accidents. However, the accident raises the question of whether emergency preparedness in the United States would be challenged if a similar-scale nuclear accident were to happen domestically when emergency responses were diverted to deal with concurrent disasters.

6.5.1 Emergency Response Planning Around Nuclear Power Plants

Because of the severe damage to the reactors at the Fukushima Daiichi plant, actions that would normally be associated with the "early phase" of a nuclear incident extended over many days to weeks, rather than the expected hours to days that inform nuclear emergency planning in the United States (see Sidebar 6.1 for definitions of accident phases). Consequently, response staff and resources were needed for an extended period of time. Furthermore, the earthquake and tsunami consumed local police and fire response resources which, similar to the United States, play an integral role in conducting orderly evacuations and providing other emergency services.[40] Medical and evacuation center staff and resources were overextended by the demands of responding to two natural disasters (the earthquake and tsunami) and the lengthy time frame of the unfolding accident.

The committee has identified three challenges from the Fukushima Daiichi accident that have the potential to compromise offsite emergency responses to Fukushima-scale events in the United States. These challenges are described below:

1. *Widespread loss of offsite electrical power and severe damage to critical infrastructure:* The impact to the communications, transportation, and electrical power infrastructure caused by the earthquake and tsunami affected the ability of the Japanese national government to communicate with prefectural and local governments. These types of impacts might also be seen in a response to a Fukushima-scale event in the United States. It is worth examining contingency planning for events that include major disruptions to communications, transportation, and the electrical power infrastructure and for which state, local, medical, and emergency reception center staff and resources are diverted by a competing disaster.

---

[40] In the United States, for example, many plans call for state or local first responders to conduct contamination screening during an event to support the movement of evacuees to an emergency reception center.

2. *Lack of real-time information about conditions at the Fukushima Daiichi plant:* Questions related to the reliability and ease of use of information from radiation monitoring equipment in Japan proved to be largely irrelevant during the accident response because onsite and offsite instrumentation was not functional, at least at full capacity.[41] As a result, information was too sparse for creation of timely and reliable analyses of the plant status and future prospects. Many emergency response decisions were made without a firm basis of situational knowledge. This led to a preference for evacuation versus sheltering in place for populations affected by the accident. It is worth examining whether, given the same set of circumstances in the United States, the information needed to select protective actions would be available—or whether failures, for example, of the Safety Parameter Display System (SPDS),[42] Emergency Response Data System (ERDS),[43] or normal telephone system would result in a similar loss of information. The USNRC is aware of this issue and believes that its ERDS modernization program, which was under way before the Fukushima Daiichi accident, is sufficient (USNRC, 2011c).[44] The Committee did not review this modernization program.

3. *Dispersion of radioactive material beyond the 10-mile EPZ resulting in doses exceeding one or more protective action guidelines (PAGs)*: Radiation doses exceeded some PAGs beyond 30 km (18.6 miles) from the Fukushima Daiichi plant. This suggests that, if a similar-scale accident were to occur in the United States, the 10-mile plume exposure pathway emergency planning zone (EPZ) currently established by the USNRC[45] may prove inadequate.[46] Given the Japanese experience, it would be worthwhile for the USNRC, FEMA, state and local entities, and industry to review and assess the scalability and effectiveness of emergency response plans for events that lead to significant potential radiation doses and radioactive contamination extending beyond the 10-mile EPZ. This would include reviewing the plans for issuing protective actions such as evacuation, sheltering in

[41] See Appendix M for the committee's opinions on how access to timely and reliable information to support decision making can be improved.

[42] The SPDS displays a set of plant parameters from which the plant operators can assess the safety status of plant operation (USNRC, 1980b).

[43] The ERDS provides electronic transmission capability of a limited set of parameters from the plant computer to the USNRC during an emergency.

[44] The aim of the USNRC's ERDS modernization program is to ensure that the agency can receive data from all affected reactor units during a multiunit event.

[45] Emergency response plans are in place for the plume exposure pathway EPZ to avoid or reduce dose from potential radiation exposure from the release of radioactive materials after a nuclear event. These plans provide the structure for implementing protective actions such as evacuation, sheltering in place, and the use of KI.

[46] As noted in Section 6.4.3.1, the U.S. government advised Americans within 50 miles of the Fukushima Daiichi plant to evacuate; similarly, a domestic accident could require evacuation and other protective actions in populations at a similar distance from a U.S. nuclear plant.

place, and use of KI[47] to populations that reside beyond the 10-mile EPZ and testing the effectiveness and scalability of these plans by performing regular exercises.

There is a need to ensure that emergency response plans in the United States include scalable and effective strategies, well-trained personnel, and adequate resources for responding to severe and long-duration nuclear emergencies. Elements of an effective communications plan with the affected populations are discussed separately in Section 6.5.3.

6.5.2 Principles for Formulating Protective Actions

The Fukushima Daiichi accident demonstrated that evacuation of populations at risk is problematic if not executed carefully, and it revealed challenges for evacuation of children, the ill, and elderly (see discussion in Section 6.4.3.1). It also revealed that evacuation, when used as a default protective action, is problematic when long-term consequences related to relocation, mental health impacts to the evacuated population, as well as the material impacts such as loss of business or employment are not considered.[48]

Additionally, the continuing concerns in Japan that dose levels applied for the protection of the population as a whole do not provide sufficient

---

[47] In the United States, the USNRC amended its regulations in 2001 to require that state and local emergency planners consider the use of KI to supplement other protective actions in the case of a general emergency at a nuclear plant (USNRC, 2002c). Since 2002, if a state requests it, the USNRC will provide enough KI for one or two daily doses to the population within the 10-mile EPZ (USNRC, 2002b). The USNRC expected to issue further guidance but, in 2002, decided not to pursue that effort (USNRC, 2000a, 2001, 2002a). The USNRC has continued to replenish state stockpiles in accordance with expiration dates. Since the inception of the program in 2001, the USNRC has shipped over 47,000,000 KI tablets to participating states. In 2004, the National Research Council issued a report (NRC, 2004a) in response to a congressional request for "a study to determine what is the most effective and safe way to distribute and administer potassium iodide tablets on a mass scale." (The congressional request is described in the Public Health Security and Bioterrorism Preparedness and Response Act of 2002. Section 127 is available at http://www.gpo.gov/fdsys/pkg/PLAW-107publ188/pdf/PLAW-107publ188.pdf, accessed March 20, 2014.) The National Research Council report noted the need for a strategy whereby local planning agencies could develop geographic boundaries for a KI distribution plan based on site-specific considerations. These geographic boundaries would be decoupled from the planning boundaries of the 10-mile EPZ (NRC, 2004a). The USNRC is currently planning to consider prestaging KI outside the 10-mile EPZ (available at http://www.nrc.gov/reactors/operating/ops-experience/japan-dashboard/priorities.html#tier-03, accessed March 20, 2014).

[48] ICRP notes that decision makers need to justify disruptive protective actions from the perspective of the radiation exposure saved but also consider other issues that are beyond the scope of radiation protection (González et al., 2013).

protection to children could also arise during the response to a severe nuclear accident in the United States: PAGs in the United States are generally based on average risks for the total population and do not provide separate guidelines for children.[49]

In view of the Fukushima Daiichi accident, the nuclear industry and organizations with emergency management responsibilities in the United States should assess the balance of protective actions for offsite populations affected by severe nuclear accidents. The analysis should specifically address protective actions for special populations (children, ill, and elderly) and their caregivers and the long-term impacts of sheltering in place, evacuation, and/or relocation, including social, psychological, and economic impacts. It does not appear that USEPA explicitly informed its analysis in the recent PAG draft manual (USEPA, 2013) with lessons learned from the Fukushima Daiichi accident.[50]

6.5.3 Plans for Communicating with the Public

The importance of a good communication strategy during a crisis is recognized by U.S. government agencies and internationally. The USNRC, FEMA, and the IAEA, for example, emphasize in their guidelines (Persensky et al., 2004; USNRC, 2011a,c; FEMA, 2013a; IAEA, 2013a) the need to deliver understandable, accurate, and timely information while acknowledging uncertainty when communicating with the public. What needs to be evaluated within the United States is how agencies and organizations

---

[49] There are at least two exceptions: The Food and Drug Administration provides guidance on the threshold for taking KI, which includes a separate limit for children. It also recommends that state and local agencies should consider applying the threshold for children to the entire population to simplify decision making in an emergency. Also, FDA's Derived Intervention Levels (DILs) include consideration of children in the guidance on food and water interdiction (USFDA, 1998). This same approach is not taken with doses to other organs and therefore does not apply to evacuation or shelter-in-place orders.

[50] In a November 13, 2013, conference call with the committee, a representative of the USEPA indicated that the agency had taken the experiences from the Fukushima Daiichi accident into account when updating the PAG manual. However, the committee did not find explicit evidence for this in that draft (USEPA, 2013). The draft manual recommends evacuating an area if the expected dose for the first few days of the accident is within the 10- to 50-mSv range. However, informed by the Katrina and other domestic natural disasters, which showed that emergency evacuation plans in nursing homes and hospitals were inadequate in many parts of the United States (Wise, 2006; Blanchard and Dosa, 2009; OIG, 2012; Fink, 2013), the draft manual notes that sheltering in place may be a preferred protective action for special populations (e.g., the elderly and the sick who are not readily mobile) at projected doses of up to 50 mSv over 4 days. When evacuations are deemed difficult because of weather conditions or other hazards, sheltering in place may be justified for those populations for projected doses up to 100 mSv. Additionally, any decisions about sheltering in place for these vulnerable populations would also apply to their caregivers, typically young and healthy individuals.

with emergency management responsibilities coordinate their efforts to effectively deliver informed and concise messages to the public.

Announcements about radioactive material releases from the Fukushima Daiichi plant triggered public health concerns in the United States, especially on the West Coast and Pacific Islands (Tupin et al., 2012). As suggested elsewhere (Salame-Alfie et al., 2012), the Fukushima Daiichi accident could have been used as a test scenario for how communications among the responders and the public would play out if an accident were to occur in the United States. The U.S. National Response Framework (see Sidebar 6.2) was not followed, so there was no declaration of a lead federal agency for the response, and a Joint Information Center with collocated group of representatives from agencies and organizations with the responsibility to handle public information needs was not established.[51]

During the accident, authorities in U.S. states received a number of inquiries from members of the public regarding potential health effects from radioactive material releases from the Fukushima Daiichi plant; the safety of milk, water, and food; and need to take KI (Salame-Alfie et al., 2012). Thyroid dose projections for U.S. populations were well below levels that would trigger health concerns[52] and therefore were not high enough to meet USEPA guidelines for taking KI. Despite this fact, the U.S. Surgeon General's office issued a statement indicating that that it was appropriate for West Coast residents to take KI. This statement was viewed as incorrect by many (Fitzgerald et al., 2012; Salame-Alfie et al., 2012).

The USNRC informed the public that "no radiation at harmful levels would reach the United States" (USNRC, 2011e) and the USEPA announced that any radioactivity detected in the United States was "well below any level of public health concern."[53] However, little authoritative information was available about the human health impacts of radiation exposures; as a result, the fear of radiation exposure and public perceptions of exposure risks were not consistent with the messaging from government agencies (Haggerty, 2011; Payne, 2011).

The experience of the United States during the Fukushima Daiichi accident highlights the need to review existing plans for communicating with the public during a nuclear emergency. It is important that such plans deliver clear, timely messages about the status of the emergency and notifications of planned actions; recommendations regarding actions that could be taken by affected individuals; frank discussions of uncertainties and unavailable but necessary information; and clarification or correction

[51] However, as stated in the Joint Information Center (JIC) manual (NRT, 2000), the structure of the JIC could be useful in coordinating multiagency events internationally.

[52] Atmospheric dispersion of the radioactive materials released from the Fukushima Daiichi plant greatly reduced their concentrations by the time they reached the United States.

[53] http://www.epa.gov/japan2011/.

of alarming information and rumors originating from various sources.[54] These communication capabilities need to span all phases and activities related to an accident.

Communicating with the public about the meaning of radiation dose limits during a nuclear emergency is also important. The public confusion about dose limits that occurred in Japan (see Section 6.4.4) would most likely also occur in the United States: the United States has established a variety of radiation dose and radioactive contamination limits for different purposes; these limits are enforced by different means and different agencies. As seen in Table 6.2, dose standards applicable to the general public in the United States range from 1.0 mSv in 1 year from normal nuclear operations to 100.0 mSv in an emergency. This is a factor of 100 difference. There are additional standards, which are not included in the table, that are specific to individual organs (e.g., the thyroid).

Not all of the public confusion originates from the existence of too many standards; another source of confusion is the lack of a separate standard for children (González et al., 2013). As noted in Section 6.4.4, the concerns in Japan that dose levels applied for the protection of the population as a whole do not provide sufficient protection to children suggests that similar concerns could also arise in the United States.

6.5.4 Decision Making for Recovery

The ongoing offsite response to the Fukushima Daiichi Accident demonstrates that cleanup and resettlement of evacuated populations (collectively described here as "recovery") are complex processes. Many aspects of recovery, including issuing predetermined protective action criteria, cannot be planned in detail before an accident occurs; indeed, such criteria depend on the accident scenario, its consequences, and stakeholder preferences. However, the current situation in Japan, where about half of the evacuees (WNA, 2014) continue to live in shelters or temporary locations with uncertainty about their future plans, emphasizes the need for the United States to conduct advance planning for recovery from a nuclear plant accident.

---

[54] Reviewing the communication efforts during the 1979 Three Mile Island (TMI) accident could offer useful insights. In that instance, considerable trust was established by government leaders when Harold Denton, Director of the USNRC's Office of Nuclear Reactor Regulation and President Carter's personal adviser for the TMI accident, took over as spokesperson. He was well equipped to answer many of the questions that the public has been found to worry about in a crisis: What happened? What is being done about it? What should we do? What is likely to happen next? What is your credible worst-case scenario? What are you doing to prevent it? The USNRC's Special Inquiry Group tasked to investigate the TMI accident describes Denton as a person who if he does not have the answers, "will be willing to look for them and to share them once they are found" (Rogovin et al., 1980).

TABLE 6.2 Selected U.S. Radiation Dose Guidelines for Members of the Public

Circumstance or Pathway	Standard (mSv)	Agency
Drinking water (per year)	0.04	USEPA[a]
Air effluents (per year)	0.1	USEPA[b]
Decommissioned site (per year)	0.25	USNRC[c]
Normal nuclear operations (per year)	1.0	USNRC/USDOE[d]
Ingestion	5.0	FDA[e]
Relocation (standard per year after year 1)	5.0	USEPA[f]
Lower evacuation threshold (early-phase nuclear power plant accident—first 4 days)	10.0	USEPA[b]
Relocation (first-year dose)	20.0	USEPA[b]
Upper evacuation threshold (early-phase nuclear power plant accident—first 4 days)	50.0	USEPA[b]
Evacuation with serious adverse external conditions for special populations (during one incident)	100.0	USEPA[b]

[a]40 CFR § 141.66(d); from beta and gamma dose.
[b]40 CFR §§ 61.92 and 61.102 and 10 CFR § 20.1101(d).
[c] 10 CFR § 20.1402.
[d] 10 CFR §§ 20.1301 and 835.208.
[e] USFDA (2004).
[f] USEPA (2013).

The 1992 USEPA PAG manual (USEPA, 1992) did not address recovery following a nuclear plant accident. USEPA's recently updated PAG manual (USEPA, 2013), which is still labeled as a draft, minimally addresses recovery. It recommends that resettlement criteria should be established *after* a contamination event has occurred and notes that the process for establishing such criteria could take months to years. The draft PAG manual also recommends that the process to determine acceptable criteria for a given community should include input from community members and other stakeholders. However, no guidance is given on how to address stakeholder concerns that would likely arise in a Fukushima Daiichi-scale accident and how they might be minimized.

The USEPA draft PAG also does not provide specific recommendations for dose thresholds for long-term cleanup. It references the 1 in 10,000 to 1 in 1,000,000 acceptable lifetime risk criterion for cancer incidence, a range that is generally used for cleanup of contaminated sites under the Comprehensive Environmental Response, Compensation, and Liability Act (CER-

CLA) and the USNRC's process for decommissioning and decontamination of nuclear facilities. Assuming that the risk of developing cancer increases in proportion to dose received with no threshold (i.e., the LNT model), this risk range translates to an approximate dose to the whole body of 0.009-0.9 mSv over a lifetime.[55] The Fukushima Daiichi accident recovery has demonstrated that attaining cleanup goals in this range (i.e., a small fraction of the radiation dose received from natural background in a lifetime) may be impractical when contaminated areas are large.

International radiation protection agencies, such as the ICRP and IAEA, advocate for the principle of *optimization* when it comes to protection of populations living in an existing exposure situation such as in the areas contaminated by radioactive material releases from the Fukushima Daiichi plant (ICRP, 2007). This approach is a departure from conventional cleanup guidelines under CERCLA or decontamination of nuclear sites, both of which are based on either radiation dose or health-risk levels. The intent of these international recommendations is to take into account not only risk of developing cancer in the future, but also competing factors, for example, the local economy, future land use, cleanup options, and, ultimately, public acceptance. The NCRP, consistent with the ICRP recommendations, is currently (June 2014) finalizing a study that establishes the framework of an approach to optimizing decision making for recovery.[56]

Deciding on recovery strategies for severe nuclear accidents and their implementation should be part of the U.S. government's advance planning. The U.S. government should be able to develop and articulate guidance for state and local authorities in dealing with radiation contamination recovery. Issues for which needed policies and decision criteria are required include resettlement and decontamination, including disposal, reduction of volume, or storage of removed contaminated materials. In cases where resettlement may not be desirable, policies will also need to be developed for redirection of (and assistance to) evacuated populations to alternative permanent homes in new locations.

---

[55] The committee derived this accumulated dose range estimate as follows: Using the LNT model, the risk of cancer incidence (all cancers) for a dose equal to 1 mSv/yr over a lifetime is 621 per 100,000 for men and 1,019 per 100,000 for women (NRC, 2006a, Table 12D-3). Assuming a 50:50 gender ratio within a population, the risk for the population as a whole is 820 per 100,000 or else 8,200 per 1,000,000. For USEPA's reference to the 1 in 10,000 to 1 in 1,000,000 acceptable lifetime risk criteria for cancer incidence, the effective dose would be 0.012 mSv/yr to 0.00012 mSv/yr. Assuming a 75-year average life span, the lifetime dose would be equal to 0.009 to 0.9 mSv over a lifetime. For comparison, the annual average effective dose from background radiation to populations in the United States is 3.1 mSv annually (NCRP, 2009).

[56] Presentation by S. Y. Chen, http://www.ncrponline.org/Annual_Mtgs/2014_Ann_Mtg/PROGRAM_2-10.pdf. Last accessed March 20, 2014. The study will be published as NCRP Report No. 175.

7

Lessons Learned: Nuclear Safety Culture

The final chapter of this report focuses on the nuclear safety culture in Japan and lessons learned for the United States. Safety culture is not an explicit element of the statement of task for this study (see Box 1.1 in Chapter 1). Nevertheless, the committee quickly came to understand that the lack of a strong nuclear safety culture was an important contributing factor to the Fukushima Daiichi accident. The committee also came to appreciate the important role that nuclear safety culture plays in nuclear plant operations and regulations in the United States.

This chapter is organized into four sections: Section 7.1 describes the nuclear safety culture concept. Sections 7.2 and 7.3 describe and discuss the nuclear safety cultures in Japan and the United States, respectively. Section 7.4 provides two committee recommendations.

7.1 BACKGROUND ON NUCLEAR SAFETY CULTURE

The term *nuclear safety culture* combines two concepts: *safety* and *culture*:

- *Safety* is protection from harm and can be defined in terms of risk: an activity is considered to be safe when its associated risks are being controlled to acceptable levels.
- Culture comprises the collective beliefs, values, and behaviors of individuals belonging to an organization (e.g., a company). It includes behavioral norms, shared attitudes, shared traditions, and mechanisms for incentivizing and reinforcing desired behaviors.

Safety is considered to be an inviolable constraint and part of the social contract under which nuclear plants are allowed to operate. The shared responsibilities for nuclear plant safety are described in Sidebar 7.1.

For purposes of this report, safety culture is perhaps best understood as those organizational processes that ultimately influence and reinforce an organizational culture that emphasizes safety. Taken together, these processes create a continuous desire for improvement that is fueled by individuals who, in turn, find motivation from the organization's safety culture (Guldenmund, 2010).

The safety culture concept was first applied to the nuclear power industry by the International Atomic Energy Agency's (IAEA's) International Nuclear Safety Advisory Group (INSAG, 1986). The term was used to explain how the lack of knowledge about risk and safety and failure to act appropriately contributed to the Chernobyl accident. According to this group (INSAG, 1992, pp. 23-24), the Chernobyl accident was caused by a "deficient safety culture at Chernobyl and throughout the Soviet design, operating and regulatory organizations."

The use of the term by the U.S. Nuclear Regulatory Commission (USNRC) developed from a 1989 policy statement issued in response to unprofessional conduct and operator inattentiveness in nuclear plant control rooms (USNRC, 1989). The statement stresses that management at nuclear power plants

> has a duty and obligation to foster the development of a "safety culture" at each facility and provide a professional work environment in the control room and throughout the facility. (p. 3425)

The USNRC published a formal safety culture policy statement in 2011. That statement defines a nuclear safety culture as the

> core values and behaviors resulting from a collective commitment by leaders and individuals to emphasize safety over competing goals to ensure protection of people and the environment. (USNRC, 2011b)

The USNRC has taken the position that safety culture applies to all licensees, including nuclear plant operators.

The Institute of Nuclear Power Operations (INPO; see Sidebar 7.2) has published guidance on the nuclear safety culture for the U.S. power industry (INPO, 2013). That guidance notes that

> nuclear safety is a collective responsibility. The concept of nuclear safety culture applies to every employee in the nuclear organization, from the

SIDEBAR 7.1
Who Is Responsible for Nuclear Plant Safety?

Nuclear plant safety begins with a plant's design and construction and extends through its full life cycle including operation, maintenance, and, inevitably, decommissioning. Consequently, a number of organizations are responsible for plant safety: plant and equipment designers and manufacturers; constructors; plant owners/operators, from upper management through reactor operator and plant maintenance staff; and regulators who set, oversee, and enforce the standards and requirements for plant design, construction, and operation. These organizations have a shared responsibility to protect public safety and the environment during both normal and off-normal plant operations.

These organizations play different but complementary roles in meeting their shared responsibilities:

- Regulators are independent institutional bodies whose focus is on protection of the public and the environment, not for the promotion of nuclear technology or protection of investment in assets.
- Design, manufacturing, and construction firms are responsible for building as much inherent safety and environmental protection into the plants as can be reasonably achieved.
- Plant owners/operators are responsible for operating their plants so that safety and environmental protection goals are achieved.

Other organizations also contribute to nuclear plant safety. These include national authorities who appoint regulators' leadership and appropriate regulators' funding; governmental organizations such as the USNRC's inspector general and

> board of directors to the individual contributor. No one in the organization is exempt from the obligation to ensure safety first. (INPO, 2013, p. 6)

In its final safety culture policy statement (USNRC, 2011b), the USNRC notes that assessments of incidents involving U.S. civilian uses of nuclear materials demonstrate that significant mistakes occur when safety culture is weak. To prevent accidents from developing into severe core-damage events, and to prevent large-scale, long-term contamination, the importance of maintaining high safety culture standards cannot be overemphasized (Hogberg, 2013).

The IAEA promotes the development of a nuclear safety culture through workshops, written guidance, and peer review. The IAEA has also published guidance on enhancing the safety culture in nuclear installations (IAEA, 2002). The agency cautions that the Government Accountability Office, which conduct independent investigations of USNRC and industry actions; and public interest organizations, which offer technical critiques and advice.

The plant owner's/operator's first, foremost, and overriding responsibility is to ensure the safe operation of its plants. The owner/operator has other responsibilities, of course, including the provision of a reliable supply of electric power and protection of plant investments. A major accident can challenge the continued viability of an operating company, and so owners/operators could elect to adopt stricter safety standards and management practices than required by regulations.

The ultimate responsibility for nuclear plant safety and environmental protection resides with the plant's owners, managers, and operating staff and the agencies that regulate them. Five decades of nuclear plant operating experience demonstrate clearly that it is not possible to anticipate all combinations and permutations of operating conditions that can occur at a nuclear plant. Consequently, safety cannot be achieved only through rules, regulations, hardware design, and operating procedures. It also requires onsite intelligence, learning, and decision making by plant operating staff. More importantly, it requires "commitment by leaders and individuals to emphasize safety over competing goals to ensure protection of people and the environment" (INPO, 2013, p. iv).

The nuclear industry and its regulators can work together to promote nuclear safety, especially through the development of common understandings of problems and potential solutions. However, safety can be compromised if plant owners/operators adopt a compliance-only operating philosophy. On the other hand, regulators can become ineffective or even captured by the nuclear industry if independence is lost. Both of these situations can weaken the industry and the regulator's responsibilities to protect the public interest.

> The biggest danger in trying to understand culture is to oversimplify it in our minds. It is tempting to say that culture is just "the way we do things around here", or "our basic values", or "our rituals", and so on. These are all manifestations of the culture, but none is the culture at the level that culture matters. A better way to think about culture is to realize that it exists at several "levels" and that we must endeavor to understand the different levels, but especially the deeper levels. (IAEA, 2002, p. 3)

There is international acceptance by the nuclear power community that a strong nuclear safety culture needs to be adopted universally: by senior management of organizations operating nuclear power plants, by individuals who work in those plants, and by regulatory bodies and other organizations that set nuclear power policies. Indeed, this commitment to safety is

SIDEBAR 7.2
Institute of Nuclear Power Operations

The Institute of Nuclear Power Operations (INPO), a not-for-profit organization headquartered in Atlanta, Georgia, was established by the nuclear power industry after the 1979 Three Mile Island nuclear accident. It has instituted several important efforts to foster and improve a safety culture at U.S. nuclear power plants.

One of INPO's safety culture activities involves linking its evaluation of nuclear power plants to the ability of plant owners to obtain liability insurance. INPO evaluations are carried out at each plant every 2 years. INPO evaluation teams spend approximately 2 weeks at each plant, interviewing plant personnel and watching their actions on the job. The evaluation team meets daily, prepares a report, and presents that report to plant management and to the utility's chief nuclear and chief executive officers. The report rates the plant's performance using a numerical scale. If a plant receives a low rating, its chief executive officer and chief nuclear officer may be asked to make a presentation to INPO to explain what steps have been taken to correct deficiencies. Plants that receive a high rating are eligible for a discount on their liability insurance provided by an industry insurance organization.

INPO also maintains and/or sponsors training for nuclear power plant personnel at all levels. For example, training is given to first- and second-line supervisors, potential plant managers, as well as members of the board of directors of nuclear power companies. INPO also provides technical consulting to its member companies on an as-needed basis.

INPO collects, analyzes, and publishes "lessons learned" from events that occur at nuclear plants in the United States and abroad. When appropriate, INPO requires its members to implement enhancements in response to these lessons. A number of such have been made based on lessons learned from the Fukushima Daiichi accident.

The results of INPO's inspection program are shared among INPO members, but such information is not made available to the public. INPO judges that this limited sharing encourages candor and places the decision about what information to release to the public in the hands of company managements. The reports are also available at INPO for review by the USNRC. Additionally, INPO encourages each plant to allow USNRC regional staff to review the reports onsite. The USNRC does not review INPO inspection reports in detail, but the agency is aware of the overall results of these inspections. Lochbaum et al. (2014, pp. 151-152) note that because these inspection results are not made public, the public cannot determine how serious the identified problems are or whether, or to what extent, the identified problems have been addressed. Releases of summaries of these inspections by management to the public would help increase transparency.

an international priority, as evidenced by treaties such as the Convention on Nuclear Safety.[1]

7.1.1 Regulatory Independence

To establish a strong nuclear safety culture, it is not enough for nuclear plant operators to adopt a safety culture: The establishment, implementation, and maintenance of a robust nuclear safety culture are also dependent on a strong and independent regulator. Noggerath et al. (2011, p. 45) notes that

> A well established national safety culture depends not only on nuclear operators to meet the highest standards, but also on a nuclear authority to keep the national requirements updated and to require modernization of plants when necessary.

A Nuclear Energy Agency report asserts that

> The nature of the relationship between the regulator and the operator can influence the operator's safety culture at a plant either positively or negatively. In promoting safety culture, a regulatory body should set a good example in its own performance. This means, for example, the regulatory body should be technically competent, set high safety standards for itself, conduct its dealings with operators in a professional manner and show good judgment in its regulatory decisions. (NEA, 1999, p. 11)

The principle of "effective independence," as explained by the IAEA, defines the international nuclear communities' commitment to strong and effective regulation:

> The government shall ensure that the regulatory body is effectively independent in its safety related decision making and that it has functional separation from entities having responsibilities or interests that could unduly influence its decision making.
>
> To be effectively independent, the regulatory body shall have sufficient authority and sufficient staffing and shall have access to sufficient financial resources for the proper discharge of its assigned responsibilities. The regulatory body shall be able to make independent regulatory judgments and decisions, free from any undue influences that might compromise safety, such as pressures associated with changing political circumstances

[1] See especially Articles 8, 10–14. The treaty text is available at http://www.iaea.org/Publications/Documents/Infcircs/Others/inf449.shtml. The United States and Japan have ratified this treaty.

> or economic conditions, or pressures from government departments or from other organizations. Furthermore, the regulatory body shall be able to give independent advice to government departments and governmental bodies on matters relating to the safety of facilities and activities.
>
> No responsibilities shall be assigned to the regulatory body that might compromise or conflict with its discharging of its responsibility for regulating the safety of facilities and activities. (IAEA, 2010, pp. 6-7)

Effective independence means that the nuclear regulatory body must be able to make decisions and perform its duties without undue pressure or constraints from the government, organizations that promote nuclear power, or organizations opposed to nuclear power (Bacon-Dussault, 2013). While regulators need to be independent of the organizations they regulate, they must exercise their regulatory authority in ways that support robust programs at nuclear power plants to identify and correct problems before they become significant safety issues.

For example, a nuclear power plant in the United States can log over 100 "problems" daily requiring some sort of corrective action, but many of these problems typically have low safety significance. A system that encourages problem identification, reporting, and correction will operate most effectively when regulatory agencies use sound judgment to prioritize reported problems according to their safety significance. Plant operators will be more willing to disclose small problems—which can be caught and corrected before they become significant—when they understand that regulators will exercise their regulatory authority fairly.

7.2 NUCLEAR SAFETY CULTURE IN JAPAN

FINDING 7.1: While the Government of Japan acknowledged the need for a strong nuclear safety culture prior to the Fukushima Daiichi accident, TEPCO and its nuclear regulators were deficient in establishing, implementing, and maintaining such a culture. Examinations of the Japanese nuclear regulatory system following the Fukushima Daiichi accident concluded that regulatory agencies were not independent and were subject to regulatory capture.

The Government of Japan acknowledged the need for a strong nuclear safety culture by entering into the Convention on Nuclear Safety. Preamble Clause iv and Article 10 of the Convention note that

> Each Contracting Party shall take the appropriate steps to ensure that all

> organizations engaged in activities directly related to nuclear installations shall establish policies that give due priority to nuclear safety.

The Government of Japan also confirmed the priority of safety in its reporting to the IAEA on implementation of the Convention (Government of Japan, 2004, 2007, 2010). These reports are a rich source of information about the status of Japanese efforts to implement a safety culture prior to the March 11, 2011, Fukushima Daiichi accident.

The safety culture deficiencies at TEPCO and its regulator that contributed to the Fukushima Daiichi accident have been explicitly acknowledged in Japanese government reports (e.g., Government of Japan, 2011a,b; NAIIC, 2012). For example, NAIIC chairman Dr. Kiyoshi Kurokawa concluded that the

> accident at the Fukushima Daiichi Nuclear Power Plant cannot be regarded as a natural disaster. It was a profoundly manmade disaster—that could and should have been foreseen and prevented. (NAIIC, 2012, p. 9)

Dr. Kurokawa also commented on the mindset that led to the accident:

> [N]uclear power became an unstoppable force, immune to scrutiny by civil society. Its regulation was entrusted to the same government bureaucracy responsible for its promotion. . . . Only by grasping this mindset can one understand how Japan's nuclear industry managed to avoid absorbing the critical lessons learned from Three Mile Island and Chernobyl; and how it became accepted practice to resist regulatory pressure and cover up small-scale accidents. It was this mindset that led to the disaster at the Fukushima Daiichi Nuclear Plant. (NAIIC, 2012, p. 9)

TEPCO has acknowledged that it was ill-prepared for the March 11, 2011, earthquake and tsunami-induced flooding that occurred at the Fukushima Daiichi and Daini plants[2]:

> Top management of [the] nuclear division did not show strong willingness in enhancing plant safety against external events even in a step-by-step manner. . . . They were stuck on probability of risk and did not have [a] clear idea to take practically effective countermeasures against external events in a timely manner. . . . Top management of nuclear division and safety experts did not try to face [the] regulatory body and the public squarely.

Chapters 3 and 4 of this report describe TEPCO's preparation for and response to the earthquake and tsunami.

---

[2] The quoted material is taken from slides presented to the committee by Mr. Akira Kawano (TEPCO) on November 26, 2012 (Kawano, 2012).

On paper, TEPCO and its nuclear regulator were committed to a nuclear safety culture prior to the Fukushima Daiichi accident. However, there is strong evidence for a deficient safety culture in both of these organizations:

- As noted in Chapter 3, for example, TEPCO and its nuclear regulator failed to take strong and timely action to implement improved seismic and tsunami safety standards for the Fukushima plants.
- As noted in Chapters 2 and 4, Japanese regulatory agencies did not inform utilities of the USNRC's B.5.b requirements for responding to beyond-design-basis events even after the USNRC made them public. TEPCO also failed to inform itself of these B.5.b requirements after they became public.
- TEPCO has admitted to falsifying reports to its regulator in 29 cases between 1988 and 1998 and to frauds in safety-related inspections at the Fukushima Daiichi plant in 1993-1994.

Taken together, these examples provide evidence of a continuing lack of safety focus in the period prior to the Fukushima Daiichi accident.

7.2.1 Regulatory Capture

The term *regulatory capture* refers to the processes by which regulated entities manipulate regulators to put their interests ahead of public interests (see Bratton and McCahery, 1995; Dal Bó, 2006; Helm, 2006). In the context of this report, regulatory capture refers specifically to the manipulation of the Nuclear and Industrial Safety Agency (NISA) before the accident and therefore before regulatory restructuring.

The problem with regulatory capture of the NISA was highlighted by NAIIC (2012, p. 20):

> The [Japanese] regulators did not monitor or supervise nuclear safety. The lack of expertise resulted in "regulatory capture," and the postponement of the implementation of relevant regulations. They avoided their direct responsibilities by letting operators apply regulations on a voluntary basis.

The report also noted that TEPCO "manipulated the cozy relationship with the regulators to take the teeth out of regulations" (2012, p. 20).

A commissioner of Japan's new nuclear regulator, the Nuclear Regulation Authority (NRA; see Chapter 2), confirmed to U.S. Government Accountability Office investigators that regulatory capture existed prior to the Fukushima Daiichi accident:

> An NRA commissioner told us that Japan's restructuring of its nuclear regulatory system is necessary to address the issue of "regulatory capture"—the collusion between NISA and the nuclear industry—that compromised the nation's nuclear safety prior to the accident and to regain the public trust, which the commissioner told us was NRA's biggest challenge. (USGAO, 2014, p. 16)

Prior to the Fukushima Daiichi accident, NISA was part of the Ministry of Economy, Trade and Industry[3] (METI), an aggressive advocate for promotion of nuclear power in Japan and abroad. The Japanese government contended that this association did not affect NISA's independence:

> NISA has clear responsibilities for safety regulations pursuant to the Atomic Energy Basic Law and the Reactor Regulation Law and the functions of NISA are substantially separated, by the law, from those of other bodies or organizations concerned with the promotion or utilization of nuclear energy. (Government of Japan, 2004, p. 8-1)

Nevertheless, analysts who have studied Japan's regulatory structure have shown that Japanese nuclear safety regulators were subject to regulatory capture prior to the Fukushima Daiichi accident. These analysts have noted that METI's dual and conflicting interests seem at odds with NISA's mission to regulate nuclear power reactors (Dorfman, 2012; Wang and Chen, 2012; Aoki and Rothwell, 2013). METI was ultimately in charge of issuing licenses to Japanese nuclear plants.

Kaufmann and Penciakova (2011) suggest that "[t]o a significant extent, it appears that regulatory capture of NISA by Japan's nuclear industry turned the regulator into a caretaker of industry rather than one for public safety." NISA's lack of regulatory independence has been described as a significant problem in regulatory practice (Dorfman, 2012; Wang and Chen, 2012; Benz, 2013).

Analysts have described two practices that hindered effective regulatory control and impeded the implementation of a strong nuclear safety culture (Dorfman, 2012; Wang and Chen, 2012; Aoki and Rothwell, 2013). These are referred to as *amakudari* and *amaagari* (Wang and Chen, 2012; Wang et al., 2013; see also Schaede, 1995):

- *Amakudari* means "descent from heaven" and it refers to the practice of hiring retired, high-profile public officials for private-sector jobs (Horiuchi and Shimizu, 2001; Dorfman, 2012; Wang and Chen, 2012). It also refers to the practice of maintaining a rigid hierarchy in nuclear utili-

[3] Formerly the Ministry of International Trade and Industry.

ties and regulatory agencies whereby when a senior-level person retires, his junior would take his place (Wang and Chen, 2012).

- *Amaagari* means "ascent to heaven" and is the movement of experts from the private sector into government or government advisory positions (Wang and Chen, 2012).

Expertise in the nuclear energy technologies is difficult to obtain, so it was frequently necessary for the Japanese government and industry to take advantage of each other's technical knowledge. It was not unusual for nuclear experts to move between the nuclear industry and its regulator during the course of their careers. However, the practices of *amakudari* and *amaagari* worked together to create a system that integrated the interests of the Japanese industry and regulators to produce a system that was insular, lacking in transparency, and difficult to improve.

Wang and Chen (2012, p. 2613) assert that the nuclear regulator placed an overreliance on the technical expertise of the nuclear industry in designing and evaluating regulations:

> Japan's safety rulemaking is deeply flawed. Because NISA lacks full-time technical experts to draw up comprehensive regulations, it depended largely on retired or active engineers from nuclear-industry-related companies to set rulemaking.

Prior to the Fukushima Daiichi accident, Japanese government officials, the nuclear power industry, and regulators consistently argued that nuclear power was completely safe. This "safety myth" stifled an honest and open discussion about risks (Noggerath et al., 2011).

7.2.2 Changes Following the Fukushima Daiichi Accident

The Japanese government called for a stronger emphasis on safety culture following the Fukushima Daiichi accident:

> All those involved with nuclear energy should be equipped with a safety culture . . . . Learning this message and putting it into practice is a starting point, duty and responsibility of those who are involved with nuclear energy. *Without a safety culture, there will be no constant improvement of nuclear safety.* (Government of Japan, 2011b, p. XII-13, emphasis added)

The government has taken a series of actions to improve its regulatory institutions and its commitment to nuclear safety. Most notably, the government established[4] a new regime for regulating civilian nuclear power (see Chapter 2, especially Figure 2.12). This regime includes the NRA as an extra-ministerial organization of the Environment Ministry. This agency combines the roles of the former Nuclear Safety Commission and NISA as well as the monitoring functions of MEXT.

The NRA is responsible for promulgating rules and regulations for nuclear plants and is also charged with evaluating whether current Japanese plants can resume operations (Bacon-Dussault, 2013; Ferguson and Jansson, 2013; Geller, 2014). The Authority has been established as an "Article 3" organization under Japanese law, which means that it has greater independence than NISA (Shiroyama, 2012).

The Japanese government is taking at least two additional steps to improve the effective independence of Japanese regulation of nuclear power:

- Not allowing senior-level regulators from the Nuclear Regulation Authority to assume jobs in METI or MEXT; and
- Limiting the ability of regulators to seek jobs in the nuclear industry.

The committee was not tasked to evaluate the effectiveness of this new regulatory structure. Nevertheless, past history suggests that Japan's new regulatory organizations are unlikely to be effective unless they establish and closely adhere to good safety culture practices. Discussions involving the new Japanese regulatory structure and its effectiveness continue as Japan considers the restart of some of its nuclear reactors (Geller, 2014).

7.3 NUCLEAR SAFETY CULTURE IN THE UNITED STATES

FINDING 7.2: The establishment, implementation, maintenance, and communication of a nuclear safety culture in the United States are priorities for the U.S. nuclear power industry and the U.S. Nuclear Regulatory Commission. The U.S. nuclear industry, acting through the Institute of Nuclear Power Operations, has voluntarily established nuclear safety culture programs and mechanisms for evaluating their implementation at nuclear plants. The U.S. Nuclear Regulatory Commission has published a policy statement on nuclear safety culture, but that statement does not contain implementation steps or specific requirements for industry adoption.

[4] Act for Establishment of Nuclear Regulation Authority (Act No. 47 of June 27, 2012).

7.3.1 U.S. Nuclear Regulator

The USNRC regulates the commercial uses of nuclear material, including nuclear power, to protect people and the environment. The agency has documented its expectations for the nuclear safety culture in a series of policy pronouncements, including a 1989 Policy Statement on the Conduct of Nuclear Power Plant Operation. The policy statement declares that

> [e]ach individual licensed by the [US]NRC to operate the controls of a nuclear power reactor must be keenly aware that he or she holds the special trust and confidence of the American people, conferred through the [US]NRC license, and that his or her first responsibility is to assure that the reactor is in a safe condition at all times. (USNRC, 1989, p. 3425)

In 2011, after a public input process, the USNRC published a Final Safety Culture Policy Statement (USNRC, 2011b) that establishes nine traits of a positive safety culture:

1. *Leadership safety values and actions*: Leaders demonstrate a commitment to safety in their decisions and behaviors.
2. *Problem identification and resolution*: Issues potentially impacting safety are promptly identified, fully evaluated, and promptly addressed and corrected commensurate with their significance.
3. *Personal accountability*: All individuals take personal responsibility for safety.
4. *Work processes*: The process of planning and controlling work activities is implemented so that safety is maintained.
5. *Continuous learning*: Opportunities to learn about ways to ensure safety are sought out and implemented.
6. *Environment for raising concerns*: A safety-conscious work environment is maintained where personnel feel free to raise safety concerns without fear of retaliation, intimidation, harassment, or discrimination.
7. *Effective safety communication*: Communications maintain a focus on safety.
8. *Respectful work environment*: Trust and respect permeate the organization.
9. *Questioning attitude*: Individuals avoid complacency and continuously challenge existing conditions and activities in order to identify discrepancies that might result in error or inappropriate action.

A *safety-conscious work environment* is an important element of a strong nuclear safety culture (see point 6, above). The USNRC defines a safety-conscious work environment as an "environment in which employees feel free to raise safety concerns, both to their management and to the [US]NRC, without fear of retaliation" (USNRC, 2005, p. 2).

The safety culture policy statement is not a regulation. Moreover, licensees are not required to adopt it or modify inconsistent practices. The policy statement also does not contain specific implementation steps. It leaves implementation to licensees and recommends that implementation begin immediately.

The statement clearly sets out the USNRC's expectation "that individuals and organizations performing regulated activities establish and maintain a positive safety culture commensurate with the safety and security significance of their activities and the nature and complexity of their organizations and functions" (USNRC, 2005, p. 14). The USNRC views its policy statement as a living document and closely monitors actual nuclear power plant events that occur both domestically and internationally.

7.3.2 U.S. Nuclear Industry

The U.S. nuclear industry has also demonstrated a clear and strong commitment to nuclear safety. INPO has taken the lead in promoting a strong nuclear safety culture in the U.S. nuclear industry through training and evaluation programs (Sidebar 7.2). The Nuclear Energy Institute, an industry advocacy group, supports INPO's activities.

INPO was established to promote excellence, safety, and reliability in nuclear plant operations (see Sidebar 7.2). The organization strongly endorses the nuclear safety culture as a key operating feature, and philosophy, of its membership (INPO, 2004) and also asserts that every nuclear power station needs a strong safety culture.

INPO has established eight key principles that apply to a healthy nuclear safety culture (INPO, 2004, p. 1; 2013, p. 31[5]):

1. Everyone is personally responsible for nuclear safety.
2. Leaders demonstrate commitment to safety.
3. Trust permeates the organization.
4. Decision making reflects safety first.
5. Nuclear technology is recognized as special and unique.
6. A questioning attitude is cultivated.

[5] INPO (2013) contains two addendums: Addendum I: "Behaviors and Actions That Support a Healthy Nuclear Safety Culture, by Organizational Level" describes nuclear safety behaviors and actions that contribute to a healthy nuclear safety culture by organizational level-executive/senior manager, manager, supervisor, and individual. Addendum II: "Cross-References" provides cross-references from Traits of a Healthy Nuclear Safety Culture to the safety culture guidance developed by the Department of Energy and the Energy Facility Contractors Group.

7. Organizational learning is embraced.
8. Nuclear safety undergoes constant examination.

INPO's key principles are slightly different from the USNRC key traits, which were described earlier. This difference is not surprising given that the USNRC traits apply to all of its licensees, whereas INPO is speaking for its membership, which comprises nuclear plant operators.

The INPO principles show that implementation of the nuclear safety culture is an organizational obligation that begins at the top of the corporate ladder and applies to every worker at nuclear plants. The principles make clear that the special nature of nuclear power production demands an enhanced level of diligence and that continuous improvement is the expected norm. Organizational learning through continuous training, communications, and discussion is imperative, because highly complex technologies such as nuclear power generation can fail in unexpected and unique ways (INPO, 2004, pp. 4-6). The World Association of Nuclear Operators[6] has joined INPO in recognizing the centrality of the nuclear safety culture for nuclear plant operations worldwide.[7]

Nuclear plant owners evaluate their safety cultures using various means. INPO provides biannual evaluations of nuclear plant operations (see Sidebar 7.2). Additionally, plant owners have established safety review groups, usually as a requirement of their USNRC licenses. These groups typically consist of the plant manager, other plant personnel, and members who are independent of the plant or utility. The groups meet at regular intervals to review plant operations from a safety perspective and report their findings to the plant's senior vice president and other plant management.

Some utilities have also voluntarily established high-level independent review groups that visit the plant and report to the utility's senior management and/or board of directors (INPO, 2005). These groups consist of people who are independent of the plant and utility and typically include people who have served in high-level positions in the industry and the USNRC.

Efforts are also being made to develop safety culture metrics and relate them to nuclear safety. For example, INPO has developed a questionnaire instrument to measure safety culture at U.S. nuclear plants. It administered the survey to 63 nuclear plants (97 percent of operating plants) with an average of 46 respondents per plant (48 percent response rate). Morrow

---

[6] WANO is an international not-for-profit organization comprising nuclear power companies and associated organizations with a mission to promote nuclear safety.

[7] WANO Guideline 2006-02, Principles for a Strong Nuclear Safety Culture, January 2006. This report has limited distribution. However, the report's contents are described in a paper by a WANO staff member (Brumfield, 2012).

and Barnes (2012) evaluated the survey results to assess how the safety culture factors identified from the INPO survey relate to safety performance at nuclear plants.

They note that

> [t]he overall safety culture survey results were significantly correlated with concurrent unplanned scrams, forced outage hours, inspection findings, and cross-cutting aspects (Morrow and Barnes, 2012, p. 48)

but that

> [a]dditional, ongoing research would be necessary to determine whether the relationships observed are consistent over time, whether the same factors consistently emerge in subsequent survey administrations within the nuclear power industry, and whether different safety culture factors are uniquely related to different aspects of performance. (p. 49)

7.3.3 Discussion

Committee members have a range of views about the current status of the nuclear safety culture in the United States. A selection of committee views is provided in this section to frame the committee's recommendations in Section 7.4. The committee did not undertake a formal assessment of the status of the U.S. nuclear safety culture because that was not part of its study charge.

7.3.3.1 Independence of the U.S. Regulator

Prior to the Fukushima Daiichi accident, there were some clear differences between the nuclear regulatory system in Japan and the United States. Prior to 1974, for example, the U.S. Atomic Energy Commission was responsible for both promoting and regulating the use of nuclear power. The U.S. Congress found it in the public interest to segregate these functions into separate agencies. The Energy Reorganization Act of 1974 abolished the U.S. Atomic Energy Commission and reorganized its functions into two new agencies: the USNRC became responsible for the regulation of civilian nuclear activities and the Energy Research and Development Administration (which subsequently became part of the Department of Energy) became responsible for nuclear energy research and promotion.

As an independent federal agency, the USNRC is not part of the executive branch of the federal government—although as a matter of policy it generally follows the laws, regulations, and guidance that apply to executive agencies. The USNRC's authority comes from the statutes enacted by Congress; moreover, the agency is ultimately answerable to Congress, particularly to its authorizing and appropriation committees (Gutierrez and Polonsky, 2007). A recent letter[8] from the chair of a Senate congressional committee makes this point clear:

> The United States Constitution gives Congress broad authority over Executive Branch agencies like the [US]NRC. As an "independent agency" [US]NRC is independent from the Executive Branch—not Congressional oversight.

The USNRC must also answer to a number of other stakeholders including industry, public interest groups, and communities that host USNRC-regulated facilities. All of these stakeholders seek to influence USNRC actions, which is traditional in the U.S. system of government. The USNRC must take into consideration the preferences of its stakeholders and the broader public while maintaining its independence as a regulator.

A recent letter from a House congressional committee[9] stressed the importance of balance in USNRC regulatory decisions:

> In the Atomic Energy Act, Congress declared that nuclear energy should "make the maximum contribution to the general welfare (Section 1 (a))" which recognizes nuclear energy's vital role in contributing to our nation's energy security. In choosing such language, Congress endeavored to balance the benefits of nuclear energy with protection of public health and safety. Our goal as legislators and yours as regulators should be to preserve that balance.

The USNRC has had to navigate carefully among competing interests to preserve its regulatory independence.

Committee members hold a range of views about whether the USNRC is being successful in maintaining appropriate independence and balance in its regulatory decision making. Some members note that there is a natural tension between the regulator, which ultimately answers to the public and its representatives in Congress, and the regulated industry, which answers to its shareholders. This situation is not unique to the nuclear industry.

---

[8] Letter from Senator Barbara Boxer, Chair of the Senate Committee on Environment and Public Works, to USNRC Chairman Allison Macfarlane, November 26, 2013, concerning a USNRC decision to withhold certain information requested by the committee. Available at http://www.epw.senate.gov/public/index.cfm?FuseAction=PressRoom.PressReleases&ContentRecord_id=94f17a8e-bf47-43f0-5627-96a1508794b7.

[9] Letter from the House Committee on Energy and Commerce to USNRC Chairman Allison Macfarlane, January 15, 2013. Available at http://energycommerce.house.gov/sites/republicans.energycommerce.house.gov/files/letters/20130115NRC.

Regulatory independence necessarily involves the continuous balancing of interests between the public and private spheres.

Other committee members point to two specific examples as evidence for the possible erosion of independence: filtered vents and industry participation in the rulemaking process. These examples are described in the following paragraphs.

The USNRC is currently involved in a rulemaking to determine whether filtered vents should be added to nuclear plants with Mark I and Mark II containments (see Sidebar 5.5, Appendix F, and Appendix L). In early 2013, Senate and House committees sent letters[10] to the USNRC complaining that the agency was moving too quickly with costly post-Fukushima safety upgrade requirements. The letters criticized a USNRC staff recommendation that the agency require owners of nuclear plants with Mark I and Mark II containments to install filtered vents to reduce radioactive releases in the event of an accident. Some committee members view these letters as an effort to weaken the agency's regulatory independence.

Other committee members view the congressional letters as a normal part of the give and take in the U.S. regulatory process. They point out that the Union of Concerned Scientists also sent a letter to the USNRC[11] urging it to reject requests to weaken critical post-Fukushima safety reforms or slow down their implementation.

As another example, in the late 1990s, the USNRC came under pressure from Congress[12] to reduce the regulatory burden on the nuclear industry by moving to risk-informed, performance-based regulations. At about the same time, the Center for Strategic and International Studies recommended (CSIS, 1999) that the USNRC and the industry should strive to work in a more informal and constructive atmosphere and conduct an open dialogue with the public to arrive at regulatory procedures.

The USNRC committed to using risk information and risk analysis as part of a policy framework and initiated a policy of increased industry participation in regulatory activities. Some committee members perceive that, as a result of this USNRC commitment, the industry began to participate

---

[10] The letters can be found at http://www.epw.senate.gov/public/index.cfm?FuseAction=Minority.PressReleases&ContentRecord_id=a79c7514-cf71-9bab-769a-0f4d16587726&Region_id=&Issue_id= and http://energycommerce.house.gov/sites/republicans.energycommerce.house.gov/files/letters/20130115NRC.pdf.

[11] Available at http://www.ucsusa.org/news/press_release/nrc-should-reject-calls-to.html.

[12] For example, Senator Pete Domenici (see Domenici et al., 2004) states that he met privately with the then-chair of the USNRC, Dr. Shirley Jackson, to directly address what some nuclear industry representatives saw as an "adversarial attitude" toward the nuclear industry. Jackson had been aggressively pursuing design-basis flaws and the USNRC had issued a series of significant fines based on these problems. Senator Domenici alleges that he threatened to reduce the USNRC's budget unless greater cooperation with industry was seen.

more actively in USNRC activities such as rulemakings and implementing guidance development, including initiating voluntary industry programs in lieu of USNRC regulatory action. This resulted, for example, in a USNRC decision to allow the industry to voluntarily implement severe accident management guidelines at nuclear plants (see Chapter 5).

These voluntary industry programs have not always been successful. Following the Fukushima Daiichi accident, for example, the USNRC's Near-Term Task Force (see Chapter 5) examined the implementation of severe accident management guidelines at U.S. nuclear plants. It found inconsistent implementation of these guidelines by licensees. The Task Force recommended that the USNRC initiate a rulemaking on severe accident management guidelines (see Chapter 5) to replace the voluntary program.

Other committee members note that the committee does not have enough information to determine whether industry participation in regulatory processes has increased since the late 1990s or whether voluntary initiatives are being substituted for regulatory actions. These members also note that industry, including the Electric Power Research Institute and vendor organizations such as BWR and PWR owners' groups have been active participants in the regulatory process and the development of voluntary initiatives since before the late 1990s.[13] Industry is well organized and has a deep resource base to support a high level of participation in the regulatory process.

Vigorous involvement of outside parties occurs in other U.S. regulatory agencies and is anticipated by the laws that govern federal rulemaking. Indeed, it is important for the USNRC to carefully consider the advice it receives from outside parties when it makes regulatory decisions. It is also essential that the USNRC balance the interests of those outside parties with those of the broader public. This requires independent (and wise) technical and policy judgments by USNRC staff and commissioners.

The importance of regulatory independence was highlighted in a recent speech[14] by USNRC Chairman Allison Macfarlane:

> A nuclear regulator must be independent, but simply being separated from promotional activities on an organization chart isn't enough. The regulator must be adequately funded and staffed with highly-competent subject matter experts. It must have the authority to stop an activity if it identifies a safety concern, even if it means that a project is delayed. It must be able to

[13] For example, the industry responded to the USNRC's safety culture policy statement (USNRC, 2011b) through a voluntary initiative. It also voluntarily proposed the FLEX initiative in response to the USNRC's Mitigation Strategies Order (EA-12-049) for beyond-design-basis external events (see Chapter 5 and Appendix F).

[14] Available at http://www.nrc.gov/reading-rm/doc-collections/commission/speeches/2014/s-14-002.pdf. Accessed August 28, 2014.

> shut down a plant that's not operating safely, even if it means a population is temporarily deprived of electricity.
>
> To have this authority, a regulator must have the ability to make truly independent safety decisions, with the confidence that those decisions won't be overturned for political reasons. Put another way, safety and security must be the entire government's priorities.

Adequate funding and highly competent staff are necessary, but not sufficient, conditions for regulatory independence. Strong leadership that maintains a laser focus on safety and does not allow itself to become distracted by outside pressures is also required. The President and Senate of the United States also play important roles in helping to maintain the USNRC's regulatory independence by nominating and appointing highly qualified agency leaders (i.e., commissioners) and working to ensure that the agency is free from undue influences.

The loss of regulatory independence is often hard to identify and in fact it may go undetected until a tragic accident occurs. See, for example, the April 2010 Deepwater Horizon accident (NAE and NRC, 2011) and the September 2010 rupture of a Pacific Gas and Electric (PG&E) Company natural gas transmission pipeline in San Bruno, California (NTSB, 2011).

7.3.3.2 Regulatory Capture

Some committee members point to specific incidents as evidence for the possible capture of the USNRC by industry. A well-documented example is the near-accident at the Davis-Besse nuclear plant in 2002. On February 16, 2002, during a refueling outage, the Davis-Besse plant conducted a routine inspection of the nozzles entering the head of the reactor pressure vessel. These inspections indicated that three control-rod drive-mechanism nozzles had indications of cracking, which had resulted in leakage through the reactor's pressure boundary. During repairs of the nozzles it was discovered that the

> [reactor pressure vessel] head material adjacent to the nozzle had disintegrated and that the affected (or "wastage") area was approximately 5 inches long, up to 4 to 5 inches wide, and 6 inches deep. The remaining thickness of the [reactor pressure vessel] head in the wastage area was found to be approximately 3/8 inch which was the stainless steel cladding on the inside surface of the RPV head. This was the only material preventing a breach of the reactor coolant pressure boundary and leak of radioactive coolant into the containment building. (OIG, 2002, p. 14)

An investigation of the incident by the USNRC's Office of the Inspector General noted that

> [US]NRC appears to have informally established an unreasonably high burden of requiring absolute proof of a safety problem, versus lack of reasonable assurance of maintaining public health and safety, before it will act to shut down a power plant.
>
> The USNRC staff had articulated this standard to the Office of the Inspector General (OIG) as a rationale for allowing Davis-Besse to operate until February 16, 2002, even in light of information that strongly indicated Davis-Besse was not in compliance with USNRC regulations and plant technical specifications and may have operated with reduced safety margins. (OIG, 2002, p. 23)

Committee members agree that the Davis-Besse incident was a serious safety violation. Some committee members also note that this incident took place over 10 years ago and is not necessarily indicative of current conditions. Moreover, the USNRC took several steps to address this problem once it was discovered: the Davis-Besse plant was shut down for repair, the company was fined and subjected to more intensive regulatory scrutiny, and the USNRC took several steps to strengthen the safety culture components of its reactor oversight process (USNRC, 2011f). For example, the USNRC's resident inspector training was augmented to include safety culture, and inspection procedures were developed to assess safety culture at plants with degraded performance.

The USNRC has also been criticized for failing to enforce fire regulations at U.S. nuclear plants. The USNRC issued prescriptive fire safety regulations following a fire at the Browns Ferry nuclear plant (located in Alabama) in 1975. Some nuclear plants have had difficulties in meeting these regulations and have sought exemptions (USGAO, 2008). The Union of Concerned Scientists (UCS), arguably the most technically informed public interest stakeholder on nuclear power issues in the United States, has criticized the USNRC's regulatory performance on this issue:

> The NRC has for many years turned a blind eye to the broad use of unapproved manual actions and long-term use of compensatory measures. It has known for two decades about substandard insulation widely used to protect electric cables but has not corrected the situation. (UCS, 2013a, p. 5)

The USGAO (2012) noted that the USNRC is making progress in resolving this issue but that some challenges remain.

UCS has been issuing annual reports on the performance of the USNRC (UCS, 2011b, 2012, 2013b; Lochbaum, 2014). These reports include discussions of recent incidents at U.S. nuclear plants and the USNRC's responses. The most recent UCS report (Lochbaum, 2014) praised the USNRC's performance:

> The Nuclear Regulatory Commission (NRC) demonstrated it can be an effective watchdog in 2013 . . . . In many cases, the agency does an admirable job protecting the public and industry workers by enforcing safety regulations.

The report also offers criticisms of the USNRC's performance:

> But the agency too often turns into Mr. Hyde, and that kind of behavior could lead to a serious accident.

Some committee members note that the USNRC exhibits independence from the U.S. nuclear industry in many matters. For example, the USNRC ordered that the vents in Mark I and II BWRs be hardened and severe accident capable (Order EA-13-109[15]; see Appendix F) even though it did not pass the backfit rule's cost-effectiveness test; the USNRC noted that

> These modifications are needed to protect health and to minimize danger to life or property because they will give licensees greater capabilities to respond to severe accidents and limit the uncontrolled release of radioactive materials. (Order, p. 7)

The requirement to extend station blackout capabilities through Order EA-12-049 (see Appendix F) is a similar example.

7.4 RECOMMENDATIONS

RECOMMENDATION 7.2A: The U.S. Nuclear Regulatory Commission and the U.S. nuclear power industry must maintain and continuously monitor a strong nuclear safety culture in all of their safety-related activities. Additionally, the leadership of the U.S. Nuclear Regulatory Commission must maintain the independence of the regulator. The agency must ensure that outside influences do not compromise its nuclear safety culture and/or hinder its discussions with and disclosures to the public about safety-related matters.

---

[15] Available at http://pbadupws.nrc.gov/docs/ML1314/ML13143A321.pdf.

RECOMMENDATION 7.2B: The U.S. nuclear industry and the U.S. Nuclear Regulatory Commission should examine opportunities to increase the transparency of and communication about their efforts to assess and improve their nuclear safety cultures.

The Fukushima Daiichi accident demonstrates that statements in support of a strong nuclear safety culture are no guarantee that one exists. In fact, the development and maintenance of a strong nuclear safety culture requires a focused and sustained commitment from all involved parties:

- Nuclear plant operators,
- Nuclear plant management,
- Nuclear industry organizations,
- Nuclear regulators—both staff and leadership, and
- Executive and legislative branches of government.

The committee sees opportunities to improve the transparency of U.S. industry and regulator efforts to assess and improve their nuclear safety cultures. This would require that the industry and regulators disclose additional information to the public about their efforts to assess safety culture effectiveness, remediate deficiencies, and implement improvements. The committee fully recognizes that any such disclosures need to be carefully planned and implemented so that they do not inhibit the full and prompt reporting of safety problems. The committee also recognizes that some types of information, for example, personnel- and security-related information, should not be disclosed to the public.

The committee judges that there would be several tangible benefits from increased communication with stakeholders and disclosures: It would help to demonstrate the nuclear industry's commitment to safety in both word and deed and demonstrate the USNRC's commitment to safety and regulatory independence. Public feedback from such disclosures might also improve the quality of safety culture assessment and improvement activities.

There are tangible benefits associated with a more frank and direct relationship between the nuclear industry, nuclear plants, and host communities (Richardson et al., 2013, p. 266). Continuing public support for nuclear power depends on the safe operation of nuclear plants. Nuclear plants must be—and must also be seen by the public to be—safe and well regulated. Many U.S. nuclear plants have been granted 20-year license renewals[16] and spent fuel is stored at all operating plants and is likely to remain onsite for an indeterminate period of time. Consequently, nuclear plants by necessity will have long-term relationships with their communi-

[16] See http://www.nrc.gov/reactors/operating/licensing/renewal/applications.html.

ties and would likely benefit from strengthened community relationships and communication efforts. Indeed, open and transparent communication is an important component of the nuclear safety culture and essential to maintaining confidence in nuclear power (Macfarlane, 2012). Including the public by extending communication and engagement is consistent with the principles that underlie a strong nuclear safety culture.

References

ACRS (Advisory Committee on Reactor Safeguards). 2011. Initial ACRS Review of: (1) The NRC Near-Term Task Force Report on Fukushima and (2) Staff's Recommended Actions to Be Taken Without Delay. Memorandum to Gregory B. Jaczko, U.S. Nuclear Regulatory Commission. October 13. Available at http://pbadupws.nrc.gov/docs/ML1128/ML11284A136.pdf.

ACRS. 2012. Draft Interim Staff Guidance Documents in Support of Tier 1 Orders. Memorandum to R. W. Borchardt, U.S. Nuclear Regulatory Commission. July 17. Available at http://pbadupws.nrc.gov/docs/ML1219/ML12198A196.pdf.

AEC (U.S. Atomic Energy Commission). 1957. Theoretical Possibilities and Consequences of Major Accidents in Large Nuclear Power Plants. WASH-740. Prepared by Brookhaven National Laboratory. Washington, DC: AEC.

AEC. 1962. Calculation of Distance Factors for Power and Test Reactors. TID-14844. Washington, DC: AEC. Available at http://pbadupws.nrc.gov/docs/ML0217/ML021720780.pdf.

AEC. 1974. Ultimate Heat Sink for Nuclear Power Plants. Regulatory Guide 1.27, Rev. 1. Washington, DC: U.S. Atomic Energy Commission. Available at http://pbadupws.nrc.gov/docs/ML1303/ML13038A084.pdf.

AESJ (Atomic Energy Society of Japan). 2011a. Estimated Current Status of Fukushima-Daiichi Nuclear Power Plant Units 1-3. April 18. Available at http://www.aesj.or.jp/en/release/chousacom_EN_01_110418.pdf.

AESJ. 2011b. Lessons Learned from the Accident at the Fukushima Daiichi Nuclear Power Plant. May 9. Available at http://www.aesj.or.jp/en/release/gbcom_kyokun_EN_20110530.pdf.

Alvarez, R., J. Beyea, K. Janberg, J. Kang, E. Lyman, A. Macfarlane, G. Thompson, and F.N. von Hippel. 2003. Reducing the hazards from stored spent power-reactor fuel in the United States. Science and Global Security 11: 1-51.

ANS (American Nuclear Society) Special Committee on Fukushima. 2012. Fukushima Daiichi: ANS Committee Report. March (Revised June 2012). La Grange, IL: ANS.

Aoki, M., and G. Rothwell. 2013. A comparative institutional analysis of the Fukushima nuclear disaster: Lessons and policy implications. Energy Policy 53: 240-247.

Asahi Shimbun. 2013. State considers loaning TEPCO additional 3 trillion yen. November 11. Available at http://ajw.asahi.com/article/business/AJ201311110038.

Atlantic and Gulf of Mexico Tsunami Hazard Assessment Group. 2008. Evaluation of Tsunami Sources with the Potential to Impact the U.S. Atlantic and Gulf Coasts: An Updated Report to the Nuclear Regulatory Commission. U.S. Geological Survey Administrative Report.

Bacon-Dussault, M. 2013. Independence of nuclear regulators in the aftermath of the Fukushima Daiichi nuclear accident: A comparative approach. Presentation at IAEA International Conference on Effective Nuclear Regulatory Systems, Ottawa, Ontario, April 8-12.

Barkan, R., U. S. ten Brink, and J. Lin. 2009. Far field tsunami simulations of the 1755 Lisbon earthquake: Implications for tsunami hazard to the U.S. East Coast and the Caribbean. Marine Geology 264: 109-122.

Benz, E. 2013. Lessons learned from Fukushima: Strengthening the international regulation of nuclear energy. William and Mary Environmental Law and Policy Review 37(3/8): 845-880.

Beyea, J., E. Lyman, and F. N. von Hippel. 2013. Accounting for long-term doses in "worldwide health effects of the Fukushima Daiichi nuclear accident." Energy and Environmental Science 6: 1042-1045.

Blanc, P.-L. 2009. Earthquakes and tsunami in November 1755 in Morocco: A different reading of contemporaneous documentary sources. Natural Hazards and Earth Systems Sciences 9: 725-738.

Blanchard, G., and D. Dosa. 2009. A comparison of the nursing home evacuation experience between Hurricanes Katrina (2005) and Gustav (2008). Journal of the American Medical Directors Association 10(9): 639-643.

Bley, D., S. Kaplan, and D. Johnson. 1992. The strengths and limitations of PSA: Where we stand. Reliability Engineering and System Safety 38: 3-26.

Bolduc, L. 2002. GIC observations and studies in the Hydro-Québec power system. Journal of Atmospheric and Solar-Terrestrial Physics 64: 1793-1802.

Bondevik, S., F. Løvholt, C. Harbitz, J. Mangerud, A. Dawson, and J. Inge Svendsen. 2005. The Storegga Slide tsunami—comparing field observations with numerical simulations. Marine and Petroleum Geology 22(1-2): 195-208.

Borchardt, R. W. 2012a. Policy Issue (Notation Vote) Memorandum to the Commissioners, Subject: Consideration of Additional Requirements for Containment Venting Systems for Boiling Water Reactors with Mark I and Mark II Containments. SECY-12-0157. Rockville, MD: U.S. Nuclear Regulatory Commission. Available at http://www.nrc.gov/reading-rm/doc-collections/commission/secys/2012/2012-0157scy.pdf.

Borchardt, R. W. 2012b. Policy Issue (Notation Vote) Memorandum to the Commissioners, Subject: State-of-the-Art Reactor Consequence Analyses—Recommendation for Limited Additional Analysis. SECY-12-0092. Rockville, MD: U.S. Nuclear Regulatory Commission. Available at http://www.nrc.gov/reading-rm/doc-collections/commission/secys/2012/2012-0092scy.pdf.

Bratton, W. W., and J. A. McCahery. 1995. Regulatory competition, regulatory capture, and corporate self-regulation. North Carolina Law Review 73: 1861-1948.

BRC (Blue Ribbon Commission on America's Nuclear Future). 2012. Report to the Secretary of Energy. January. Washington, DC: U.S. Department of Energy. Available at energy.gov/sites/prod/files/2013/04/f0/brc_finalreport_jan2012.pdf.

Breitung, W., C. Chan, S. Dorofeev, A. Eder, B. Gelfand, M. Heitsch, R. Klein, A. Malliakos, E. Shepherd, E. Studer, and P. Thibault. 2000. Flame Acceleration and Deflagration-to-Detonation Transition in Nuclear Safety: State-of-the-Art Report. Issy-les-Moulineaux, France: OECD Nuclear Energy Agency. Available at https://www.oecd-nea.org/nsd/docs/2000/csni-r2000-7.pdf.

Bromet, E. J. 2014. Emotional consequences of nuclear power plant disasters. Health Physics 106(2): 206-210.

Brumfiel, G. 2011. Fukushima reaches cold shutdown. Nature, doi:10.1038/nature.2011.9674.

Brumfield, T. 2012. Evaluation of Safety Culture in WANO Pre-Startup Reviews. Atlanta, GA: World Association of Nuclear Operators. Available at https://gnssn.iaea.org/NSNI/SC/SCPoP/Papers%20prepared%20for%20meeting/Todd%20Brumfield_Evaluation%20of%20Safety%20Culture%20in%20WANO%20Pre-Startup%20Reviews_Paper.pdf.

Bye, A., E. Lois, V. N. Dang, G. Parry, J. Forester, S. Massaiu, R. Boring, P. Ø. Braarud, H. Broberg, J. Julius, I. Männistö, and P. Nelson. 2011. International Human Reliability Analysis (HRA) Empirical Study—Phase 2 Report: Results from Comparing HRA Method Predictions to Simulator Data from SGTR Scenarios. NUREG/IA-0216, Vol. 2. Rockville, MD: U.S. Nuclear Regulatory Commission. Available at http://pbadupws.nrc.gov/docs/ML1125/ML11250A010.pdf.

Cabinet Secretariat, Government of Japan. 1999. Act on Special Measures Concerning Nuclear Emergency Preparedness (Act No. 156 of December 17, 1999). Tokyo: Government of Japan.

Camp, A. L., J. C. Cummings, M. P. Sherman, C. F. Kupiec, R. J. Healy, J. S. Caplan, J. R. Sandhop, and J. H. Saunders. 1983. Light Water Reactor Hydrogen Manual. NUREG/CR-2726, SAND82-1137, Rev. 3. Prepared by Sandia National Laboratories and General Physics Corp. Washington, DC: U.S. Nuclear Regulatory Commission. Available at http://www.nrc.gov/reading-rm/doc-collections/nuregs/contract/cr2726/.

Cannon-Bowers, J. A., and E. Salas. 1998. Making Decisions Under Stress: Implications for Individual and Team Training. Washington, DC: American Psychological Association.

Carafano, J. J. 2011. The Great Eastern Japan Earthquake: Assessing Disaster Response and Lessons for the United States. Special Report No. 94. Washington, DC: The Heritage Foundation.

Cardis, E., D. Krewski, M. Boniol, V. Drozdovitch, S. C. Darby, E. S. Gilbert, S. Akiba, J. Benichou, J. Ferlay, S. Gandini, C. Hill, G. Howe, A. Kesminiene, M. Moser, M. Sanchez, H. Storm, L. Voisin, and P. Boyle. 2006. Estimates of the cancer burden in Europe from radioactive fallout from the Chernobyl accident. International Journal of Cancer 119(6): 1224-1235.

Carlowicz, M. J., and R. E. Lopez. 2002. Storms from the Sun: The Emerging Science of Space Weather. Washington, DC: Joseph Henry Press.

Chaytor, J. D., U. S. ten Brink, A. R. Solow, and B. D. Andrews. 2009. Size distribution of submarine landslides along the U.S. Atlantic margin. Marine Geology 264(1): 16-27.

CENTRA (CENTRA Technology, Inc.). 2011. OECD/IFP Futures Project on "Future Gobal Shocks": Geomagnetic Storms. IFP/WKP/FGS(2011)4. Burlington, MA: CENTRA Technology, Inc.

Chang, Y. J., D. Bley, L. Criscione, B. Kirwan, A. Mosleh, T. Madary, R. Nowell, R. Richards, E. Roth, S. Sieben, and A. Xoulis. 2014. The SACADA database for human reliability and human performance. Reliability Engineering & System Safety 126(SI): 117-133.

Chen, S. Y., and T. S. Tenforde. 2012. Optimizing decision making for late-phase recovery one year after the Fukushima nuclear accident [Letter to Editor]. Journal of Radiological Protection 32: 191-192.

Chester, F. M., C. Rowe, K. Ujiie, J. Kirkpatrick, C. Regalla, F. Remitti, J. C. Moore, V. Toy, M. Wolfson-Schwehr, S. Bose, J. Kameda, J. J. Mori, E. E. Brodsky, N. Eguchi, S. Toczko, and the Expedition 343 and 343T Scientists. 2013. Structure and composition of the plate-boundary slip zone for the 2011 Tohoku-Oki earthquake. Science 342(6163): 1208-1211.

CNSC (Canadian Nuclear Safety Commission). 2011. CNSC Fukushima Task Force Report. INFO-0824. Ottawa, Ontario: Canadian Nuclear Safety Commission. October.

Commonwealth Edison (Commonwealth Edison Company of Chicago). 1981. Zion Probabilistic Safety Study. Prepared by Pickard, Lowe and Garrick Inc., Westinghouse Electric Corporation, and Fauske and Associates, Inc.

Conklin, C., and J. Edwards. 2000. Selection of protective action guides for nuclear incidents. Journal of Hazardous Materials 75(2-3): 131-144.

Consolidated Edison (Consolidated Edison Company of New York and the New York Power Authority). 1982. Indian Point Probabilistic Safety Study. Prepared by Pickard, Lowe and Garrick Inc., Westinghouse Electric Corporation, and Fauske and Associates, Inc.

Cook, D. H., S. R. Greene, R. M. Harrington, S. A. Hodge, and D. D. Yue. 1981. Station Blackout at Browns Ferry Unit One—Accident Sequence Analysis. NUREG/CR-2182, Vol. 1; ORNL/NUREG/TM-455/V1. Prepared for the U.S. Department of Energy. Oak Ridge, TN: Oak Ridge National Laboratory. Available at http://web.ornl.gov/info/reports/1981/3445600211884.pdf.

CSIS (Center for Strategic and International Studies), Nuclear Regulatory Process Review Steering Committee. 1999. The Regulatory Process for Nuclear Power Reactors: A Review. Washington, DC: CSIS.

Cyranoski, D. 2011. Japan faces up to failure of its earthquake preparations. Nature 471: 556-557.

Dal Bó, E. 2006. Regulatory capture: A review. Oxford Review of Economic Policy 22(2): 203-225.

Dogariu, A., and R. B. Miles. 2011. Detecting localized trace species in air using radar resonance-enhanced multi-photon ionization. Applied Optics 50(4): A68-A73.

Domenici, P. V., B. J. Lyons, and J. J. Steyn. 2004. A Brighter Tomorrow: Fulfilling the Promise of Nuclear Energy. Lanham, MD: Rowman and Littlefield.

Dorfman, D. 2012. The changing perspectives of U.S. and Japanese nuclear energy policies in the aftermath of the Fukushima Daiichi disaster. Pace Environmental Law Review 30(1): 255-283.

Driscoll, N. W., J. K. Weissel, and J. A. Goff. 2000. Potential for large-scale submarine slope failure and tsunami generation along the U.S. mid-Atlantic coast. Geology 28: 407-410.

ENSI (Swiss Federal Nuclear Safety Inspectorate). 2011. Lessons: Fukushima 11032011, Lessons Learned and Checkpoints Based on the Nuclear Accidents at Fukushima. ENSI-AN-8001. Aargau: Swiss Federal Nuclear Safety Inspectorate.

EPRI (Electric Power Research Institute). 2012a. Fukushima Daiichi Accident—Technical Causal Factor Analysis: 2012. Technical Report 1024946. Palo Alto, CA: EPRI. Available at http://www.epri.com/search/Pages/results.aspx?k=1024946&r=mpkeyword%3AAQdUc3VuYW1pCW1wa2V5d29yZAEBXgEk,mptabresults%3AARABUmVzZWFyY2ggUmVzdWx0cwxtcHRhYnJlc3VsdHMBAl4iAiIk.

EPRI. 2012b. Severe Accident Management Guidance Technical Basis Report. Palo Alto, CA: Electric Power Research Institute. Available at http://www.epri.com/abstracts/Pages/ProductAbstract.aspx?ProductId=000000000001025295.

EPRI. 2012c. Summary of the EPRI Early Event Analysis of the Fukushima Daiichi Spent Fuel Pools Following the March 11, 2011 Earthquake and Tsunami in Japan: 2012. Technical Update 1025058. Palo Alto, CA: EPRI. May.

EPRI. 2013. Fukushima Technical Evaluation: Phase 1—MAAP5 Analysis. Technical Report 1025750. Prepared by ERIN Engineering and Research, Inc. Palo Alto, CA: EPRI. April.

EPRI and NERC (Electric Power Research Institute and North American Electric Reliability Corporation). 2011. Geo-magnetic Disturbances (GMD): Monitoring, Mitigation, and Next Steps (A Literature Review and Summary of the 2011 NERC GMD Workshop). Technical Report 1024629. Prepared by Hoffman Power Consulting. Palo Alto, CA: Electric Power Research Institute. November.

Evangeliou, N., Y. Balkanski, A. Cozic, and A. P. Møller. 2014. Global and local cancer risks after the Fukushima Nuclear Power Plant accident as seen from Chernobyl: A modeling study for radiocaesium (^{134}Cs & ^{137}Cs). Environment International 64: 17-27.

Exelon Generation Company. 2013. Peach Bottom Atomic Power Station, Units 2 and 3 Mitigation Strategies (MS) Overall Integrated Plan. Available at http://pbadupws.nrc.gov/docs/ML1305/ML13059A305.pdf.

Farmer, F. R. 1967. Siting criteria—a new approach. Pp. 303-329 in Proceedings of a Symposium on the Containment and Siting of Nuclear Power Plants Held by the International Atomic Energy Agency in Vienna, 3-7 April 1967, Vienna, Austria (SM-89/34). Available at http://www.iaea.org/inis/collection/NCLCollectionStore/_Public/44/070/44070762.pdf.

Federal Business Opportunities. 2013. Solicitations: Defense Advanced Research Projects Agency—Solicitation Number: DARPA-BAA-14-11. Arlington, VA: Defense Advanced Research Projects Agency. December 11.

FEMA (Federal Emergency Management Agency). 2013a. Communicating During and After a Nuclear Power Plant Incident. Washington, DC: U.S. Department of Homeland Security. June 1.

FEMA. 2013b. Program Manual: Radiological Emergency Preparedness. Washington, DC: U.S. Department of Homeland Security. June.

FERC (Federal Energy Regulatory Commission). 2013. Reliability Standards for Geomagnetic Disturbances, FERC Docket No. RM12-22-000; Order No. 779. Washington, DC: FERC. May 16.

Ferguson, C., and M. Jansson. 2013. Regulating Japanese Nuclear Power in the Wake of the Fukushima Daiichi Accident. FAS Issue Brief. Washington, DC: Federation of American Scientists. May.

Figueroa, P. 2013. Risk communication surrounding the Fukushima nuclear disaster: An anthropological approach. Asia Europe Journal 11: 53-64.

Fine, I. V., A. B. Rabinovich, B. D. Bornhold, R. E. Thomson, and E. A. Kilikov. 2005. The Grand Banks landslide-generated tsunami of November 18, 1929: Preliminary analysis and numerical modeling. Marine Geology 215(1): 45-57.

Fink, S. 2013. Five Days at Memorial: Life and Death in a Storm-Ravaged Hospital. New York: Crown.

Fitzgerald, J. E., S. B. Wollner, A. A. Adalja, R. C. Morhard, A. J. Cero, and T. V. Inglesby. 2012. After Fukushima: Managing the Consequences of a Radiological Release (Final Report). Baltimore, MD: Center for Biosecurity of University of Pittsburgh Medical Center. March.

Frankel, A. 2013a. Comment on "Why earthquake hazard maps often fail and what to do about it" by S. Stein, R. Geller, and M. Liu. Tectonophysics 592: 200-206.

Frankel, A. 2013b. Rupture history of the 2011 M 9 Tohoku Japan earthquake determined from strong-motion and high-rate GPS recordings: Subevents radiating energy in different frequency bands. Bulletin of the Seismological Society of America 103(2B): 1290-1306.

FSC (Food Safety Commission of Japan). 2011. Emergency Report on Radioactive Nuclides in Foods. Tokyo: Government of Japan. March.

Fujii, Y., K. Satake, S. Sakai, M. Shinohara, and T. Kanazawa. 2011. Tsunami source of the 2011 off the Pacific coast of Tohoku Earthquake. Earth Planets Space 63(7): 815-820.

Fujiwara, T., S. Kodaira, T. No, Y. Kaiho, N. Takahashi, and Y. Kaneda. 2011. The 2011 Tohoku-Oki Earthquake: Displacement reaching the trench axis. Science 334(6060): 1240.

Furumura, T., S. Takemura, S. Noguchi, T. Takemoto, T. Maeda, K. Iwai, and S. Padhy. 2011. Strong ground motions from the 2011 off-the Pacific-Coast-of-Tohoku, Japan (Mw=9.0) earthquake obtained from a dense nationwide seismic network. Landslides 8: 333-338.

Garrick, B. J. 1968. Unified Systems Safety Analysis for Nuclear Power Plants. Ph.D. Dissertation. University of California, Los Angeles.

Garrick, B. J. 2008. Quantifying and Controlling Catastrophic Risks. Burlington, MA: Academic Press.

Gauntt, R., D. Kalinich, J. Cardoni, and J. Phillips. 2012a. MELCOR simulations of the severe accident at the Fukushima IF2 reactor. Presentation at the American Nuclear Society Winter Meeting and Nuclear Technology Expo, San Diego, CA.

Gauntt, R., D. Kalinich, J. Cardoni, J. Phillips, A. Goldmann, S. Pickering, M. Francis, K. Robb, L. Ott, D. Wang, C. Smith, S. St. Germain, D. Schwieder, and C. Phelan. 2012b. Fukushima Daiichi Accident Study (Status as of April 2012). SAND2012-6173. Albuquerque, NM, and Livermore, CA: Sandia National Laboratories. July. Available at energy.sandia.gov/wp/wp-content/gallery/uploads/Fukushima_SAND2012-6173.pdf.

Gavrilin, Y., V. Khrouch, S. Shinkarev, V. Drozdovitch, V. Minenko, E. Shemiakina, A. Ulanovsky, A. Bouville, L. Anspaugh, P. Voillequé, and N. Luckyanov. 2004. Individual thyroid dose estimation for a case-control study of Chernobyl-related thyroid cancer among children of Belarus—Part I: ^{131}I, short-lived radioiodines (^{132}I, ^{133}I, ^{135}I), and short-lived radiotelluriums (^{131}mTe and ^{132}Te). Health Physics 86(6): 565-585.

Geist, E. L., and T. Parsons. 2009. Assessment of source probabilities for potential tsunamis affecting the U.S. Atlantic coast. Marine Geology 264(1-2): 98-108.

Gelfand, B. E., M. V. Silnikov, S. P. Medvedev, and S. V. Khomik. 2012. Thermo-Gas Dynamics of Hydrogen Combustion and Explosion. Berlin: Springer-Verlag.

Geller, R. J. 2014. Back to the future: Restarting Japan's nuclear power plants. Nikkei Asian Review, March 7. Available at http://asia.nikkei.com/print/article/20248.

Gertman, D. I., B. P. Hallbert, M. W. Parrish, M. B. Sattision, D. Brownson, and J. P. Tortorelli. 2002. Review of Findings for Human Error Contribution to Risk in Operating Events. NUREG/CR-6753, INEEL/EXT-01-01166. Prepared by Idaho National Engineering and Environmental Laboratory. Rockville, MD: U.S. Nuclear Regulatory Commission Available at http://pbadupws.nrc.gov/docs/ML0209/ML020930077.pdf.

Ghofrani, H., G. M. Atkinson, K. Goda, and K. Assatourians. 2013. Stochastic Finite-Fault Simulations of the 2011 Tohoku, Japan, Earthquake. Bulletin of the Seismological Society of America 103(2B): 1307-1320.

Goldberg, S. M., and R. Rosner. 2011. Nuclear Reactors: Generation to Generation. Cambridge, MA: American Academy of Arts & Sciences. Available at amacad.org/pdfs/nuclearreactors.pdf.

González A. J. 2012. The recommendations of the ICRP vis-à-vis the Fukushima Dai-ichi NPP accident aftermath. Journal of Radiological Protection 32(1): N1-7.

González, A. J., M. Akashi, J. D. Boice, Jr., M. Chino, T. Homma, N. Ishigure, M. Kai, S. Kusumi, J. Lee, H. Menzel, O. Niwa, K. Sakai, W. Weiss, S. Yamashita, and Y. Yonekura. 2013. Radiological protection issues arising during and after the Fukushima nuclear reactor accident. Journal of Radiological Protection 33: 497-571.

Government of Japan. 2004. Convention on Nuclear Safety National Report of Japan for the Third Review Meeting. Tokyo: Government of Japan. August. Available at http://www.nsr.go.jp/archive/nisa/english/internationalcooperation/conventions/cns/pdf/3rd_NationalReport.pdf.

Government of Japan. 2007. Convention on Nuclear Safety National Report of Japan for the Fourth Review Meeting. Tokyo: Government of Japan. September. Available at http://www.nsr.go.jp/archive/nisa/english/internationalcooperation/conventions/cns/pdf/4th_NationalReport.pdf.

Government of Japan. 2010. Convention on Nuclear Safety National Report of Japan for the Fifth Review Meeting. Tokyo: Government of Japan. September. Available at http://www.nsr.go.jp/archive/nsc/NSCenglish/documents/conventions/2011.pdf.

Government of Japan. 2011a. Additional Report of the Japanese Government to the IAEA: The Accident at TEPCO's Fukushima Nuclear Power Stations (Second Report) (Summary). September. Tokyo: Government of Japan NERHQ (Nuclear Emergency Response Headquarters).

Government of Japan. 2011b. Report of Japanese Government to the IAEA Ministerial Conference on Nuclear Safety: The Accident at TEPCO's Fukushima Nuclear Power Stations. June. Tokyo: Government of Japan NERHQ (Nuclear Emergency Response Headquarters). Available at www.kantei.go.jp/foreign/kan/topics/201106/iaea_houkokusho_e.html.

Greene, S. R. 1986. The role of BWR MK I secondary containments in severe accident mitigation. Proceedings of the U.S. Nuclear Regulatory Commission 14th Water Reactor Safety Information Meeting, Bethesda, MD.

Greene, S. R. 1987. The impact of BWR MK I primary containment failure dynamics on secondary containment integrity. Presentation at the 15th Water Reactor Safety Meeting, Gaithersburg, MD, October 19. ORNL Report CONF-8710111-6. Available at http://www.osti.gov/scitech/servlets/purl/5835351.

Greene, S. R. 1988. The role of BWR secondary containments in severe accident mitigation: Issues and insights from recent analyses. Proceedings of the 4th Workshop on Containment Integrity, Arlington, VA, June 14-17. NUREG/CR-0095, ORNL Report CONF-8806153-1.

Greene, S. R. 1990. The role of BWR secondary containments in severe accident mitigation: Issues and insights from recent analyses. Nuclear Engineering and Design 120: 75-86.

Greene, S. R. 2014. The canary, the ostrich, and the black swan: A historical perspective on our understanding of BWR severe accidents and their mitigation. Nuclear Technology 186(2): 115-138.

Greene, S. R., and S. A. Hodge. 1986. The role of BWR MK I secondary containments in severe accident mitigation. Presentation at the 14th Water Reactor Safety Meeting, Gaithersburg, MD, October 27-31. ORNL Report CONF-8610135-41.

Guldenmund, F. 2010. (Mis)understanding safety culture and its relationship to safety management. Risk Analysis 30(10): 1466-1480.

Gusman, A. R., Y. Tanioka, S. Sakai, and H. Tsushima. 2012. Source model of the great 2011 Tohoku earthquake estimated from tsunami waveforms and crustal deformation data. Earth and Planetary Science Letters 341: 234-242.

Gutierrez, J. M., and A. S. Polonsky. 2007. Fundamentals of Nuclear Regulation in the United States, 2nd Ed. Washington, DC: Morgan, Lewis and Bockius LLP.

Hache, G., and H. M. Chung. 2001. The history of LOCA embrittlement criteria. Pp. 205-237 in Proceedings of the Twenty-Eighth Water Reactor Safety Information Meeting, Bethesda, MD, October 23-25, 2000. NUREG/CP-0172. Prepared by Brookhaven National Laboratory for the U.S. Nuclear Regulatory Commission.

Haggerty, R. 2011. Potassium iodide and Geiger counter sales spike after Japan disaster. Chicago Tribune, March 15. Available at articles.chicagotribune.com/2011-03-15/news/ct-met-local-iodide-sales-20110315_1_potassium-iodide-radiation-geiger.

Hamada, N., and H. Ogino. 2012. Food safety regulations: What we learned from the Fukushima nuclear accident. Journal of Environmental Radioactivity 111: 83-99.

Hamada, N., H. Ogino, and Y. Fujimichi. 2012. Safety regulations of food and water implemented in the first year following the Fukushima nuclear accident. Journal of Radiation Research 53(5): 641-671.

Harbitz, C. B. 1991. Model simulations of tsunami generated by the Storegga Slide. Institute of Mathematics, University of Oslo Preprint Series No. 5.

Hardie, S. M. L., and I. G. McKinley. 2013. Fukushima remediation: Status and overview of future plans. Journal of Environmental Radioactivity 133: 75-85.

Hasegawa, R. 2013. Disaster Evacuation from Japan's 2011 Tsunami Disaster and the Fukushima Nuclear Accident. Study No. 05/13. Paris: Institut du Développement Durable et des Relations Internationales. Available at http://www.iddri.org/Publications/Collections/Analyses/STUDY0513_RH_DEVAST%20report.pdf.

Hayano, R. S., and R. Adachi. 2013. Estimation of the total population moving into and out of the 20 km evacuation zone during the Fukushima NPP accident as calculated using "Auto-GPS" mobile phone data. Proceedings of the Japan Academy, Series B: Physical and Biological Sciences 89(5): 196-199.

Hayashi, Y. 2011. Japan officials failed to hand out radiation pills in quake's aftermath. Wall Street Journal, September 29. Available at http://online.wsj.com/news/articles/SB10001424052970204010604576596321581004368.

HERP (Headquarters for Earthquake Research Promotion), Earthquake Research Committee. 2002. Long-term evaluation of seismic activity from offshore Sanriku to offshore Bousou (in Japanese). Available at http://www.jishin.go.jp/main/chousa/kaikou_pdf/sanriku_boso.pdf.

HERP, Earthquake Research Committee. 2005. National Seismic Hazard Maps for Japan (2005)–March 23. Tokyo: Headquarters for Earthquake Research Promotion, Government of Japan. Available at jishin.go.jp/main/chousa/06mar_yosoku-e/NationalSeismicHazardMaps.pdf.

Heezen, B., and M. Ewing. 1952. Turbidity currents and submarine slumps, and the 1929 Grand Banks Earthquake. American Journal of Science 250: 849-873.

Helm, D. 2006. Regulatory reform, capture, and the regulatory burden. Oxford Review of Economic Policy 22(2): 169-185.

Helton, D., J. Chang, N. Siu, K. Coyne, and M. Leonard. 2010. The role of severe accident management in the advancement of level 2 PRA modeling techniques. In Implementation of Severe Accident Management Measures, ISAMM 2009: Workshop Proceedings. NEA/CSNI/R(2010)10/PART1. Issy-les-Moulineaux, France: OECD Nuclear Energy Agency. Available at http://www.oecd-nea.org/html/nsd/docs/2010/csni-r2010-10-vol1.pdf.

Hogberg, L. 2013. Root causes and impacts of severe accidents at large nuclear plants. AMBIO 42: 267-284.

Hollnagel, E., D. D. Woods, and N. Leveson. 2006. Resilience Engineering: Concepts and Precepts. Burlington, VT: Ashgate.

Horiuchi, A., and K. Shimizu. 2001. Did *amakudari* undermine the effectiveness of regulator monitoring in Japan? Journal of Banking and Finance 25: 573-596.

Hosono, H., Y. Kumagai, and T. Sekizaki. 2013. Development of an information package of radiation risk in beef after the Fukushima Daiichi Nuclear Power Plant accident. Pp. 187-204 in Agricultural Implications of the Fukushima Nuclear Accident, T. M. Nakanishi and K. Tanoi, eds. Tokyo: Springer Japan.

IAEA (International Atomic Energy Agency). 1994. PSA for the Shutdown Mode for Nuclear Power Plants: Proceedings of a Technical Committee Meeting in Stockholm, November 30-December 3, 1992. IAEA-TECDOC-751. Vienna, Austria: IAEA.

IAEA. 2002. Safety Culture in Nuclear Installations: Guidance for Use in the Enhancement of Safety Culture. IAEA-TECDOC-1329. Vienna, Austria: International Atomic Energy Agency. Available at www-pub.iaea.org/MTCD/publications/PDF/te_1329_web.pdf.

IAEA. 2003. Method for Developing Arrangements for Response to a Nuclear or Radiological Emergency (Updating IAEA-TECDOC-953). Vienna, Austria: International Atomic Energy Agency. Available at http://www-pub.iaea.org/books/IAEABooks/6750/Method-for-Developing-Arrangements-for-Response-to-a-Nuclear-or-Radiological-Emergency-Updating-IAEA-TECDOC-953.

IAEA. 2007. IAEA Safety Glossary: Terminology Used in Nuclear Safety and Radiation Protection. STI/PUB/1290. Vienna, Austria: International Atomic Energy Agency. Available at http://www-pub.iaea.org/MTCD/publications/PDF/Pub1290_web.pdf.

IAEA. 2010. Governmental, Legal and Regulatory Framework for Safety: General Safety Requirements, Part 1. Safety Standards Series No. GSR Part 1. Vienna, Austria: International Atomic Energy Agency. Available at www-pub.iaea.org/MTCD/publications/PDF/Pub1465_web.pdf.

IAEA. 2011. IAEA International Fact Finding Expert Mission of the Fukushima Dai-ichi NPP Accident Following the Great East Japan Earthquake and Tsunami—Tokyo, Fukushima Dai-ichi NPP, Fukushima Dai-ni NPP and Tokai Dai-ni NPP, Japan, 24 May-2 June 2011. Vienna, Austria: International Atomic Energy Agency. Available at http://www-pub.iaea.org/MTCD/meetings/PDFplus/2011/cn200/documentation/cn200_Final-Fukushima-Mission_Report.pdf.

IAEA. 2012. IAEA Mission to Onagawa Nuclear Power Station to Examine the Performance of Systems, Structures and Components Following the Great East Japanese Earthquake and Tsunami—Onagawa and Tokyo, Japan, 30 July-11 August 2012. Vienna, Austria: International Atomic Energy Agency. Available at http://www.iaea.org/newscenter/focus/actionplan/reports/onagawa0413.pdf.

IAEA. 2013a. Actions to Protect the Public in an Emergency Due to Severe Conditions at a Light Water Reactor: Emergency Preparedness and Response Report. Vienna, Austria: International Atomic Energy Agency.

IAEA. 2013b. Preliminary Summary Report: The Follow-up IAEA International Mission on Remediation of Large Contaminated Areas Off-Site the Fukushima Daiichi Nuclear Power Plant, Tokyo and Fukushima Prefecture, Japan, 14-21 October. Vienna, Austria: International Atomic Energy Agency. Available at http://www.iaea.org/newscenter/focus/fukushima/remediation-report-211013.pdf.

IAEA. 2014a. Nuclear Power Reactors in the World IAEA-RDS-2/34. Vienna, Austria: International Atomic Energy Agency.

IAEA. 2014b. The Follow-up IAEA International Mission on Remediation of Large Contaminated Areas Off-Site the Fukushima Daiichi Nuclear Power Plant, Tokyo and Fukushima Prefecture, Japan, 14-21 October 2013. Final Report. Vienna, Austria: International Atomic Energy Agency.

ICRP (International Commission on Radiological Protection). 1991. 1990 Recommendations of the International Commission on Radiological Protection. ICRP Publication 60. Annals of the ICRP 21(1-3).

ICRP. 2007. The 2007 Recommendations of the International Commission on Radiological Protection. ICRP Publication 103. Annals of the ICRP 37(2-4).

ICRP. 2011. Application of the Commission's Recommendations to the Protection of People Living in Long-Term Contaminated Areas After a Nuclear Accident or a Radiation Emergency. Annals of the ICRP 39(3): ICRP Publication 111.

Imtihani, N., and Y. Mariko. 2013. Media coverage of Fukushima nuclear power station accident 2011 (A case study of NHK and BBC WORLD TV stations). Procedia Environmental Sciences 17: 938-946.

INPO (Institute of Nuclear Power Operations). 2004. Principles for a Strong Nuclear Safety Culture. Atlanta, GA: INPO. Available at www.nrc.gov/about-nrc/regulatory/enforcement/INPO_PrinciplesSafetyCulture.pdf.

INPO. 2005. Performance Objectives and Criteria. INPO 05-003. Atlanta, GA: Institute of Nuclear Power Operations.

INPO. 2011. Special Report on the Nuclear Accident at the Fukushima Daiichi Nuclear Power Station, Revision 0. INPO 11-005. Atlanta, GA: Institute of Nuclear Power Operations.

INPO. 2012. Special Report on the Lessons Learned from the Nuclear Accident at the Fukushima Daiichi Nuclear Power Station, Revision 0. INPO 11-005 Addendum. Atlanta, GA: Institute of Nuclear Power Operations.

INPO. 2013. Traits of a Healthy Nuclear Safety Culture. INPO 12-012, Rev. 1. Atlanta, GA: Institute of Nuclear Power Operations. Available at nuclearsafety.info/wp-content/uploads/2010/07/Traits-of-a-Healthy-Nuclear-Safety-Culture-INPO-12-012-rev.1-Apr2013.pdf.

INSAG (International Nuclear Safety Group). 1986. Report by the International Nuclear Safety Advisory Group: Summary Report on the Post-Accident Review Meeting on the Chernobyl Accident. Safety Series No. 75-INSAG-1. Vienna, Austria: International Atomic Energy Agency. Available at ilankelman.org/miscellany/chernobyl.pdf.

INSAG. 1992. Pp. 23-24 in The Chernobyl Accident, Updating of INSAG-1. Safety Series No. INSAG-7: A Report by the International Safety Advisory Group. Vienna, Austria: International Atomic Energy Agency. Available at www-pub.iaea.org/MTCD/publications/PDF/Pub913e_web.pdf.

INSAG. 2011. The Fukushima Accident. Letter to International Atomic Energy Agency Director General Yukiya Amano from INSAG Chairman Richard A. Meserve. Available at http://www-ns.iaea.org/committees/files/insag/743/INSAGLetterReport20117-26-11.pdf.

INSAG. 2012. Letter to International Atomic Energy Agency Director General Yukiya Amano from INSAG Chairman Richard A. Meserve. Available at http://www-ns.iaea.org/committees/files/insag/743/gc56inf-11_enINSAGlettertoDG.pdf.

INSAG. 2013. Letter to International Atomic Energy Agency Director General Yukiya Amano from INSAG Chairman Richard A. Meserve. Available at http://www-ns.iaea.org/committees/files/insag/743/2013AnnualAssessmentLetter.pdf.

Investigation Committee (Investigation Committee on the Accident at Fukushima Nuclear Power Stations of Tokyo Electric Power Company). 2011. Interim Report (Main Text). Tokyo: Government of Japan. December 26.

Investigation Committee. 2012. Final Report on the Accident at Fukushima Nuclear Power Stations of Tokyo Electric Power Company. Tokyo: Government of Japan. July 23.

IRSN (Institut de Radioprotection et de Sûreté Nucléaire). 2011. Assessment on the 66th Day of Projected External Doses for Populations Living in the North-West Fallout Zone of the Fukushima Nuclear Accident: Outcome of Population Evacuation Measures. Report DRPH/2011-10. Paris: Directorate of Radiological Protection and Human Health. Available at www.irsn.fr/EN/news/Documents/IRSN-Fukushima-Report-DRPH-23052011.pdf.

IRSN. 2012a. Fukushima, One Year Later: Initial Analyses of the Accident and Its Consequences. Report IRSN/DG/2012-003. Paris: Directorate of Radiological Protection and Human Health. March 12.

IRSN. 2012b. Summary of Information Available in Japan Concerning the Contamination of Foodstuffs by the Radionuclides Persisting in the Environment After the Fukushima Accident. Paris: Directorate of Radiological Protection and Human Health. July 13. Available at www.irsn.fr/EN/publications/thematic/fukushima/Documents/IRSN_Fukushima-Accident_Foodstuffs-contamination_13072012.pdf.

JANTI (Japan Nuclear Technology Institute). 2011. Examination of Accident at Tokyo Electric Power Co., Inc.'s Fukushima Daiichi Nuclear Power Station and Proposal of Countermeasures. Tokyo: JANTI. October.

JNES (Japan Nuclear Energy Safety Organization). 2013. Initial Operations in the Local Nuclear Emergency Response Headquarters. JNES-RE-2013-0002. Tokyo: JNES. April.

Jo, J., C. C. Lin, and L. Neymotin. 1995. Evaluation of Potential Severe Accidents During Low Power and Shutdown Operations at Surry, Unit 1: Evaluation of Severe Accident Risk During Mid-Loop Operation, Volume 6, Part 1. NUREG/CR-6144, BNL-NUREG-52399, ON: TI95013650, TRN: 95:017227. Washington, DC: U.S. Nuclear Regulatory Commission. Available at http://www.osti.gov/scitech/biblio/83860.

JSCE (Japan Society of Civil Engineers). 2006. Tsunami Assessment Method for Nuclear Power Plants in Japan (originally published in Japanese, February 2002). Tokyo: Japan Society of Civil Engineers. Available at http://committees.jsce.or.jp/ceofnp/system/files/JSCE_Tsunami_060519.pdf.

Kagan, Y. Y., and D. D. Jackson. 2013. Tohoku Earthquake: A surprise? Bulletin of the Seismological Society of America 103(2B): 1181-1194.

Kai, M. 2012. Some lessons on radiological protection learnt from the accident at the Fukushima Dai-ichi nuclear power plant. Journal of Radiological Protection 32(1): N101-N105.

Kanamori, H. 1977. The energy release in great earthquakes. Journal of Geophysical Research 82(20): 2981-2987.

Kaplan, S., and B. J. Garrick. 1981. On the quantitative definition of risk. Risk Analysis 1(1): 11-27.

Kappenman, J. 2010. Geomagnetic Storms and Their Impacts on the U.S. Power Grid. Meta-R-319. Prepared for Oak Ridge National Laboratory. Available at http://web.ornl.gov/sci/ees/etsd/pes/pubs/ferc_Meta-R-319.pdf.

Kappenman, J. G. 2012. A perfect storm of planetary proportions. IEEE Spectrum 49(2): 26-31.

Kataoka, J., A. Kishimoto, T. Nishiyama, T. Fujita, K. Takeuchi, T. Kato, T. Nakamori, S. Ohsuka, S. Nakamura, M. Hirayanagi, S. Adachi, T. Uchiyama, and K. Yamamoto. 2013. Handy Compton camera using 3-D position-sensitive scintillators coupled with large-area monolithic MPPC arrays. Nuclear Instruments and Methods in Physics Research, Section A: Accelerators, Spectrometers, Detectors and Associated Equipment 732: 403-407.

Kaufmann, D., and V. Penciakova. 2011. Preventing nuclear meltdown: Assessing regulatory failure in Japan and the United States. Brookings, Opinion, April 1. Available at http://www.brookings.edu/research/opinions/2011/04/01-nuclear-meltdown-kaufmann.

Kawabe, R. 2012. Evaluation of the depth of concrete ablation by molten core. Presentation at the Nuclear and Industrial Safety Agency Technical Workshop on the Accident of TEPCO's Fukushima Dai-ichi NPS, Tokyo, Japan, July 23-24, 2012. Tokyo: Japan Nuclear Energy Safety Organization.

Kawamura, H., T. Kobayashi, A. Furuno, T. In, Y. Ishikawa, T. Nakayama, S. Shima, and T. Awaji. 2011. Preliminary numerical experiments on oceanic dispersion of ^{131}I and ^{137}Cs discharged into the ocean because of the Fukushima Daiichi nuclear power plant disaster. Journal of Nuclear Science and Technology 48(11): 1349-1356.

Kawano, A. 2012. Lessons learned from the Fukushima accident and challenge for nuclear reform. Presentation to the Committee on Lessons Learned from the Fukushima Nuclear Accident for Improving Safety and Security of U.S. Nuclear Plants, Tokyo, Japan, November 26, 2012. Tokyo: Tokyo Electric Power Company, Inc.

Keller, W., and M. Modarres. 2005. A historical overview of probabilistic risk assessment development and its use in the nuclear power industry: A tribute to the late Professor Norman Carl Rasmussen. Reliability Engineering and System Safety 89: 271-285.

Kirwan, B. 1994. A Guide to Practical Human Reliability Assessment. Bristol, PA: Taylor and Francis.

Kozdon, J. E., and E. M. Dunham. 2013. Rupture to the trench: Dynamic rupture simulations of the 11 March 2011 Tohoku Earthquake. Bulletin of the Seismological Society of America 103(2B): 1275-1289.

Kubo, H., and Y. Kakehi. 2013. Source process of the 2011 Tohoku Earthquake estimated from the joint inversion of teleseismic body waves and geodetic data including seafloor observation data: Source model with enhanced reliability by using objectively determined inversion settings. Bulletin of the Seismological Society of America 103(2B): 1195-1220.

Kurihara, J. 2013. Crisis communication amid the Fukushima tragedy: A postmortem analysis and recommendations. Presentation at a Workshop at the Canon Institute for Global Studies (CIGS), Tokyo, Japan, July 18, 2013. Tokyo: Canon Institute for Global Studies.

Lay, T., and H. Kanamori. 2011. Insights from the Great 2011 Japan Earthquake. Physics Today 64(12): 33-39.

Lay, T., Y. Fujii, E. Geist, K. Koketsu, J. Rubinstein, T. Sagiya, and M. Simons. 2013. Introduction to the Special Issue on the 2011 Tohoku Earthquake and Tsunami. Bulletin of the Seismological Society of America 103(2B): 1165-1170.

Lee, J. C., and N. J. McCormick. 2011. Risk and Safety Analysis of Nuclear Systems. Hoboken, NJ: John Wiley and Sons, Inc.

Leonard, L. J., G. C. Rogers, and S. Mazzotti. 2013. Tsunami hazard assessment of Canada. Natural Hazards 70(1): 237-274.

Levenson, N. 1995. Safeware: System Safety and Computers. Boston, MA: Addison Wesley.

Levy, S. 2012. How Fukushima Daiichi severe accidents could be avoided. Presentation at the 2012 American Nuclear Society Winter Meeting Embedded Topical Meeting on Severe Accidents, November 11-15, 2012, San Diego, CA. San Jose, CA: S. Levy and Associates, Inc.

Leyse, M. 2014. Preventing Hydrogen Explosions in Severe Nuclear Accidents: Unresolved Safety Issues Involving Hydrogen Generation and Mitigation. NRDC Report R:14-02-B. New York: Natural Resources Defense Council.

Locat, J., H. Lee, U. S. ten Brink, D. Twichell, E. Geist, and M. Sansoucy. 2009. Geomorphology, stability and mobility of the Currituck Slide. Marine Geology 264(1): 28-40.

Lochbaum, D. 2014. The NRC and Nuclear Power Plant Safety in 2013—More Jekyll, Less Hyde. Cambridge, MA: Union of Concerned Scientists. Available at http://www.ucsusa.org/assets/documents/nuclear_power/NRC-Nuclear-Safety-Report-2013.pdf.

Lochbaum, D., E. Lyman, S. Q. Stranahan, and the Union of Concerned Scientists. 2014. Fukushima: The Story of a Nuclear Disaster. New York: The New Press.

Lois, E., V. N. Dang, J. Forester, H. Broberg, S. Massaiu, M. Hildebrandt, P. Ø. Braarud, G. Parry, J. Julius, R. Boring, I. Männistö, and A. Bye. 2009. International Human Reliability Analysis (HRA) Empirical Study—Phase 1 Report: Description of Overall Approach and Pilot Phase Results from Comparing HRA Methods to Simulator Performance Data. NUREG/IA-0216, Vol. 1. Rockville, MD: U.S. Nuclear Regulatory Commission. Available at http://pbadupws.nrc.gov/docs/ML0933/ML093380283.pdf.

Lyons, C., and D. Colton. 2012. Aerial measuring system in Japan. Health Physics 102(5): 509-515.

Maeda, T., T. Furumura, S. Noguchi, S. Takemura, S. Sakai, M. Shinohara, K. Iwai, and S. J. Lee. 2013. Seismic- and tsunami-wave propagation of the 2011 off the Pacific coast of Tohoku Earthquake as inferred from the tsunami-coupled finite-difference simulation. Bulletin of the Seismological Society of America 103(2B): 1456-1472.

Macfarlane, A. 2012. Assessing Progress in Worldwide Nuclear Safety. Remarks at the International Nuclear Safety Group (INSAG) Forum, International Atomic Energy Agency, Vienna, Austria, September 17. Rockville, MD: U.S. Nuclear Regulatory Commission. Available at http://pbadupws.nrc.gov/docs/ML1226/ML12261A373.pdf.

Mainichi. 2013. Fukushima decontamination needs 5 tril. yen budget: nat'l institute. July 24.

Masson, D. G., C. B. Harbitz, R. B. Wynn, G. Pedersen, and F. Løvholt. 2006. Submarine landslides: Processes, triggers and hazard prediction. Philosophical Transactions of the Royal Society A, Mathematical, Physical, and Engineering Sciences 364(1845): 2009-2039.

Matsuo, Y., and Y. Yamaguchi. 2013. The rise in cost of power generation in Japan after the Fukushima Daiichi accident and its impact on the finances of the electric power utilities. November 18. Institute of Energy Economics, Japan. Available at http://eneken.ieej.or.jp/data/5252.pdf.

McCaffrey, R. 2008. Global frequency of magnitude 9 earthquakes. Geology 36: 263-266.

Melgar, D., and Y. Bock. 2013. Near-field tsunami models with rapid earthquake source inversions from land- and ocean-based observations: The potential for forecast and warning. Journal of Geophysical Research: Solid Earth 118(11): 5939-5955.

METI (Ministry of Economy, Trade and Industry). 2013. Areas to which evacuation orders have been issued. Tokyo: METI, Government of Japan. Available at www.meti.go.jp/english/earthquake/nuclear/roadmap/pdf/20130807_01.pdf.

MHLW (Ministry of Health, Labour and Welfare of Japan). 2012. New Standard Limits for Radionuclides in Foods. Presentation from the Department of Food Safety Pharmaceutical and Food Safety Bureau, Tokyo, Government of Japan, June 7. Available at www.mhlw.go.jp/english/topics/2011eq/dl/new_standard.pdf.

Minoura, K., F. Imamura, D. Sugawara, Y. Kono, and T. Iwashita. 2001. The 869 Jōgan tsunami deposit and recurrence interval of large-scale tsunami on the Pacific coast of northeast Japan. Journal of Natural Disaster Science 23(2): 83-88.

MITRE. 2011. Impacts of Severe Space Weather on the Electric Grid. JSR-11-320. McLean, VA: MITRE Corp., JASON Program Office. November.

Mizushima, A. 2012. The Japan-US "military" response to the earthquake, and the strengthening of the military alliance as a result. Fukushima on the Globe, December 12. Available at http://fukushimaontheglobe.com/the-earthquake-and-the-nuclear-accident/whats-happened/the-japan-us-military-response.

Mori, N., T. Takahashi, T. Yasuda, and H. Yanagisawa. 2011. Survey of 2011 Tohoku earthquake tsunami inundation and run-up. Geophysical Research Letters 38(7): L00G14, doi: 10.1029/2011GL049210.

Mori, N., T. Takahashi, and the 2011 Tohoku Earthquake Tsunami Joint Survey Group. 2012. Nationwide post event survey and analysis of the 2011 Tohoku Earthquake tsunami. Coastal Engineering Journal 54(01): 1-27.

Morino, Y., T. Ohara, and M. Nishizawa. 2011. Atmospheric behavior, deposition, and budget of radioactive materials from the Fukushima Daiichi nuclear power plant in March 2011. Geophysical Research Letters 38(7): L00G11, doi: 10.1029/2011GL048689.

Morrow, S., and V. Barnes. 2012. Independent Evaluation of INPO's Nuclear Safety Culture Survey and Construct Validation Study. Technical Paper. Rockville, MD: U.S. Nuclear Regulatory Commission.

Mumaw, R. J., D. Swatzler, E. M. Roth, and W. A. Thomas. 1994. Cognitive Skill Training for Nuclear Power Plant Operational Decision Making. NUREG/CR-6126. Prepared for the U.S. Nuclear Regulatory Commission. Available at http://pbadupws.nrc.gov/docs/ML1115/ML11151A201.pdf.

NAE and NRC (National Academy of Engineering and National Research Council). 2011. Macondo Well *Deepwater Horizon* Blowout: Lessons for Improving Offshore Drilling Safety. Washington, DC: The National Academies Press. Available at http://www.nap.edu/openbook.php?record_id=13273.

NAIIC (Nuclear Accident Independent Investigation Commission). 2012. The Official Report of the Fukushima Nuclear Accident Independent Investigation Commission. Tokyo: National Diet of Japan.

Nakahara, H. 2013. Envelope inversion analysis for high-frequency seismic energy radiation from the 2011 M_w 9.0 off the Pacific coast of Tohoku earthquake. Bulletin of the Seismological Society of America 103(2B): 1348-1359.

Nakamura, A., and M. Kikuchi. 2011. What we know, and what we have not yet learned: Triple disasters and the Fukushima nuclear fiasco in Japan. Public Administration Review 71(6): 893-899.

Nakamura, Y., S. Kodaira, S. Miura, C. Regalla, and N. Takahashi. 2013. High-resolution seismic imaging in the Japan trench axis area off Miyagi, northeastern Japan. Geophysical Research Letters 40(9): 1713-1718.

Narabayashi, T. 2012. Lessons learned from the Fukushima Daiichi nuclear power plant accident. Keynote address to the 7th International Symposium on Turbulence, Heat and Mass Transfer, Palermo, Italy, September 24-27, 2012.

National Police Agency of Japan. 2014. Damage situation and police countermeasures associated with 2011 Tohoku district—off the Pacific Ocean Earthquake. Presentation from Japan Emergency Disaster Countermeasures Headquarters, National Police Agency of Japan, Tokyo. Available at www.npa.go.jp/archive/keibi/biki/higaijokyo_e.pdf. Accessed on June 3, 2013.

NCRP (National Council on Radiation Protection and Measurements). 2009. Ionizing Radiation Exposure of the Population of the United States. Report No. 160. Bethesda, MD: NCRP.

NEA (Nuclear Energy Agency of the Organisation for Economic Co-operation and Development). 1999. The Role of the Nuclear Regulator in Promoting and Evaluating Safety Culture. Paris: Organisation for Economic Co-operation and Development. June.

NEA. 2000. OECD Flame Acceleration and Deflagration to Detonation Transition in Nuclear Safety: State-of-the-Art Report by a Group of Experts. Nuclear Safety NEA/CSNI/R(2000)7, August. Issy-les-Moulineaux, France: Organisation for Economic Co-operation and Development.

NEA. 2002. Chernobyl: Assessment of Radiological and Health Impacts, 2002 Update of Chernobyl: Ten Years On. Paris: Organisation for Economic Co-operation and Development.

NEA. 2013. The Fukushima Daiichi Nuclear Power Plant Accident: OECD/NEA Nuclear Safety Response and Lessons Learnt. Paris: Nuclear Energy Agency, Organisation for Economic Co-Operation and Development. Available at http://www.oecd-nea.org/pub/2013/7161-fukushima2013.pdf.

NEA. 2014. Status Report on Hydrogen Management and Related Computer Codes. NEA/CSNI/R(2013)8 (JT03359912). Available at http://www.oecd-nea.org/nsd/docs/2014/csni-r2014-8.pdf.

NEI (Nuclear Energy Institute). 1994. Severe Accident Issue Closure Guidelines. NEI 91-04, Rev. 1. Washington, DC: NEI. Available at http://pbadupws.nrc.gov/docs/ML0728/ML072850981.pdf.

NEI. 2002. B.S.b Phase 2 & 3 Submittal Guideline. NEI 06-12, Rev. 2. Washington, DC: NEI. Available at http://pbadupws.nrc.gov/docs/ML0700/ML070090060.pdf.

NEI. 2012. Diverse and Flexible Coping Strategies (FLEX) Implementation Guide. NEI 12-06 Revision 0. Washington, DC: Nuclear Energy Institute. Available at http://pbadupws.nrc.gov/docs/ML1214/ML12143A232.pdf.

NEI. 2013. Industry Support and Use of PRA and Risk-Informed Regulation, Letter to USNRC Chair Allison M. Macfarlane, December 19. Available at http://pbadupws.nrc.gov/docs/ML1335/ML13354B997.pdf.

NEI, INPO, and EPRI (Nuclear Energy Institute, Institute of Nuclear Power Operations, and Electric Power Research Institute). 2012. The Way Forward: U.S. Industry Leadership in Response to Events at the Fukushima Daiichi Nuclear Power Plant. Washington, DC: Nuclear Energy Institute. Available at http://www.nei.org/corporatesite/media/filefolder/Way_Forward_2_23_12.pdf.

NERHQ-TEPCO (Nuclear Emergency Response Headquarters-Tokyo Electric Power Company). 2011. Report of Japanese Government to the IAEA Ministerial Conference on Nuclear Safety—The Accident at TEPCO's Fukushima Nuclear Power Stations. Tokyo: Government of Japan.

Nettles, M., G. Ekström, and H. C. Koss. 2011. Centroid-moment-tensor analysis of the 2011 off the Pacific coast of Tohoku Earthquake and its larger foreshocks and aftershocks. Earth, Planets and Space 63(7): 519-523.

Ng, K. H., and M. L. Lean. 2012. The Fukushima nuclear crisis reemphasizes the need for improved risk communication and better use of social media. Health Physics 103(3): 307-310.

NISA (Nuclear and Industrial Safety Agency). 2012. Technical Knowledge of the Accident at Fukushima Dai-ichi Nuclear Power Station of Tokyo Electric Power Co., Inc. (Provisional Translation). Tokyo: Ministry of Economy, Trade and Industry. Avalable at http://www.nsr.go.jp/archive/nisa/english/press/2012/06/en20120615-1-1.pdf.

Noggerath, J., R. Geller, and V. Gusiakov. 2011. Fukushima: The myth of safety, the reality of geoscience. Bulletin of the Atomic Scientists 67(5): 37-46.

Nomura, S., S. Gilmour, M. Tsubokura, D. Yoneoka, A. Sugimoto, T. Oikawa, M. Kami, and K. Shibuya. 2013. Mortality risk amongst nursing home residents evacuated after the Fukushima nuclear accident: A retrospective cohort study. PLOS ONE 8(3): e60192.

Norman, D. 1990. The problem of automation: Inappropriate feedback and interaction, not over-automation. Philosophical Transactions of the Royal Society of London Series B 327: 585-593.

NRA (Nuclear Regulation Authority). 2013a. FY2012 Nuclear Regulation Authority Annual Report (provisional English translation). Tokyo: NRA.

NRA. 2013b. Practical measures for evacuees to return to their homes. Tokyo: Nuclear Regulation Authority. November 20.

NRC (National Research Council). 1987. Technical Aspects of Hydrogen Control and Combustion in Severe Light-Water Reactor Accidents. Washington, DC: National Academy Press.

NRC. 1998. Peer Review in Environmental Technology Development Programs: The Department of Energy's Office of Science and Technology. Washington, DC: National Academy Press. Available at http://www.nap.edu/catalog.php?record_id=6408.

NRC. 2004a. Distribution and Administration of Potassium Iodide in the Event of a Nuclear Incident. Washington, DC: The National Academies Press. Available at http://www.nap.edu/catalog.php?record_id=10868.

NRC. 2004b. Safety and Security of Commercial Spent Nuclear Fuel Storage (U). Washington, DC: The National Academies Press. An abbreviated public version of this report was issued in 2006 and is available at http://www.nap.edu/catalog.php?record_id=11263.

NRC. 2006a. Health Risks from Exposure to Low Levels of Ionizing Radiation: BEIR VII Phase 2. Washington, DC: The National Academies Press. Available at http://www.nap.edu/catalog.php?record_id=11340.

NRC. 2006b. Safety and Security of Commercial Spent Nuclear Fuel Storage: Public Report. Washington, DC: The National Academies Press. Available at http://www.nap.edu/catalog.php?record_id=11263.

NSC (Nuclear Safety Commission). 2013. Emergency Preparedness for Nuclear Facilities (May 2007), Chapter 2: Sections 2.1, 2.2, 2.6; Chapter 3: Sections 3.1, 3.2; Chapter 5; and Appendixes D, F, J, L, and N (Technical Language Service, Trans.). Tokyo: Government of Japan. Available at www.nsr.go.jp/archive/nsc/anzen/sonota/houkoku/bousai220823.pdf.

NRT (National Response Team). 2000. Joint Information Center Model: Collaborative Communications During Emergency Response. Available at http://www.au.af.mil/au/awc/awcgate/nrt/jic-model.pdf.

NTSB (U.S. National Transportation Safety Board). 2011. Accident Report: Pacific Gas and Electric Company Natural Gas Transmission Pipeline Rupture and Fire, San Bruno, California, September 9, 2010. NTSB: PAR-11-01, NTIS: PB2011-916501. Washington, DC: NTSB.

NSCJ (Nuclear Safety Commission of Japan, Special Committee on Nuclear Safety Standards and Guides). 2012. Proposed Requirements for Incorporation into Safety Design Guide and Other Related Regulatory Guides, Annex 1 (March 14) (provisional translation). Available at http://www.nsr.go.jp/archive/nsc/NSCenglish/geje/20120322review_1_1.pdf.

Ogino, H., and T. Hattori. 2013. Operational level for unconditional release of contaminated property from affected areas around Fukushima Daiichi nuclear power plant. Radiation Protection Dosimetry 157(3): 446-454.

OIG (Office of the Inspector General). 2002. Office of the Inspector General Event Inquiry: NRC's Regulation of Davis-Besse Regarding Damage to the Reactor Vessel Head. Case No. 02-03S. Washington, DC: Nuclear Regulatory Commission. December 30. Available at www.nrc.gov/reading-rm/doc-collections/insp-gen/2003/02-03s.pdf.

OIG. 2012. Gaps Continue to Exist in Nursing Home Emergency Preparedness and Response During Disasters: 2007-2010. OEI-06-09-00270. Washington, DC: Nuclear Regulatory Commission.

Okrent, D. 1981. Nuclear Reactor Safety: On the History of the Regulatory Process. Madison: University of Wisconsin Press.

ONR (Office of Nuclear Regulation). 2011. Japanese Earthquake and Tsunami: Implications for the UK Nuclear Industry Final Report. ONR-FR-REP-11-002 Revision 2. Bootle, UK: Health and Safety Executive.

Pacific Gas and Electric Company. 2010. Methodology for probabilistic tsunami hazard analysis: Trial application for the Diablo Canyon Power Plant site. Presentation to the PEER Workshop on Tsunami Hazard Analyses for Engineering Design Parameters, Berkeley, CA.

Pariès, J. 2011. Lessons from the Hudson. Chapter 2 in Resilience Engineering in Practice: A Guidebook, E. Hollnagel, J. Pariès, D. Woods, and J. Wreathall, eds. Farnham, UK: Ashgate.

Pascucci-Cahen, L., and P. Momal. 2012. Massive radiological releases profoundly differ from controlled releases. Eurosafe Forum. November 6. Available at http://www.irsn.fr/FR/Actualites_presse/Actualites/Documents/EN_Eurosafe-2012_Massive-releases-vs-controlled-releases_Cost_IRSN-Momal.pdf.

Payne, E. 2011. Radiation fears spark run on West Coast pharmacies, health food stores. Cable News Network (CNN), March 16.

Pereira, A. S. 2006. The opportunity of a disaster: The economic impact of the 1755 Lisbon Earthquake. CHERRY DP 03/06. Centre for Historical Economics and Related Research at York, York University. Available at http://www.york.ac.uk/media/economics/documents/cherrydiscussionpapers/0603.pdf.

Persensky, J., S. Browde, A. Szabo, L. Peterson, E. Specht, and E. Wight. 2004. Effective Risk Communication: The Nuclear Regulatory Commission's Guidelines for External Risk Communication. NUREG/BR-0308. Rockville, MD: U.S. Nuclear Regulatory Commission. Available at http://pbadupws.nrc.gov/docs/ML0406/ML040690412.pdf.

Philadelphia Electric Company. 1981. Probabilistic Risk Assessment, Limerick Generating Station, Philadelphia, PA.

Phillips, J., J. Cardoni, R. Gauntt, D. Kalinich, and A. Goldmann, 2012. MELCOR simulations of the severe nuclear accident at the Fukushima 1F2 reactor. Presentation at American Nuclear Society Winter Meeting, November 15, 2012.

Pöllänen, R., H. Toivonen, K. Peräjärvi, T. Karhunen, T. Ilander, J. Lehtinen, K. Rintala, T. Katajainen, J. Niemelä, and M. Juusela. 2009. Radiation surveillance using an unmanned aerial vehicle. Applied Radiation and Isotopes 67: 340-344.

Povinec, P. P., M. Gera, K. Holý, K. Hirose, G. Lujaniené, M. Nakano, W. Plastino, I. Sýkora, J. Bartok, and M. Gažák. 2013. Dispersion of Fukushima radionuclides in the global atmosphere and the ocean. Applied Radiation and Isotopes 81: 383-392.

Progress Energy. 2008. Application for Combined License for Levy Nuclear Power Plant Units 1 and 2, NRC Project No. 756, attached to a letter to the U.S. Nuclear Regulatory Commission from James Scarola. NPD-NRC-2008-022. July 28. Available at http://pbadupws.nrc.gov/docs/ML0822/ML082260277.pdf.

Rasmussen, J. 1983. Skills, rules, and knowledge; signals, signs, and symbols, and other distinctions in human performance models. IEEE Transactions on Systems, Man, and Cybernetics 13(3): 257-266.

Reason, J. 2008. The Human Contribution: Unsafe Acts, Accidents, and Heroic Recoveries. Farnham, UK: Ashgate.

Rempe, J., M. Farmer, M. Corradini, L. Ott, R. Gauntt, and D. Powers. 2012. Revisiting insights from Three Mile Island Unit 2 postaccident examinations and evaluations in view of the Fukushima Daiichi accident. Nuclear Science and Engineering 172(3): 223-248.

Richardson, P., K. Rickwood, and P. Rickwood. 2013. Public involvement as a tool to enhance nuclear safety. Energy Strategy Reviews 1(4): 266-271.

Riley, P. 2012. On the probability of occurrence of extreme space weather events. Space Weather 10(2): S02012, doi: 10.1029/2011SW000734.

RJIF (Rebuild Japan Initiative Foundation's Independent Investigation Commission on the Fukushima Nuclear Accident). 2014. The Fukushima Daiichi Nuclear Power Station Disaster: Investigating the Myth and Reality, M. K. Bricker, ed. London: Routledge.

Robb, K. R., M. W. Francis, and M. T. Farmer. 2013. Enhanced Ex-Vessel Analysis for Fukushima Daiichi Unit 1: Melt Spreading and Core-Concrete Interaction Analyses with MELTSPREAD and CORQUENCH. ORNL/TM-2012/455. Prepared by Oak Ridge National Laboratory, Oak Ridge, TN. Available at http://info.ornl.gov/sites/publications/files/Pub39382.pdf.

Rogovin, M., G. T. Frampton, and the Nuclear Regulatory Commission Special Inquiry Group. 1980. Three Mile Island: A Report to the Commissioners and to the Public, Vols. I-II. NUREG/CR-1250. Washington, DC: Rogovin, Stern and Huge.

Roth, E. M., A. Mosleh, Y. J. Chang, R. Richards, D. Bley, S. H. Shen, and A. Zoulis. 2012. Model-based framework for characterizing contextual factors for HRA applications. Presentation at the 11th International Probabilistic Safety Assessment and Management Conference and the annual European Safety and Reliability Conference, Helsinki, Finland, June 25-29, 2012.

Sakai, T., T. Takeda, H. Soraoka, K. Yanagisawa, and T. Annaka. 2006. Development of a probabilistic tsunami hazard analysis in Japan. In Proceedings of ICONE14 International Conference on Nuclear Engineering, July 17-20, Miami, FL. ICONE14-89183. New York: American Society of Mechanical Engineers. Available at http://jsmillerdesign.com/FukushimaPapers/Development%20of%20a%20Probalistic%20Tsunami%20 Hazard%20Analysis%20in%20Japan.pdf.

Salame-Alfie, A., P. Mulligan, F. Fisher-Tyler, R. McBurney, and E. Fordham. 2012. Fukushima disaster response—the states' perspective. Health Physics 102(5): 580-583.

Sarter, N. B., D. D. Woods, and C. E. Billings. 1997. Automation surprises. Pp. 1926-1943 in Handbook of Human Factors and Ergonomics, 2nd Ed., G. Salvendy, ed. New York: John Wiley and Sons, Inc.

Sasakawa, Y. 2012. International expert symposium in Fukushima: Radiation and Health Risks. Journal of Radiological Protection 32(1): E7-E8.

Satake, K., Y. Namegaya, and S. Yamaki. 2008. Numerical simulation of the AD 869 Jogan tsunami in Ishinomaki and Sendai plains (in Japanese with English abstract). Annual Report of Active Fault and Paleoearthquake Researches 8: 71-89.

Satake, K., Y. Fujii, T. Harada, and Y. Namegaya. 2013. Time and space distribution of coseismic slip of the 2011 Tohoku Earthquake as inferred from tsunami waveform data. Bulletin of the Seismological Society of America 103(2B): 1473-1492.

Schaede, U. 1995. The "Old Boy" network and government-business relationships in Japan. Journal of Japanese Studies 21(2): 293-317. Available at http://www.jstor.org/stable/133010.

SCJ (Science Council of Japan). 2011. Report to the Foreign Academies from Science Council of Japan on the Fukushima Daiichi Nuclear Power Plant Accident. Tokyo: SCJ. May 2.

Sehgal, B. R., ed. 2012. Nuclear Safety in Light Water Reactors: Severe Accident Phenomenology. Waltham, MA: Academic Press.

Shiroyama, H. 2012. Regulatory Failures of Nuclear Safety in Japan—The Case of Fukushima Accident. Available at http://tokyo2013.earthsystemgovernance.org/wp-content/uploads/2013/01/0202-SHIROYAMA_Hideaki-.pdf.

Shneider, M. N., and R. B. Miles. 2005. Microwave diagnostics of small plasma objects. Journal of Applied Physics 98: 033301.

Silva, K., Y. Ishiwatari, and S. Takahara. 2013. Cost per severe accident as an index for severe accident consequence assessment and its applications. Reliability Engineering and System Safety 123(2014): 110-122.

Siu, N., D. Marksberry, S. Cooper, K. Coyne, and M. Stutzke. 2013. PSA technology challenges revealed by the Great East Japan Earthquake. Presentation at the Probabilistic Safety Assessment and Management Topical Conference in Light of the Fukushima Dai-Ichi Accident, Tokyo, Japan, April 15-17, 2013. Available at http://pbadupws.nrc.gov/docs/ML1303/ML13038A203.pdf.

Smith, D. E., S. Shi, R. A. Cullingford, A. G. Dawson, S. Dawson, C. R. Firth, I. D. L. Foster, P. T. Fretwell, B. A. Haggart, L. K. Holloway, and D. Long. 2004. The Holocene Storegga Slide tsunami in the United Kingdom. Quaternary Science Reviews 23(23-34): 2291-2321.

Stein, S., and E. A. Okal. 2007. Ultralong period seismic study of the December 2004 Indian Ocean earthquake and implications for regional tectonics and the subduction process. Bulletin of the Seismological Society of America 97: S279-S295.

Stein, S., R. J. Geller, and M. Liu. 2012. Why earthquake hazard maps often fail and what to do about it. Tectonophysics 562-563: 1-25.

Stein, S., R. J. Geller, and M. Liu. 2013. Reply to comment by Arthur Frankel on "Why earthquake hazard maps often fail and what to do about it." Tectonophysics 592: 207-209.

Steinhauser, G., A. Brandl, and T. E. Johnson. 2014. Comparison of the Chernobyl and Fukushima nuclear accidents: A review of the environmental impacts. Science of the Total Environment 470-471: 800-817.

Sugawara, D., F. Imamura, K. Goto, H. Matsumoto, and K. Minoura. 2013. The 2011 Tohoku-oki Earthquake Tsunami: Similarities and differences to the 869 Jogan Tsunami on the Sendai Plain. Pure and Applied Geophysics 170(5): 831-843.

Suzuki, W., S. Aoi, H. Sekiguchi, and T. Kunugi. 2011. Rupture process of the 2011 Tohoku-Oki mega-thrust earthquake (M9.0) inverted from strong-motion data. Geophysical Research Letters 38(7): L00G16, doi: 10.1029/2011GL049136.

Tajima, F., J. Mori, and B. L. N. Kennett. 2013. A review of the 2011 Tohoku-Oki earthquake (Mw 9.0): Large-scale rupture across heterogeneous plate coupling. Tectonophysics 586: 15-34.

Tanigawa, K., Y. Hosoi, N. Hirohashi, Y. Iwasaki, and K. Kamiya. 2012. Loss of life after evacuation: Lessons learned from the Fukushima accident. Lancet 379(9819): 889-891.

Task Force for the Reform of Nuclear Safety Regulations and Organisations. n.d. Reform of Nuclear Regulation Organisation and System in Japan. Tokyo: Cabinet Secretariat, Government of Japan. Available at http://www.nsr.go.jp/archive/nisa/shingikai/700/14/240724/AT-6-1.pdf.

Tateno, S., and H. M. Yokoyama. 2013. Public anxiety, trust, and the role of mediators in communicating risk of exposure to low dose radiation after the Fukushima Daiichi Nuclear Plant explosion. Journal of Science Communication 12(2): A03.

Taylor, J. M. 1996. Memorandum from James M. Taylor, Executive Director of Operations, dated April 29, 1996, to the Commissioners Regarding: Status of the Integration Plan for Closure of Severe Accident Issues and the Status of Severe Accident Research. SECY-96-088. Rockville, MD: U.S. Nuclear Regulatory Commission. Available at http://www.nrc.gov/reading-rm/doc-collections/commission/secys/1996/secy1996-088/1996-088scy.pdf.

ten Brink, U. 2009. Tsunami hazard along the U.S. Atlantic Coast. Marine Geology 264(1-2): 1-3.

ten Brink, U. S., H. J. Lee, E. L. Geist, and D. Twichell. 2009. Assessment of tsunami hazard to the U.S. East Coast using relationships between submarine landslides and earthquakes. Marine Geology 264(1-2): 65-73.

Ten Hoeve, J. E., and M. Z. Jacobson. 2012. Worldwide health effects of the Fukushima Daiichi nuclear accident. Energy and Environmental Science 5: 8743-8757.

TEPCO. 2011a. Fukushima Nuclear Accident Analysis Report (Interim Report). Tokyo: Tokyo Electric Power Company, Inc. December 2. Available at http://www.tepco.co.jp/en/press/corp-com/release/betu11_e/images/111202e14.pdf.

TEPCO. 2011b. Fukushima Nuclear Accident Investigation Report: Interim Report—Supplemental Volume. Tokyo, Japan: Tokyo Electric Power Company, Inc. December 2.

TEPCO. 2011c. Mid-and-Long-Term Roadmap Towards the Decommissioning of Fukushima Daiichi Nuclear Power Units 1-4 (December 2011). Available at http://www.meti.go.jp/english/earthquake/nuclear/decommissioning/.

TEPCO. 2011d. The Nuclear Safety and Quality Assurance Meeting's Accident Investigation Examination Committee's Opinion of the Tokyo Electric Power Company's "Fukushima Nuclear Accident Investigation Report (Midterm Report)." Available at http://www.tepco.co.jp/en/press/corp-com/release/betu11_e/images/111202e15.pdf.

TEPCO. 2012a. Estimation of the Released Amount of Radioactive Materials into the Atmosphere as a Result of the Accident in the Fukushima Daiichi Nuclear Power Station. Tokyo: Tokyo Electric Power Company, Inc. March.

TEPCO. 2012b. Fukushima Nuclear Accident Analysis Report. Available at http://www.tepco.co.jp/en/nu/fukushima-np/interim/index-e.html. Tokyo: Tokyo Electric Power Company, Inc. June 20.

TEPCO. 2012c. MAAP Code-Based Analysis of the Development of the Events at the Fukushima Daiichi Nuclear Power Station. Tokyo: Tokyo Electric Power Company, Inc. March.

TEPCO. 2013. Evaluation of the Situation of Cores and Containment Vessels of Fukushima Daiichi Nuclear Power Station Units-1 to 3 and Examination into Unsolved Issues in the Accident Progression: Progress Report No. 1 (December). Available at http://www.tepco.co.jp/en/press/corp-com/release/betu13_e/images/131213e0101.pdf.

TEPCO. 2014. New Comprehensive Special Business Plan. January 15. Available at http://www.tepco.co.jp/en/press/corp-com/release/betu14_e/images/140115e0205.pdf.

Tohoku (Tohoku Electric Power Company). 2012. Safe shutdown of Onagawa Nuclear Power Station: The closest BWRs to the 3/11/11 epicenter. Presentation to the Committee on Lessons Learned from the Fukushima Nuclear Accident for Improving Safety and Security of U.S. Nuclear Plants, Tokyo, Japan, November 29, 2012. Honcho, Japan: Tohoku Electric Power Company.

Tominaga, T., M. Hachiya, H. Tatsuzaki, and M. Akashi. 2014. The accident at the Fukushima Daiichi Nuclear Power Plant in 2011. Health Physics 106(6): 630-637.

Tupin, E. A., M. A. Boyd, J. E. Mosser, and J. S. Wieder. 2012. U.S. EPA response to the Fukushima Daiichi Nuclear Power Plant accident. Health Physics 102(5): 563-569.

UCS (Union of Concerned Scientists). 2011a. Statement by David Lochbaum—Director of Nuclear Safety Project, before the U.S. Senate Energy and Natural Resources Committee, March 29. Available at http://www.ucsusa.org/assets/documents/nuclear_power/lochbaum-senate-energy-3-29-2011.pdf.

UCS. 2011b. The NRC and Nuclear Power Plant Safety in 2010: A Brighter Spotlight Needed. March. Cambridge, MA: UCS Publications. Available at http://www.ucsusa.org/assets/documents/nuclear_power/nrc-2010-full-report.pdf.

UCS. 2012. The NRC and Nuclear Power Plant Safety in 2011: Living on Borrowed Time. Cambridge, MA. UCS Publications. Available at http://www.ucsusa.org/assets/documents/nuclear_power/nrc-nuclear-safety-2011-full-report.pdf.

UCS. 2013a. NRC's Failure to Enforce Reactor Fire Regulations. Cambridge, MA: Union of Concerned Scientists. Available at www.ucsusa.org/assets/documents/nuclear_power/ucs-nrc-fire-regulations-5-2-13.pdf.

UCS. 2013b. The NRC and Nuclear Power Plant Safety in 2012: Tolerating the Intolerable. March. Cambridge, MA. UCS Publications. Available at http://www.ucsusa.org/assets/documents/nuclear_power/NRC-nuclear-safety-2012-report.pdf.

UNSCEAR (United Nations Scientific Committee on the Effects of Atomic Radiation). 2011. Sources and Effects of Ionizing Radiation: UNSCEAR 2008 Report to the General Assembly, Volume II with Scientific Annexes C, D, and E. New York: United Nations. Available at www.unscear.org/docs/reports/2008/11-80076_Report_2008_Annex_D.pdf.

UNSCEAR. 2013a. Effects of radiation exposure of children. Annex B in Sources, Effects and Risks of Ionizing Radiation, UNSCEAR 2013 Report, Volume II. New York: United Nations.

UNSCEAR. 2013b. Levels and effects of radiation exposure due to the nuclear accident after the 2011 Great East-Japan Earthquake and tsunami. Scientific Annex A in Sources, Effects, and Risks of Ionizing Radiation, UNSCEAR 2013 Report, Volume I, Report to the General Assembly. New York: United Nations. Available at http://www.unscear.org/docs/reports/2013/13-85418_Report_2013_Annex_A.pdf.

USDHS (U.S. Department of Homeland Security). 2013. National Response Framework, 2nd ed. Washington, DC: USDHS. May.

U.S. Embassy (Embassy of the United States, Tokyo, Japan). 2011. Travel Warning. Available at japan.usembassy.gov/e/acs/tacs-travel20110317.html. Washington, DC: U.S. Department of State.

USEPA (U.S. Environmental Protection Agency). 1992. Manual of Protective Action Guides and Protective Actions for Nuclear Incidents (Revised 1991). EPA 400-R-92-001. Washington, DC: USEPA.

USEPA. 2013. PAG Manual: Protective Action Guides and Planning Guidance for Radiological Incidents (Draft for Interim Use and Public Comment). Washington, DC: USEPA.

USFDA (U.S. Food and Drug Administration). 1998. Accidental Radioactive Contamination of Human Food and Animal Feeds: Recommendations for State and Local Agencies. UCM094513. Washington, DC: U.S. Department of Health and Human Services. August 13.

USFDA. 2004. Supporting Document for Guidance Levels for Radionuclides in Domestic and Imported Foods. UCM078341. Washington, DC: U.S. Department of Health and Human Services.

USGAO (U.S. General Accounting Office). 1983. Response to Specific Questions on the Indian Point Probabilistic Safety Study. GAO/RCED-83-158. Washington, DC: U.S. General Accounting Office. Available at http://gao.gov/assets/150/140084.pdf.

USGAO (U.S. Government Accountability Office). 2008. Report to Congressional Requesters—Nuclear Safety: NRC's Oversight of Fire Protection at U.S. Commercial Nuclear Reactor Units Could Be Strengthened. GAO-08-747. Washington, DC: U.S. Government Accountability Office. Available at www.gao.gov/new.items/d08747.pdf.

USGAO. 2012. Report to Congressional Requesters—Nuclear Regulatory Commission: Oversight and Status of Implementing a Risk-Informed Approach to Fire Safety. GAO-13-8. Washington, DC: U.S. Government Accountability Office. Available at www.gao.gov/assets/650/649658.pdf.

USGAO. 2014. Report to the Chairman, Subcommittee on Transportation and Infrastructure, Committee on Environment and Public Works, U.S. Senate—Nuclear Safety: Countries' Regulatory Bodies Have Made Changes in Response to the Fukushima Daiichi Accident. GAO-14-109. Washington, DC: U.S. Government Accountability Office. Available at www.gao.gov/assets/670/661464.pdf.

USNRC (U.S. Nuclear Regulatory Commission). 1975. Reactor Safety Study: An Assessment of Accident Risks in U.S. Commercial Nuclear Power Plants, WASH-1400, NUREG-75/014. Washington, DC: U.S. Nuclear Regulatory Commission. Available at http://www.nrc.gov/reading-rm/doc-collections/nuregs/staff/sr75-014/.

USNRC. 1980a. Clarification of TMI Action Plan Requirements: Final Report. NUREG-0737. Washington, DC: U.S. Nuclear Regulatory Commission. Available at www.nrc.gov/reading-rm/doc-collections/nuregs/staff/sr0737/final/index.html.

USNRC. 1980b. Functional Criteria for Emergency Response Facilities. NUREG-0696. Washington, DC: U.S. Nuclear Regulatory Commission. Available at http://www.nrc.gov/reading-rm/doc-collections/nuregs/staff/sr0696/sr0696.pdf.

USNRC. 1980c. NRC Action Plan Developed as a Result of the TMI-2 Accident, Vols. 1 and 2. NUREG-0660. Washington, DC: U.S. Nuclear Regulatory Commission. Available at http://pbadupws.nrc.gov/docs/ML0724/ML072470526.pdf (Vol. 1) and http://pbadupws.nrc.gov/docs/ML0724/ML072470524.pdf (Vol. 2).

USNRC. 1980d. Proposed rule: Severe accident design criteria. Federal Register 45(Oct. 2): 65474.

USNRC. 1980e. TMI-2 Lessons Learned Task Force Status Report and Short-Term Recommendations. NUREG-0578. Washington, DC: U.S. Nuclear Regulatory Commission. Available at http://pbadupws.nrc.gov/docs/ML0900/ML090060030.pdf.

USNRC. 1982. Guidelines for the Preparation of Emergency Operating Procedures: Resolution of Comments on NUREG-0799. NUREG-0899. Washington, DC: U.S. Nuclear Regulatory Commission. Available at http://pbadupws.nrc.gov/docs/ML1025/ML102560007.pdf.

USNRC. 1983. Clarification of TMI Action Plan Requirements: Requirements for Emergency Response Capability. NUREG-0737, Supplement No. 1. Washington, DC: U.S. Nuclear Regulatory Commission. Available at http://www.nrc.gov/reading-rm/doc-collections/nuregs/staff/sr0737/sup1/sr0737sup1.pdf.

USNRC. 1985. Policy statement on severe reactor accidents regarding future designs and existing plants. Federal Register 50(153): 32138. Available at http://pbadupws.nrc.gov/docs/ML0037/ML003711521.pdf.

USNRC. 1986. Safety goals for the operation of nuclear power plants; policy statement; republication. Federal Register 51(Aug. 21): 30028. Available at http://www.nrc.gov/reading-rm/doc-collections/commission/policy/51fr30028.pdf.

USNRC. 1987. Severe Accident Risks: An Assessment for Five U.S. Nuclear Power Plants—Draft Report. NUREG-1150. Washington, DC: U.S. Nuclear Regulatory Commission.

USNRC. 1989. Policy statement on the conduct of nuclear power plant operations. Federal Register 54: 3424. Available at http://www.nrc.gov/reading-rm/doc-collections/commission/policy/54fr3424.pdf.

USNRC. 1990. Severe Accident Risks: An Assessment for Five U.S. Nuclear Power Plants. NUREG-1150. Washington, DC: U.S. Nuclear Regulatory Commission. Available at http://www.nrc.gov/reading-rm/doc-collections/nuregs/staff/sr1150/.

USNRC. 1995. Use of Probabilistic Risk Assessment Methods in Nuclear Regulatory Activities: Final Policy Statement. Federal Register 60(158): 42622-42629. Available at http://www.gpo.gov/fdsys/pkg/FR-1995-08-16/pdf/95-20237.pdf.

USNRC. 1997a. Individual Plant Examination Program: Perspectives on Reactor Safety and Plant Performance, Final Report, Vols. 1-5. NUREG-1560. Washington, DC: U.S. Nuclear Regulatory Commission.

USNRC. 1997b. Regulatory Analysis Technical Evaluation Handbook, Final Report. NUREG/BR-0184. Rockville, MD: U.S. Nuclear Regulatory Commission. Available at http://pbadupws.nrc.gov/docs/ML0501/ML050190193.pdf.

USNRC. 1998. Peer Review for High-Level Waste Repositories. NUREG-1297. Rockville, MD: U.S. Nuclear Regulatory Commission.

USNRC. 2000a. Letter from Richard A. Meserve, Chairman of U.S. Nuclear Regulatory Commission, to Edward J. Markey, U.S. House of Representatives, dated March 9, in response to letter of November 10, 1999, on potassium iodide stockpiles. Washington, DC: U.S. Nuclear Regulatory Commission. Available at http://pbadupws.nrc.gov/docs/ML0036/ML003693033.pdf.

USNRC. 2000b. Memorandum from Annette L. Vietti-Cook, Secretary of the Commission, dated December 22, 2000, Regarding: SECY-00-0037—Status of Potassium Iodide Activities and SECY-00-0040—Final Amendments to 10 CFR 50.47. Rockville, MD: U.S. Nuclear Regulatory Commission. Available at http://pbadupws.nrc.gov/docs/ML0037/ML003780292.pdf.

USNRC. 2000c. Technical Basis and Implementation Guidelines for a Technique for Human Event Analysis (ATHEANA). NUREG-1624, Rev. 1. Washington, DC: U.S. Nuclear Regulatory Commission. Available at http://www.nrc.gov/reading-rm/doc-collections/nuregs/staff/sr1624/r1/.

USNRC. 2001. Commission Voting Record, Decision Item: SECY-02-0069—Status of Potassium Iodide Activities. Rockville, MD: U.S. Nuclear Regulatory Commission. June 29. Available at http://pbadupws.nrc.gov/docs/ML0118/ML011860293.pdf.

USNRC. 2002a. Commission Voting Record, Decision Item: SECY-02-0089—Revised Draft NUREG-1633 and Public Information Brochure on Potassium Iodide (KI) for the General Public. Rockville, MD: U.S. Nuclear Regulatory Commission. November 26. Available at http://www.nrc.gov/reading-rm/doc-collections/commission/secys/2002/secy2002-0089/2002-0089scy.pdf.

USNRC. 2002b. Federal policy on use of potassium iodide. Federal Register 67(7): 1355.

USNRC. 2002c. Perspectives Gained from the Individual Plant Examination of External Events (IPEEE) Program—Final Report, Volume 1. NUREG-1742. Rockville, MD: U.S. Nuclear Regulatory Commission. Available at http://pbadupws.nrc.gov/docs/ML0212/ML021270070.pdf.

USNRC. 2005. NRC Regulatory Issue Summary 2005-18: Guidance for Establishing and Maintaining a Safety Conscious Work Environment. Rockville, MD: U.S. Nuclear Regulatory Commission. Available at http://www.nrc.gov/reading-rm/doc-collections/gen-comm/reg-issues/2005/ri200518.pdf.

USNRC. 2006. Regulatory Guide 1.97: Criteria for the Accident Monitoring Instrumentation for Nuclear Power Plants, Revision 4. Available at http://pbadupws.nrc.gov/docs/ML0615/ML061580448.pdf.

USNRC. 2007. Regulatory Guide 1.23: Meteorological Monitoring Programs for Nuclear Power Plants, Rev. 1. Rockville, MD: U.S. Nuclear Regulatory Commission. Available at http://pbadupws.nrc.gov/docs/ML0703/ML070350028.pdf.

USNRC. 2009a. Power reactor security requirements: Final rule. Federal Register 74 (Mar. 27): 13926-13993.

USNRC. 2009b. Regulatory Guide 1.200, Rev. 2: An Approach for Determining the Technical Adequacy of Probabilistic Risk Assessment Results for Risk-Informed Activities. Rockville, MD: U.S. Nuclear Regulatory Commission. Available at http://pbadupws.nrc.gov/docs/ML0904/ML090410014.pdf.

USNRC. 2010. The Evolution of Mitigating Measures for Large Fire and Explosions: A Chronological History from September 11, 2001 Through October 7, 2009 (summary and detailed chronology). Available at http://pbadupws.nrc.gov/docs/ML0929/ML092990417.pdf.

USNRC. 2011a. Developing an Emergency Risk Communication (ERC)/Joint Information Center (JIC) Plan for a Radiological Emergency. NUREG/CR-7032. Prepared by Center for Risk Communication, New York. Rockville, MD: U.S. Nuclear Regulatory Commission. Available at http://pbadupws.nrc.gov/docs/ML1104/ML110490119.pdf.

USNRC. 2011b. Final safety culture policy statement. Federal Register 76: 34773. Available at http://www.gpo.gov/fdsys/pkg/FR-2011-06-14/html/2011-14656.htm.

USNRC. 2011c. Guidance on Developing Effective Radiological Risk Communication Messages: Effective Message Mapping and Risk Communication with the Public in Nuclear Plant Emergency Planning Zones. NUREG/CR-7033. Prepared by Center for Risk Communication, New York. Rockville, MD: U.S. Nuclear Regulatory Commission. Available at http://pbadupws.nrc.gov/docs/ML1104/ML110490120.pdf.

USNRC. 2011d. NRC Provides Protective Action Recommendations Based on U.S. Guidelines. NRC News No. 11-050, March 16. Rockville, MD: U.S. Nuclear Regulatory Commission. Available at www.nrc.gov/reading-rm/doc-collections/news/2011/11-050.pdf.

USNRC. 2011e. NRC sees no radiation at harmful levels reaching U.S. from damaged Japanese nuclear power plants (revised). NRC News No. 11-046, March 13. Rockville, MD: U.S. Nuclear Regulatory Commission. Available at http://pbadupws.nrc.gov/docs/ML1107/ML110720002.pdf.

USNRC. 2011f. Supplemental Inspection for Repetitive Degraded Cornerstones, Multiple Degraded Cornerstones, Multiple Yellow Inputs, or One Red Input. NRC Inspection Procedure 95003. Rockville, MD: U.S. Nuclear Regulatory Commission. Available at http://www.nrc.gov/reading-rm/doc-collections/insp-manual/inspection-procedure/ip95003.pdf.

USNRC. 2012a. A Proposed Risk Management Regulatory Framework: A Report to NRC Chairman Gregory B. Jaczko from the Risk Management Task Force. NUREG-2150. Rockville, MD: U.S. Nuclear Regulatory Commission. Available at http://pbadupws.nrc.gov/docs/ML1210/ML12109A277.pdf.

USNRC. 2012b. Boiling water reactor (BWR) systems. Chapter 3 in Nuclear Power for Electrical Generation (Reactor Concepts Manual). USNRC Technical Training Center. Available at http://pbadupws.nrc.gov/docs/ML1209/ML120970422.pdf.

USNRC. 2012c. Compliance with Order EA-12-049, Order Modifying Licenses with Regard to Requirements for Mitigation Strategies for Beyond-Design-Basis External Events. Interim Staff Guidance (ISG) Japan Lessons-Learned Project Directorate (JLD). JLD-ISG-2012-01. Available at http://pbadupws.nrc.gov/docs/ML1214/ML12146A014.pdf.

USNRC. 2012d. Letter and Order Modifying Licenses with Regard to Requirements for Mitigation Strategies for Beyond-Design-Basis External Events. EA-12-049. Rockville, MD: U.S. Nuclear Regulatory Commission. Available at http://pbadupws.nrc.gov/docs/ML1205/ML12054A735.pdf.

USNRC. 2012e. Long-term cooling and unattended water makeup of spent fuel pools. Federal Register 77(243): 74778-74798.

USNRC. 2013a. A Comparison of U.S. and Japanese Regulatory Requirements in Effect at the Time of the Fukushima Accident. Rockville, MD: U.S. Nuclear Regulatory Commission. Available at http://pbadupws.nrc.gov/docs/ML1332/ML13326A991.pdf.

USNRC. 2013b. Information Digest, 2013-2014, Vol. 25. NUREG 1350. Washington, DC: U.S. Nuclear Regulatory Commission, Office of Public Affairs. Available at www.nrc.gov/reading-rm/doc-collections/nuregs/staff/sr1350/.

USNRC. 2013c. State-of-the-Art Reactor Consequence Analyses Project, Vol. 1: Peach Bottom Integrated Analysis. NUREG/CR-7110, Rev. 1. Prepared by Sandia National Laboratories. Rockville, MD: U.S. Nuclear Regulatory Commission. Available at http://pbadupws.nrc.gov/docs/ML1202/ML120260675.pdf.

USNRC. 2013d. State-of-the-Art Reactor Consequence Analyses Project, Vol. 2: Surry Integrated Analysis. NUREG/CR-7110, Revision 1. Prepared by Sandia National Laboratories. Rockville, MD: U.S. Nuclear Regulatory Commission. Available at http://www.nrc.gov/reading-rm/doc-collections/nuregs/contract/cr7110/v2/.

USNRC NTTF (U.S. Nuclear Regulatory Commission Near-Term Task Force). 2011. Recommendations for Enhancing Reactor Safety in the 21st Century: The Near-Term Task Force Review of Insights from the Fukushima Dai-Ichi Accident. Rockville, MD: U.S. Nuclear Regulatory Commission. Available at http://pbadupws.nrc.gov/docs/ML1118/ML111861807.pdf.

USNRC and USEPA (U.S. Nuclear Regulatory Commission and U.S. Environmental Protection Agency Task Force on Emergency Planning). 1978. Planning Basis for the Development of State and Local Government Radiological Emergency Response Plans in Support of Light Water Nuclear Power Plants. NUREG-0396/EPA 520/1-78-016. Washington, DC: USNRC and USEPA. Available at http://pbadupws.nrc.gov/docs/ML0513/ML051390356.pdf.

USNRC and EPRI (U.S. Nuclear Regulatory Commission and Electric Power Research Institute). 2012. EPRI/NRC-RES Fire Human Reliability Analysis Guidelines. NUREG-1921, EPRI 1023001. Washington, DC: U.S. Nuclear Regulatory Commission; Palo Alto, CA: Electric Power Research Institute. Available at http://pbadupws.nrc.gov/docs/ML1221/ML12216A104.pdf.

USNRC and FEMA (U.S. Nuclear Regulatory Commission and Federal Emergency Management Agency). 1980. Criteria for Preparation and Evaluation of Radiological Emergency Response Plans and Preparedness in Support of Nuclear Power Plants. NUREG-0654/FEMA-REP-1, Rev. 1. Washington, DC: USNRC and FEMA. Available at http://pbadupws.nrc.gov/docs/ML0404/ML040420012.pdf.

Walker, J. S. 2004. Three Mile Island: A Nuclear Crisis in Historical Perspective. Berkeley: University of California Press.

Wall, J., R. Elms, and D. Nock. 2004. Developing an incident management simulation for training emergency responders. Paper 1865. Presentation at the Interservice/Industry Training, Simulation, and Education Conference (I/ITSEC), Orlando, Florida, May 15, 2004.

Wang, Q., and X. Chen. 2012. Regulatory failures for nuclear safety—the bad example of Japan—implication for the rest of the world. Renewable and Sustainable Energy Reviews 16(5): 2610-2617.

Wang, Q., X. Chen, X. Yi-chong. 2013. Accident like the Fukushima unlikely in a country with effective nuclear regulation: Literature review and proposed guidelines. Renewable and Sustainable Energy Reviews 17: 126-146.

Whaley, A. M., J. Xing, R. Boring, S. Hendrickson, J. Joe, and K. Le Blanc. 2012. Building a Psychological Foundation for Human Reliability Analysis (Draft). NUREG-2114, INL/EXT-11-23898. Available at http://pbadupws.nrc.gov/docs/ML1131/ML113180490.pdf.

WHO (World Health Organization). 2013. Health risk assessment from the nuclear accident after the 2011 Great East Japan Earthquake and Tsunami based on a preliminary dose estimation. Geneva, Switzerland: WHO.

Williamson, B., J. Lyter, D. Roniger, and P. Ellison. 2013. Revising BWR emergency procedures: The BWR Owners' Group has completed a revision of emergency procedure guidelines in light of lessons from Fukushima. Nuclear Energy International, November 18. Available at www.neimagazine.com/features/featurerevising-bwr-emergency-procedures/.

Wise, C. R. 2006. Organizing for Homeland Security After Katrina: Is Adaptive Management What's Missing? Public Administration Review 66(3): 302-318.

WNA (World Nuclear Association). 2014. Fukushima Accident. London, UK: World Nuclear Association. Available at www.world-nuclear.org/info/Safety-and-Security/Safety-of-Plants/Fukushima-Accident/.

World Energy Council. 2012. World Energy Perspective: Nuclear Energy One Year After Fukushima. London, UK: World Energy Council. Available at http://www.worldenergy.org/wp-content/uploads/2012/10/PUB_world_energy_perspective__nuclear_energy_one_year_after_fukushima_2012_WEC.pdf.

World Travel and Tourism Council. 2011. The Tohoku Pacific Earthquake and Tsunami. December Update. Available at http://www.google.com/url?sa=t&rct=j&q=&esrc=s&source= web&cd=2&ved=0CCUQFjAB&url=http%3A%2F%2Fwww.wttc.org%2F~%2Fmedia%2Ffiles%2Freports%2Fspecial%2520and%2520periodic%2520reports%2Fjapan_report_dec_update_rgb_1.ashx&ei=tjXxU5e2LY3-yQS-iIHYBg&usg=AFQjCNHj0JHQEZfLtmywx2W_x4_0P0gZfg&sig2=Mkt3NyoNuwAZYXIf3IDJiw&bvm=bv.73231344,d.aWw.

Yamanaka, Y. 2012. Research plan regarding improvement of simulation code for understanding the status of fuel debris in the reactor. Presentation at the International Symposium on the Decommissioning of TEPCO's Fukushima Daiichi Nuclear Power Plant Unit 1-4, March 14, 2012, Tokyo, Japan. Tokyo: Minister of Economy, Trade, and Industry.

Yamazaki, Y., K. F. Cheung, and T. Lay. 2013. Modeling of the 2011 Tohoku near-field tsunami from finite-fault inversion of seismic waves. Bulletin of the Seismological Society of America 103(2B): 1444-1455.

Yasumura, S., M. Hosoya, S. Yamashita, K. Kamiya, M. Abe, M. Akashi, K. Kodama, and K. Ozasa. 2012. Study protocol for the Fukushima health management survey. Journal of Epidemiology 22(5): 375-383.

Yasumura, S., A. Goto, S. Yamazaki, and M.R. Reich. 2013. Excess mortality among relocated institutionalized elderly after the Fukushima nuclear disaster. Public Health 127(2):186-8.

Yasutaka, T., W. Naito, and J. Nakanishi. 2013. Cost and effectiveness of decontamination strategies in radiation contaminated areas in Fukushima in regard to external radiation dose. PLOS ONE 8(9): e75308.

Yue, H., and T. Lay. 2013. Source rupture models for the M_w 9.0 2011 Tohoku Earthquake from joint inversions of high-rate geodetic and seismic data. Bulletin of the Seismological Society of America 103(2B): 1242-1255.

Zitellini, N., F. Chierici, R. Sartori, and L. Torelli. 1999. The tectonic source of the 1755 Lisbon earthquake and tsunami. Annals of Geophysics 42(1): 49-55.

Appendix A

Biographical Sketches of Committee, Technical Advisor, and Staff

Norman P. Neureiter, Ph.D., *Chair*, is a senior advisor at the Center for Science Diplomacy and acting director of the Center for Science, Technology, and Security Policy of the American Association for the Advancement of Science (AAAS). He is also the U.S. co-chair of the Indo-U.S. Science and Technology Forum—an organization created by the two governments for furthering scientific cooperation between the United States and India. Dr. Neureiter received a B.A. in chemistry from the University of Rochester in 1952 and a Ph.D. in organic chemistry from Northwestern University in 1957. In 1957, he joined Humble Oil and Refining (now part of Exxon) in Baytown, Texas, as a research chemist, also teaching German and Russian at the University of Houston. He joined the International Affairs Office of the U.S. National Science Foundation in 1963 and managed the newly established U.S.-Japan Cooperative Science Program. Entering the U.S. Foreign Service in 1965, he was named deputy scientific attache at the U.S. Embassy in Bonn. In 1967, he was transferred to Warsaw as the first U.S. Scientific Attache in Eastern Europe with responsibility for Poland, Hungary, and Czechoslovakia. Dr. Neureiter returned to Washington in 1969 as assistant for international affairs to the President's science advisor in the White House Office of Science and Technology. He left the government in 1973 and joined Texas Instruments (TI), where he held a number of staff and management positions including vice president of TI Asia, based in Tokyo from 1989 to 1994. After retirement from TI in 1996, he worked as a consultant until being appointed as the first science and technology adviser to the U.S. Secretary of State in September 2000. Finishing the 3-year assignment in 2003, he was made a Distinguished Presidential

Fellow for International Affairs at the U.S. National Academy of Sciences. In 2008, he was elected a fellow of the American Academy of Arts and Sciences, and also received the Public Welfare Medal of the National Academy of Sciences. He received the Order of the Rising Sun–Gold and Silver Star from the Emperor of Japan in 2010, for fostering scientific cooperation between the United States and Japan. He speaks German, Russian, Polish, French, Spanish and Japanese.

B. John Garrick, Ph.D., *Vice Chair*, was appointed as chairman of the U.S. Nuclear Waste Technical Review Board by President George W. Bush in 2004. Dr. Garrick is an executive consultant on the application of the risk sciences to complex technological systems in the space, defense, chemical, marine, transportation, and nuclear fields. His areas of expertise include risk assessment and nuclear science and engineering. He served for 10 years (1994-2004), 4 years as chair, on the U.S. Nuclear Regulatory Commission's Advisory Committee on Nuclear Waste. A founder of the firm PLG, Inc., Dr. Garrick retired as president, chairman, and chief executive officer in 1997. Before PLG's acquisition and integration into a new firm, it was an international engineering, applied science, and management consulting firm. Dr. Garrick received his Ph.D. in engineering and applied science and an M.S. in nuclear engineering from the University of California, Los Angeles, and a B.S. in physics from Brigham Young University. He is past president of the Society for Risk Analysis (1989-1990) and recipient of that Society's most prestigious award, the Distinguished Achievement Award. Dr. Garrick was elected to the National Academy of Engineering in 1993. He has been a member and chair of several National Research Council committees and recently chaired the National Academy of Engineering Committee on Combating Terrorism. He is a member of the first class of lifetime National Associates of the National Academies.

Robert A. Bari, Ph.D., is senior physicist and senior advisor at Brookhaven National Laboratory. He has been involved in the design and safety assessments of complex, high-technology facilities since he joined the applied programs at the laboratory in 1974. He has worked on projects and issues regarding nuclear safety and nonproliferation technologies, nuclear waste management, and development of advanced nuclear reactors and has directed numerous studies of advanced nuclear energy concepts. Dr. Bari is currently international co-chairman of the working group that has developed a comprehensive methodology for evaluation of proliferation resistance and physical protection of all new nuclear energy concepts being proposed within the multinational Generation IV International Forum. He continues to be a frequent delegate to the International Atomic Energy Agency in Vienna and has participated in several programs for the Organisation for Economic Co-operation and Development in Paris. He received his Ph.D. in physics from Brandeis University and his B.S. in physics from Rutgers University. He has served as an adjunct faculty member and advisor to several major universities in the field of nuclear technology as well as on the board of directors of the American Nuclear Society. He is also past president of the International Association for Probabilistic Safety Assessment and Management and past chairman of the American Nuclear Society (ANS) Consensus Standards Committee for Probabilistic Risk Assessment. For his achievements in nuclear safety, Dr. Bari was awarded the Theo J. "Tommy" Thompson Award in 2003 by ANS. In 2004, he received the Brookhaven National Laboratory Award for Outstanding Achievement in Science and Technology. Dr. Bari was awarded membership in the Phi Beta Kappa, Sigma Xi, and Sigma Pi Sigma honor societies and is an elected fellow of ANS and of the American Physical Society.

Percy (Pat) M. Beard, Jr., Ph.D., is a retired nuclear utility executive. He graduated from the Naval Academy in 1958 and received a Ph.D. in nuclear physics in 1964 from Duke University under a special Navy postgraduate program. Dr. Beard then entered the nuclear submarine program and served on five submarines including command of the Francis Scott Key. Following his retirement from the Navy, he joined the Institute of Nuclear Power Operations in 1981, serving in various capacities including vice president of the Evaluation and Assistance Group and vice president and director of Government Relations. Dr. Beard was senior vice president of nuclear operations at the Florida Power Corporation from 1989 to 1997. He served on the Nuclear Energy Institute (NEI) Strategic Issues Advisory Committee (SIAC) (1994-1997); NEI SAIC Steering Group (1994-1997); University of Florida Nuclear Engineering Sciences Advisory Board (1995-1997); Virginia Power Nuclear Oversight Board (1994-1997); National Nuclear Accrediting Board (1995-2005); Executive Oversight Board for South Texas Project Nuclear Station (2003-2006); and Nuclear Safety Review Boards for three Exelon nuclear stations (2001-2003).

Jan Beyea, Ph.D., is chief scientist at Consulting in the Public Interest where he consults on energy/environmental topics for local, national, and international organizations. He has expertise in energy technologies and associated environmental and health concerns and has written numerous articles on energy and the environment, including articles on planning for reactor accidents. His current research interests are in the field of epidemiology. Dr. Beyea previously served as chief scientist and vice president of the National Audubon Society and held positions at Holy Cross College, Columbia University, and Princeton University's Center for Energy and Environmental Studies. He received a B.A. from Amherst College and a

Ph.D. in physics from Columbia University. Dr. Beyea has been a member of numerous advisory committees and panels, including the National Research Council's Board on Energy and Environmental Systems, Energy Engineering Board, Committee on Assessment of the Prospects for Inertial Fusion Energy, Committee on America's Energy Future, Committee on Alternative Energy R&D Strategies, Committee to Review DOE's Fine Particulates Research Plan, and Committee on Alternatives for Controlling the Release of Solid Materials from Nuclear Regulatory Commission-Licensed Facilities. He currently is a member of the World Trade Center Health Registry Scientific Advisory Committee. He has also served on the Secretary of Energy Advisory Board's Task Force on Economic Modeling, has been a member of the policy committee of the Recycling Advisory Council, and has advised various studies of the Office of Technology Assessment. He recently served as a guest editor for and contributor to a theme issue of the *Bulletin of the Atomic Scientists* on the subject of risks from low-level radiation. Dr. Beyea is an elected fellow of the American Physical Society.

M. Quinn Brewster, Ph.D., is currently the Hermia G. Soo Professor of Mechanical Engineering at the University of Illinois at Urbana–Champaign. He conducts fundamental scientific and engineering research in radiation heat transfer, solid propellant and metal combustion, thermophysical properties of materials, and laser-aided materials processing. Dr. Brewster was involved in the Academic Strategic Alliance Program and a multidisciplinary university research initiative whose objectives were to develop a scientific basis for understanding solid rocket motors and energetic materials combustion. Dr. Brewster holds a Ph.D. in mechanical engineering from the University of California, Berkeley. He has authored one book on thermal radiative transfer and chapters in four other books and several publications on combustion science. He is a fellow of the American Society of Mechanical Engineers and associate fellow of the American Institute of Aeronautics and Astronautics. Dr. Brewster served on the National Research Council's Committee on Safety and Security of Commercial Spent Nuclear Fuel Storage.

Michael L. Corradini, Ph.D., is chair and professor in the Department of Engineering Physics at the University of Wisconsin, Madison. Dr. Corradini's research focus is nuclear engineering and multiphase flow with specific interests that include light-water reactor safety, fusion reactor design and safety, waste management and disposal, vapor explosions research and molten core–concrete interaction research, and energy policy analysis. He received his B.S. in mechanical engineering from Marquette University and his M.S. and Ph.D. degrees in nuclear engineering from the Massachusetts Institute of Technology. He is a member of the American Institute of

Chemical Engineers, the American Society of Engineering Education, the American Society of Mechanical Engineers, and a fellow of the American Nuclear Society. Dr. Corradini has received numerous awards including the National Science Foundation's Presidential Young Investigator Award, the American Nuclear Society reactor safety best paper award, and the University of Wisconsin, Madison, campus teaching award. He has served on various technical review committees, including the research review panel of the U.S. Nuclear Regulatory Commission. He currently serves on the Nuclear Regulatory Commission's Advisory Committee on Reactor Safeguards and is past president of the American Nuclear Society. Dr. Corradini was elected to the National Academy of Engineering in 1998.

Vijay K. Dhir, Ph.D., is a distinguished professor of mechanical and aerospace engineering and has been dean of UCLA's Henry Samueli School of Engineering and Applied Science since 2003. He also leads the boiling heat transfer laboratory, which conducts pioneering work in fundamental and applied sciences involving boiling, an efficient process of heat removal. Currently, his laboratory is involved in the study of flow boiling, microgravity boiling, and nuclear reactor thermal hydraulics. Born in India, Dr. Dhir received his B.Sc. from Punjab Engineering College in Chandigarh, India, and his Master of Technology from the Indian Institute of Technology in Kanpur, India. He received his Ph.D. from the University of Kentucky. In the late 1960s, he worked for a short period in industry as an engineer, and for the past 35 years he has been a consultant for numerous organizations. Dr. Dhir served as chair of the UCLA Department of Mechanical and Aerospace Engineering from 1994 to 2000. In 2004, he was selected as an inductee into the University of Kentucky's Engineering Hall of Distinction, and in 2012 he received his alma mater's Honorary Ph.D. degree. The American Society of Mechanical Engineers (ASME) has honored him with the Heat Transfer Memorial Award and the Robert Henry Thurston Lecture Award. The American Institute of Chemical Engineers (AIChE) honored him with the Donald Q. Kern Award and the Max Jakob Memorial Award (awarded jointly with ASME). He is recipient of the Technical Achievement Award of the Thermal Hydraulics Division of the American Nuclear Society. Most recently, he received the Lifetime Achievement Award at ICCES (the International Conference on Computational & Experimental Engineering and Sciences). Dr. Dhir has more than 300 publications in archival journals and proceedings of conferences. He was elected to the National Academy of Engineering in 2006.

Michael W. Golay, Ph.D., is a professor of nuclear science and engineering at the Massachusetts Institute of Technology (MIT) where he has worked since 1971. He is director of the Reactor Technology Course for Utility

Executives and the Nuclear Operational Risk Management Course, both cosponsored by MIT and the National Academy for Nuclear Training. Most recently he has focused his research and teaching on improving nuclear power performance both in the United States and internationally, particularly through use of probabilistic and dynamic methods of analysis. He has also been an active advisor to governmental and industrial organizations, particularly concerning risk-informed regulation and nuclear nonproliferation. Dr. Golay received his Ph.D. in nuclear engineering from Cornell University in 1969 and performed postdoctoral research at Rennselaer Polytechnic Institute. In 1980, he was a visiting researcher at Électricité de France. He has served on the Institute of Nuclear Power Operations Advisory Council, the U.S. Nuclear Regulatory Commission's Research Review Committee, the Department of Energy's TOPS Committee (on nonproliferation), and national laboratory and nuclear power plant oversight committees. He is a fellow of the American Association for the Advancement of Science and the American Nuclear Society.

Barbara L. Hamrick, J.D., CHP, is the radiation safety officer at the University of California, Irvine Medical Center where she oversees the use of radiation and radioactive materials in medical and research applications, including use in radiology, nuclear medicine, radiation oncology, pathology, and neurology. Ms. Hamrick received a B.S. and an M.S. in physics from the University of California, Irvine, in 1985 and 1987, respectively. She also received a J.D. from Loyola Law School in Los Angeles and was admitted to the California State Bar in 1999. Prior to joining the staff at the Medical Center, Ms. Hamrick worked for 18 years in regulatory agencies at the local, state, and federal levels, including the Los Angeles County Office of Radiation Management, the California Department of Public Health, and the U.S. Nuclear Regulatory Commission. Her work involves the application of health physics to a diverse set of problems, including survey and remediation at decommissioning facilities, external and internal dose assessments, air and water effluent modeling and monitoring, emergency planning, population monitoring, and radioactive waste management and disposal. Ms. Hamrick also has extensive experience and knowledge related to federal and state statutes and regulations governing the use of radiation and radioactive materials. She has participated in and co-chaired interagency working groups established to develop regulation, policy, and guidance related to the use of radiation and radioactive material in coordination with the U.S. Nuclear Regulatory Commission, U.S. Department of Homeland Security, U.S. Department of Energy, U.S. Environmental Protection Agency, Organization of Agreement States, and the Conference of Radiation Control Program Directors. Ms. Hamrick is currently serving as president of the Health Physics Society. She was certified by the American Board of Health Physics in 2002.

Shelley A. Hearne, Dr.P.H., is a visiting professor at Johns Hopkins Bloomberg School of Public Health Department of Health Policy and Management and director of the Big Cities Health Coalition at the National Association of County and City Health Officials. Previously, she was managing director of the Pew Health Group at The Pew Charitable Trusts, working to improve the health and well-being of Americans by reducing unnecessary risks in food, medical, and consumer products. She is also a member of the Board of Councilors of the Radiation Effects Research Foundation (RERF). RERF collects and analyzes health information on the World War II atomic bombing survivors. Dr. Hearne holds a B.A. in chemistry and environmental studies from Bowdoin College and a Dr.P.H. in environmental health sciences from Columbia University School of Public Health. She was the founding executive director of Trust for America's Health—a nonprofit, nonpartisan organization dedicated to preventing epidemics and protecting people, and she was the national recipient of the 2004 Delta Omega Curriculum Award honoring innovative public health teaching. She has worked in various roles in government, environmental nonprofits, and philanthropy, ranging from serving as the executive director of the Pew Environmental Health Commission to acting director of the New Jersey Pollution Prevention Office. Dr. Hearne is the past chair of the American Public Health Association Executive Board and was the former vice president of the Council on Education for Public Health, which accredits graduate public health institutions.

Paul A. Locke, Dr.P.H., J.D., M.P.H., an environmental health scientist and attorney, is an associate professor at the Johns Hopkins University Bloomberg School of Public Health. He holds his primary appointment in the Department of Environmental Health Sciences and a joint appointment in the Department of Health Policy and Management. Dr. Locke directs the Doctor of Public Health Program in Environmental Health Sciences. Dr. Locke's research and practice focus on how decision makers use scientific data and research in regulation and policy making and how environmental health sciences influence the policy-making process. His areas of study include designing and evaluating radiation protection initiatives and radiation policies, especially in low-dose radiation science, radon risk reduction, safe disposal of high-level radioactive waste, and uranium mining and recovery operations. He holds an M.P.H from Yale University School of Medicine, a Dr.P.H. from the Johns Hopkins University Bloomberg School of Public Health, and a J.D. from Vanderbilt University School of Law. Dr. Locke was a member of the National Academy of Sciences Nuclear and Radiation Studies Board from 2003 to 2009, and chaired the National Academy's Committee on Uranium Mining in Virginia. He is a member of the editorial board of the *International Journal of Low Radiation* and is on the board of directors of the National Council on Radiation Protection and Measurements. Dr. Locke is admitted to practice law before the bars of New York and the District of Columbia, the Southern District Court of New York, and the U.S. Supreme Court.

James E. Matheson, Ph.D., is chairman and chief financial officer of SmartOrg, Inc. and a world-recognized leader in the development and application of decision analysis. Dr. Matheson has been a consulting professor in the Department of Management Science and Engineering at Stanford University since 1967. He received a B.S. in electrical engineering from the Carnegie Institute of Technology and an M.S. and a Ph.D. in electrical engineering from Stanford University, where he studied in the computer-coordinated systems track of the Engineering-Economic Systems Program. During and after college, Dr. Matheson worked at the Westinghouse Research Laboratories where he was on the design team of an early process control computer and co-developed the first digital solutions to magnetic and electrical field problems that revolutionized the analysis and design of many products. In the mid 1960s, Dr. Matheson created and directed SRI International's Decision Analysis Group, which for 15 years led the profession of decision analysis consulting. He also led the initial development of R&D portfolio analysis for major industrial and pharmaceutical firms. He personally co-invented influence diagrams, which have become a standard professional tool for capturing the interrelationships among decisions, uncertainties, and values. Dr. Matheson was awarded the Ramsey Medal, the highest honor in the field of decision analysis, by the Institute for Operations Research and Management Sciences. He serves on the advisory board of Right Side Capital Management, an early-stage venture capital fund. He also was a founder and a member of the board of directors of the Decision Education Foundation, a nonprofit dedicated to helping individuals learn to make better personal decisions.

Thomas G. Moser, U.S. Navy (retired), is the chief of staff of Osprey Global Solutions, where he coordinates corporate business development and project management activities and provides antiterrorism and security expertise to federal, state, and local government entities and private-sector clients. He holds a B.S. in business administration from Waynesburg College in Pennsylvania and an M.B.A. from Southern Illinois University. While on active service as a career Navy SEAL Officer, Mr. Moser served as commanding officer of the Navy's unique antiterrorism RED CELL team, commanding officer of SEAL Team FOUR, commanding officer of the Naval Special Warfare Development Group (a classified special SEAL Unit), and as chief of staff at the Joint Special Operations Command. Following his naval career, Mr. Moser served as a counterterrorism and special operations consultant and exercise planner for Department of Defense Special Operations Units. He developed plans to exercise the nation's military and first-responder units' response to incidents involving the use of chemical, biological, radiological, and nuclear weapons of mass destruction. He later worked with the Department of Energy (DOE) as the site manager of the Andrews Air Force Base laboratory facility that was responsible for one of the nation's Nuclear Emergency Search Teams. Following service at DOE, Mr. Moser was selected to serve as one of the first Department of Homeland Security (DHS) Protective Security Advisors (PSAs) and served as PSA to the State of South Carolina, representing DHS as an onsite critical infrastructure and vulnerability assessment specialist. Mr. Moser participated in comprehensive security assessments at nuclear power plants and material production facilities in North and South Carolina and served on the National Research Council Committee on Risk-Based Approaches for Securing the DOE Nuclear Weapons Complex. More recently, Mr. Moser participated in a survey and assessment of the Coast Guard Service of a Gulf Coast Cooperative member state, addressing counterpiracy and countersmuggling missions in the Persian Gulf and Arabian Sea. Mr. Moser is an American Society of Industrial Security Certified Protection Professional and Physical Security Professional.

Arthur T. Motta, Ph.D., is chair of the Nuclear Engineering Program and a professor of nuclear engineering and of materials science and engineering at Penn State University. His research focuses on the environmental degradation of materials in the reactor environment with specific emphasis on nuclear fuel cladding. His research interests include radiation damage, corrosion and hydrogen ingress, mechanical behavior of materials, and materials characterization. He holds a B.Sc. in mechanical engineering and an M.Sc. in nuclear engineering from the Federal University of Rio de Janeiro, Brazil, and a Ph.D. in nuclear engineering from the University of California, Berkeley. Before coming to Penn State, he worked for the CEA at the Centre for Nuclear Studies in Grenoble, France, and for Atomic Energy of Canada Limited at the Chalk River Laboratories in Canada. He is a member of the editorial board of the *Journal of Nuclear Materials*. He received the Penn State Engineering Society Outstanding Advising Award in 2001 and the Outstanding Research Award in 2012, and the Outstanding Research Achievement Award from the Materials Science and Technology Division of the American Nuclear Society in 2010.

John A. Orcutt, Ph.D., is a distinguished professor of geophysics at Scripps Institution of Oceanography and a secretary of the Navy/Chief of Naval Operations Oceanography Chair. He received his B.S. in mathematics and physics from the U.S. Naval Academy in Annapolis, his M.Sc. in physical chemistry as a Fulbright Scholar at the University of Liverpool, and his Ph.D. in earth sciences from the Scripps Institution of Oceanography. He served as a submariner in the U.S. Navy and was the Chief Engineer on USS *Kamehameha* including a shipyard overhaul with refueling of the nuclear plant. His research interests include the exploitation of information technology for the collection and processing of real-time environmental data as well as marine and continental seismology and geophysics. He is the principal investigator for the National Science Foundation Major Research Equipment and Facilities Construction Ocean Observatories Initiative Cyberinfrastructure Implementing Organization. He is also chair of the MEDEA Ocean Panel and recently completed a review of hydroacoustics monitoring by the UN Comprehensive Test Ban Treaty Office in the Indian Ocean. He is a charter member of the National Research Council Ocean Studies Board and just began another term nearly 25 years after his first. He is the principal investigator of a BP research institute at Scripps, which began operations in 2004. He received the Ewing Medal from the U.S. Navy and the American Geophysical Union (AGU) in 1994; the Newcomb-Cleveland Prize from the American Association for the Advancement of Science in 1983; and the Marine Technology Society's Lockheed Martin Award for Ocean Science and Technology in 2007. He chaired the National Research Council review of the NOAA Tsunami Warning System and the Ocean Panel of the Climate, Energy and National Security Committee. He served as president of the American Geophysical Union from 2004 to 2006 and was elected as an Honorary Fellow of the Royal Astronomical Society in 2005. He was elected to the American Philosophical Society in 2002 and the National Academy of Engineering in 2011.

Emilie M. Roth, Ph.D., is the owner and principal scientist of Roth Cognitive Engineering. A cognitive psychologist, Dr. Roth's work involves the analysis of human problem solving and decision making in real-world environments (e.g., military command and control, intelligence analysis, nuclear power plant emergencies, railroad operations, and surgery), and the impacts of support systems (e.g., computerized procedures, alarm systems, advanced graphical displays, and new forms of automation) on cognitive performance. Dr. Roth has conducted empirical studies of naturalistic decision making, developed and applied cognitive task analysis and cognitive work analysis techniques for understanding the cognitive demands imposed by work environments, and developed principles for effective decision support for individuals and teams. Dr. Roth has supported design of first-of-a-kind systems including the command center for a next-generation Navy ship; a next-generation nuclear power plant control room; and work-centered support systems for flight planning and monitoring for an Air Force organization. She received her Ph.D. in cognitive psychology from the University of Illinois at Urbana-Champaign. She serves on the editorial board of the journals *Human Factors* and *Journal of Cognitive Engineering and Decision Making*. She was elected a fellow of the Human Factors and Ergonomics Society. She recently participated in the National Research Council Committee on Human-System Design Support for Changing Technology.

Joseph E. Shepherd, Ph.D., is the C. L. "Kelly" Johnson Professor of Aeronautics and Mechanical Engineering and, since 2009, the dean of graduate studies at the California Institute of Technology (Caltech). His research interests are fluid and solid dynamics, combustion chemistry, thermodynamics, and dynamic structural response with applications to explosions, propulsion, high-speed flight, and energy technology. Since 1980, he has carried out research on hydrogen combustion behavior during severe accidents in nuclear power plants as well as in nuclear material processing and storage facilities. He has worked with private industry, the U.S. Nuclear Regulatory Commission, the U.S. Department of Energy, U.S. national laboratories, and international organizations to evaluate hydrogen control methodologies and assess potential hazards including the effects of explosions. He received his Ph.D. in applied physics from Caltech in 1980, served as a member of the technical staff at Sandia National Laboratories from 1980 to 1986, and was an assistant professor of mechanical engineering at Rensselaer Polytechnic Institute from 1986 to 1993. He has been on the faculty at Caltech since 1993. Dr. Shepherd served on the National Research Council Committee on Determining Basic Research Needs to Interrupt the Improvised Explosive Device Delivery Chain.

Elizabeth Q. Ten Eyck is president of ETE Consulting, Inc. She is an expert in domestic and international nuclear safeguards and security for government-owned and -licensed commercial nuclear facilities and is involved in consulting work on vulnerability assessments of U.S. critical infrastructure for the Department of Homeland Security through Argonne National Laboratory. Ms. Ten Eyck received her B.S. in electrical engineering from the University of Maryland. She has over 30 years of career federal service, first as a security engineer for the U.S. Secret Service, then as director of the Office of Safeguards and Security for the U.S. Department of Energy, and, until she retired in 2000, as director of the Division of Fuel Cycle Safety and Safeguards for the U.S. Nuclear Regulatory Commission (USNRC), where she managed the safety and safeguards regulatory program for commercial fuel cycle facilities. During her career at the USNRC, she also managed transportation activities and the safeguards program for nuclear power reactors. Ms. Ten Eyck served on the National Research Council Committee on Transportation of Radioactive Waste.

Frank N. von Hippel, Ph.D., is a senior research physicist and professor of public and international affairs emeritus at Princeton University's Program on Science and Global Security, which he co-founded. In 1989, he co-founded the journal *Science & Global Security*. He co-founded and is currently co-chair of the nongovernmental International Panel on Fissile Materials, which includes experts from 17 countries and develops proposals for initiatives to reduce global stocks of plutonium and highly enriched uranium and the numbers of locations where they can be found. He received a Ph.D. in nuclear physics from University of Oxford and a B.A. from Massachusetts Institute of Technology. As a former assistant director for national security in the White House Office of Science and Technology, Dr. von Hippel's areas of policy research include nuclear arms control and nonproliferation, energy, and checks and balances in policy making for technology. He has been involved in reactor safety issues since he served as a member of he American Physical Society's 1974-1975 Study Group on Light Water Reactor Safety. Prior to coming to Princeton, he worked for 10 years in the field of elementary-particle theoretical physics. Dr. von Hippel's awards include the American Physical Society (APS) 2010 Leo Szilard Lectureship Award for outstanding work and leadership in using physics to illuminate public policy on nuclear arms control and nonproliferation, nuclear energy, and energy efficiency; the American Association for the Advancement of Science 1994 Hilliard Roderick Prize for Excellence in Science, Arms Control and International Security; a MacArthur Foundation Prize Fellowship (1993-1998); and the 1977 APS Forum Award for Promoting the Understanding of the Relationship of Physics and Society. Dr. von Hippel recently served on the National Research Council Committee on Best Practices for Nuclear Materials Protection, Control and Accounting.

Loring A. Wyllie, Jr., M.S., is chairman emeritus of the board and senior principal at Degenkolb Engineers. He has more than 45 years of professional experience in seismic evaluations, analysis, and design of strengthening measures for improved seismic performance. He serves as consultant to several University of California campuses, various commercial and architectural clients, Department of Energy laboratories, and many others. He received his B.S. and M.S. from the University of California, Berkeley. He is a past chairman of the state historical building safety board, whose mandate is to evaluate and analyze methods for strengthening buildings that preserve their historic character. He is also the past president of the

Earthquake Engineering Research Institute (EERI). His contributions to the profession of structural engineering were recognized by his election to the National Academy of Engineering in 1990. In 2007, he was honored with the prestigious Outstanding Projects and Leaders Lifetime Achievement Award by the American Society of Civil Engineers (ASCE). He was made an honorary member of the Structural Engineers Association of Northern California and Earthquake Engineering Research Institute. In recognition of his expertise in concrete design and performance, the American Concrete Institute named him an honorary member in 2000. Mr. Wyllie was elected an honorary member of ASCE in 2001.

Technical Advisor

Najmedin Meshkati, Ph.D., CPE, is a professor of civil/environmental engineering and a professor of industrial and systems engineering at the Viterbi School of Engineering, University of Southern California (USC). For the past 25 years, he has been teaching and conducting research on risk reduction and reliability enhancement of complex technological systems, including nuclear power, aviation, and petrochemical and transportation industries. Dr. Meshkati simultaneously received a B.S. in industrial engineering and a B.A. in political science from Sharif (Arya-Meher) University of Technology and Shahid Beheshti University (National University of Iran), respectively; an M.S. in engineering management from USC; and a Ph.D. in industrial and systems engineering from USC. He was a Jefferson Science Fellow and a senior science and engineering advisor in the Office of Science and Technology Adviser to the U.S. Secretary of State (2009-2010). Dr. Meshkati has inspected many petrochemical and nuclear power plants around the world, including Chernobyl. He is an elected fellow of the Human Factors and Ergonomics Society, an AT&T faculty fellow in industrial ecology, a NASA faculty fellow (Jet Propulsion Laboratory, 2003 and 2004), and a recipient of the Presidential Young Investigator Award from the National Science Foundation (NSF). He is the 2007 recipient of the Oliver Keith Hansen Outreach Award from the Human Factors and Ergonomics Society (HFES) and was honored by the HFES for his scholarly efforts on human factors of complex, large-scale technological systems. He is also a certified professional ergonomist. He was a member of the National Academy of Engineering/National Research Council's Committee on the Analysis of Causes of the *Deepwater Horizon* Explosion, Fire, and Oil Spill to Identify Measures to Prevent Similar Accidents in the Future.

Staff

Kevin D. Crowley is senior board director of the Nuclear and Radiation Studies Board (NRSB) at the National Research Council–National Academy of Sciences in Washington, D.C. He is responsible for managing the NRSB's work on nuclear safety and security, radioactive waste management and environmental cleanup, and radiation health effects. He is also the principal investigator for a long-standing cooperative agreement between the National Academy of Sciences and the U.S. Department of Energy to provide scientific support for the Radiation Effects Research Foundation in Hiroshima, Japan. Dr. Crowley's professional interests and activities focus on safety, security, and technical efficacy of nuclear and radiation-based technologies. He has directed over 20 National Research Council studies on these and other topics, including *Safety and Security of Commercial Spent Nuclear Fuel Storage* (2004, 2006); *Going the Distance? The Safe Transport of Spent Nuclear Fuel and High-Level Radioactive Waste in the United States* (2006); *Medical Isotope Production Without Highly Enriched Uranium* (2009); *America's Energy Future: Technology and Transformation* (2009); and *Analysis of Cancer Risks in Populations near Nuclear Facilities*. Before joining the National Academies staff in 1993, Dr. Crowley held teaching/research positions at Miami University of Ohio, the University of Oklahoma, and the U.S. Geological Survey. He holds M.A. and Ph.D. degrees, both in geology, from Princeton University.

Ourania (Rania) Kosti joined the staff of the Nuclear and Radiation Studies Board in January 2011. Prior to her current appointment, Dr. Kosti was a postdoctoral fellow at the Lombardi Comprehensive Cancer Center at Georgetown University Hospital in Washington, D.C., where she conducted research on biomarker development for early cancer detection using case-control epidemiologic study designs. She focused primarily on prostate, breast, and liver cancers and trying to identify those individuals who are at high risk of developing malignancies. She contributed on hypotheses generation, study design, data analysis and management of clinical databases and biospecimen repositories. Dr. Kosti also trained at the National Cancer Institute (NCI) (2005-2007) in the Cancer and Developmental Biology Laboratory; the same period she volunteered in NCI's Division of Cancer Epidemiology and Genetics. She received a B.Sc. in biochemistry from the University of Surrey, UK, an M.Sc. in molecular medicine from the University College London, and a Ph.D in molecular endocrinology from St. Bartholomew's Hospital in London, UK.

Daniel Pomeroy joined the Nuclear and Radiation Studies Board as a Christine Mirzayan Science and Technology Policy Graduate Fellow in August 2012 before working as postdoctoral fellow from December 2012 until August 2013. He is currently serving as an AAAS Congressional Science and Engineering Fellow for the American Geophysical Union. He received his Ph.D. in experimental high-energy physics from Brandeis University in June 2012. For his graduate research he worked in experimental high-energy physics at CERN in Switzerland. There he helped construct portions of the ATLAS detector and then used it to search for new fundamental physics phenomena, focusing primarily on lepton flavor violation. He received his B.S. from the University of Massachusetts Amherst, where he spent his summers interning at Thomas Jefferson National Accelerator Facility.

Appendix B

Presentations, Breakout Sessions, and Visits

WASHINGTON, D.C., JULY 19, 2012

Plenary Presentations

- U.S. Nuclear Regulatory Commission's Response to the Fukushima Nuclear Accident and Recommendations for this NAS Study, Mike Johnson, Deputy Executive Director for Reactor and Preparedness Programs, U.S. Nuclear Regulatory Commission; Rob Taylor, Deputy Director of the Japan Lessons Learned Project Directorate, U.S. Nuclear Regulatory Commission
- U.S. Nuclear Industry Response to the Fukushima Nuclear Accident & Recommendations for This NAS Study, Marv Fertel, President and Chief Executive Officer, Nuclear Energy Institute
- Union of Concerned Scientists Views on the U.S. Response to the Fukushima Accident and Recommendations for This NAS Study, Dave Lochbaum, Director, Nuclear Safety Program, Union of Concerned Scientists; Ed Lyman, Senior Scientist, Nuclear Safety Program, Union of Concerned Scientists

WASHINGTON, D.C., SEPTEMBER 6-7, 2012

Plenary Presentations

- TEPCO Overview of Fukushima Accident, Shin Takizawa, Manager, Nuclear International Relations and Strategy Group, Nuclear Power and Plant Siting Administrative Department, Tokyo Electric Power Company; Toshiaki Sakai, General Manager, Construction Engineering Center, Construction Department, Tokyo Electric Power Company; Yasunori Yamanaka, Manager, Nuclear Safety Engineering Group, Nuclear Asset Management Department, Tokyo Electric Power Company; Kenji Tateiwa, Manager, Nuclear Power Programs, Tokyo Electric Power Company Washington Office

- INPO Overview of the Fukushima Accident Time Line, William E. Webster, Jr., Senior Vice President; Steven W. Meng, Manager, Emergency Preparedness, Institute of Nuclear Power Operations
- Comments from Ichiro Fujisaki, Ambassador of Japan to the United States

Breakout Sessions

Session 1: Accident Progression, Management, and Recovery
Moderator: B. John Garrick, committee vice chair
Rapporteur: Kevin Crowley, study director

Invited Participants:

Industry

- William Berg, Senior Licensing Engineer, GE Hitachi Nuclear Energy
- Randall Gauntt, Sandia National Laboratories, Severe Accident Analysis Department
- David Hembree, Vice President, Emergency Response, Institute of Nuclear Power Operations (INPO)
- Steven W. Meng, Manager, Emergency Preparedness, INPO
- Toshiaki Sakai, General Manager, Construction Engineering Center, Construction Department, TEPCO
- Shin Takizawa, Manager, Nuclear International Relations and Strategy Group, Nuclear Power and Plant Siting Administrative Department, TEPCO
- William E. Webster, Senior Vice President, Industry Evaluations, INPO
- William T. Williamson, Reactor Engineer Specialist, Browns Ferry Nuclear Plant

Government

- Richard Lee, Chief, Fuel and Source Term Code Development Branch, Office of Nuclear Regulatory Research, USNRC

Session 2: Response of Physical Plant During Accident
Moderator: Joseph Shepherd, committee member
Rapporteurs: Ourania Kosti, senior program officer and Micah Lowenthal, CISAC board director

Invited Participants:

- Mark Ajluni, Nuclear Licensing Manager, Southern Nuclear Operating Company
- Randy Ferrer, Senior Design Engineer, Constellation Energy Nuclear Group, LLC
- Neil Gannon, Vice President of Nuclear Operations, PPL Susquehanna
- Jeff Gasser, INPO
- Robert Paley, Senior Evaluator, Performance Improvement and Learning, INPO
- Kenji Tateiwa, Manager, Nuclear Power Programs, TEPCO
- Yasunori Yamanaka, Manager, Nuclear Safety Engineering Group, Nuclear Asset Management Department, TEPCO

TOKYO, JAPAN, NOVEMBER 26-28, 2012

Plenary Presentations

- Fukushima Nuclear Accident Independent Investigation Commission, Kiyoshi Kurokawa, Chair, Academic Fellow, National Graduate Institute for Policy Studies
- Lessons-Learned by TEPCO from the Fukushima Accident, Akira Kawano, General Manager, Tokyo Electric Power Company
- Rebuild Japan Initiative Foundation Independent Investigation Commission on the Fukushima Nuclear Accident, Koichi Kitazawa, Chair, Rebuild Japan Initiative Foundation Independent Investigation Commission on the Fukushima Nuclear Accident
- Response of the Onagawa Plant to the Great East-Japan Earthquake and Tsunami, Akiyoshi Obonai, Deputy Manager of Nuclear Power Department, Tohoku Electric Power Company; Jun Iida, Assistant Manager of Nuclear Power Department, Tohoku Electric Power Company; Kazuo Hirata, Assistant Manager of Civil & Architectural Engineering Department, Tohoku Electric Power Company
- Nuclear Regulation Authority: Overview, Time Line for Establishment, Current and Future Plans, Toyoshi Fuketa, Commissioner, Nuclear Regulation Authority

Breakout Sessions

Session 1: Accident Progression Analysis
Moderator: Michael Corradini, committee member
Rapporteur: Arthur Motta, committee member

Questions:

1. How much fuel damage/core melting occurred in the reactors?
2. Did the core penetrate and react with the pressure vessels?
3. Did the core/pressure vessel material pierce the containment?
4. Is there evidence (e.g., vessel wall temperatures), other than from severe accident codes, that could indicate whether the lower head of the reactor vessel at any unit was penetrated and whether there was some amount of core-concrete interaction?
5. Is there any evidence of steam explosions and/or core material underneath the reactor?
6. Is there any evidence of recriticality?
7. Are there measurements of airborne concentrations of radionuclides (or accumulation of ground deposition) that would enable validation of core degradation timing and extent?
8. How were total release inventories of radionuclides estimated during the accident?
9. Is there any evidence of continued release of radionuclides by an airborne pathway from the plant site (such as the release of iodine vapor or radionuclide aerosol release) subsequent to the time at which core degradation was arrested at all three units?
10. Is there evidence of damage to any systems, structures, or components (beyond the damage to the electrical power lines and towers) of Units 1-6 due directly to the seismic events?
11. What is the current status of the cooling systems? To what extent were the reactor and their PV piping systems damaged during the earthquake?
12. What is the current status of electrical equipment and diesel generators?
13. How useful were the severe accident codes (e.g., MAAP, RELAP) for assessing core damage in real time?
14. How was the lack of information and uncertainty of parameters (e.g., IC, SRVs, RCIC, HPIC, RPV integrity) managed during the accident?
15. Were reactor simulators used to assess the condition of the reactors and/or potential operator responses?
16. What lessons have been learned in the operation of safety systems (RCIC, IC) during the accident? For example, was the IC properly used in Unit 1? If not, how would it be changed in the future?

Invited Participants:

Industry

- Toshihiko Fukuda, General Manager, Nuclear Quality and Safety Management Department, Tokyo Electric Power Company (TEPCO)
- Shinichi Kawamura, General Manager, Nuclear Seismic Engineering Center, General Manager, Nuclear Safety System Engineering, TEPCO
- Rikiro Kikuchi, Group Manager, Architectural Engineering Group (Seismic Integrity), Nuclear Seismic Engineering Center, Nuclear Asset Management Department, TEPCO
- Hideaki Kiyoura, Deputy Manager, Seismic Integrity Engineering Group, Nuclear Seismic Engineering Center, Nuclear Asset Management Department, TEPCO
- Shinya Mizokami, Deputy Manager, Nuclear Reactor Safety Engineering Group, Nuclear Asset Management Department, TEPCO
- Kumiaki Moriya, Corporate Chief Engineer, Hitachi-GE Nuclear Energy, Ltd.
- Masahisa Ohtsuki, General Manager, Nuclear Power Plant Management Department, TEPCO
- Junichi Taira, Environmental Evaluation Group, Nuclear and Siting Headquarters, TEPCO
- Shin Takizawa, Manager, Nuclear International Relations and Strategy Group, Nuclear Power and Plant Siting Administrative Department, TEPCO
- Nobuyuki Ueda, Nuclear Safety Division, Japan Nuclear Safety Institute (JANSI)
- Yasunori Yamanaka, Group Manager, Nuclear Reactor Safety Engineering Group, Nuclear Asset Management Department, TEPCO

Government

- Kiyoharu Abe, Senior Technical Advisor, Japan Nuclear Energy Safety Organization (JNES)
- Toyoshi Fuketa, Commissioner, Nuclear Regulation Authority
- Haruataka Hoshi, Severe Accident Evaluation Group, Nuclear Energy System Safety Division, Incorporated Administrative Agency, JNES
- Akitoshi Hotta, Principal Officer, Severe Accident Group, Nuclear Energy System Safety, JNES
- Hiroshi Yamagata, Senior Coordinator for Severe Accident Measures, Secretariat of Nuclear Regulation Authority (NRA)

Academia/Professional Organizations and Societies

- Michio Ishikawa, Former President and CEO, Japan Nuclear Technology Institute
- Hideki Nariai, Professor Emeritus, University of Tsukuba
- Koji Okamoto, Department of Nuclear Engineering and Management, University of Tokyo
- Ayao Tsuge, President, The Japan Federation of Engineering Societies

Session 2: OffSite Emergency Response and Regulatory Oversight
Moderator: Paul Locke, committee member
Rapporteur: Shelley Hearne, committee member

Questions:

1. What offsite protection measures were taken during and after the accident? For example, was potassium iodide distributed to the population and, if so, how?
2. What was the public's reaction to these measures and their adequacy?
3. How were decisions on managing the health impacts from the accident, to either workers or the surrounding population, made and communicated?
4. What were the criteria used with respect to intervention on food? How are the Nobuyuki Hamada publications that review the food safety regulations put into place in Japan after the declaration of nuclear emergency conditions viewed by experts and the public?
5. Describe coordination of response in terms of the evacuation plan (e.g., communication with residents on explanation of the accident and evacuation directions). How well were the responses coordinated between TEPCO, government, and international organizations?
6. What lessons were learned in emergency response decision making and communication?
7. Can you explain the problems that arose in attempting to use the Network System for Prediction of Environmental Emergency Dose Information (SPEEDI)?
8. What changes do you anticipate making, if any, to improve the SPEEDI system and its use in planning the evacuation strategy?
9. Did NISA have onsite inspectors? Did they play any role in accident response?
10. Did NISA have an independent emergency control center? Did it play any role in accident response?

11. What regulatory requirements establish criteria for seismic and tsunami design requirements?

12. Has the Japanese regulator imposed special requirements for:

- Station blackout events—such as a minimum coping time without restoration of AC power?
- Emergency lighting?
- Backup station to the control room, from which critical safety functions can be managed?
- Anticipated transients without scram?
- Strainer requirements for protection of recirculation flow from the suppression pool from clogging?
- Hydrogen control?

13. Does the regulatory agency use performance indicators to assess the adequacy of safety culture of plant management?

14. Does the Japanese regulatory process incorporate risk analysis to prioritize safety requirements and regulatory oversight activities?

15. Are plants required to comply with deterministic safe shutdown requirements?

16. Are there requirements for the protection of critical safety equipment from internally initiated flooding? If so, why were they ineffective in protecting against the tsunami?

17. Under what conditions can the regulator order shutdown of a plant?

18. Under what conditions can the regulator require backfits to the plant?

19. Were the venting system and emergency management procedures implemented at the plant in response to regulatory requirements or as voluntary initiatives?

20. Did the regulator perform an independent review of the adequacy of these beyond-design-basis requirements?

Invited Participants:

Industry

- Akira Kawano, General Manager, Nuclear International Relations and Strategy Group, Nuclear Power and Plant Siting Administrative Department, TEPCO
- Akira Suzuki, Group Manager, Radiological Health and Safety Center, Nuclear Power and Plant Siting Administrative Department, TEPCO

- Tatsuya Taminami, Group Manager, Nuclear Security and Industrial Safety Management Group, Nuclear Power Plant Management Department, TEPCO

Government

- Toshihiro Funahashi, Senior Staff, Emergency Response Training Group, Nuclear Emergency Response and Preparedness Department, Incorporated Administrative Agency, JNES
- Toshimitsu Homma, Japan Atomic Energy Agency (JAEA)
- Kazumi Miyagi, Assistant Director-General, Nuclear Energy Response and Preparedness Department, Incorporated Administrative Agency, JNES
- Tatsujiro Suzuki, Vice Chairman, Japan Atomic Energy Commission Central
- Tomoho Yamada, Secretariat of Nuclear Regulation Authority
- Tetsuro Yamaguchi, Deputy Director, Nuclear Emergency Preparedness Division, Secretariat of NRA

Academia/Professional Organizations and Societies

- Takashi Sawada, Secretary General, Director, Atomic Energy Society of Japan
- Ayao Tsuge, President, The Japan Federation of Engineering Societies

Session 3: Accident Management and Operator Training
Moderator: Emilie Roth, committee member
Rapporteur: Najm Meshkati, committee member

Questions:

1. Was the organizational structure and staffing within the plant sufficient to cope with severe accidents, in particular co-incident failures or events impacting two or more plants on the same site? Was essential information about the state and operations of the units passed on efficiently from one shift to the next?
2. Can you explain the chain of decision making/command for emergency response that was specified in regulatory and TEPCO policies and procedures (e.g., Nuclear Emergency Preparedness Act, Nuclear Operator Emergency Action Plan, Nuclear Emergency Response Manual)?

3. In what ways was the chain of command that developed during the Fukushima accident different from the chain of command that was specified in documented policies and procedures? What factors do you believe contributed to this?

4. Given the lessons learned from the Fukushima accident, what changes, if any, do you anticipate making to the formal command-and-control chain for emergency response within the plant?

5. What changes do you anticipate making, if any, to the physical command-and-control sites (e.g., the plant control room, the Emergency Response Center, the offsite center, the emergency response center) and/or to the communications available among them?

6. What, if any, severe accident management guidelines were in place at the time of the accident? Were these guidelines followed in accident response?

7. What kind of engineering support was available within the plant to support decision making?

8. Do you anticipate the need for any decision aids (e.g., simplified computer decision-making codes) to support decision making during severe accidents?

9. What role did external advice (e.g., from IAEA, the U.S. Department of Energy, or U.S. Nuclear Regulatory Commission) play in managing the accident?

10. How does training account for design-specific unit characteristics?

11. How are plant personnel trained on severe accident management procedures? How often does training occur?

12. How often are operators trained on simulators for response to design-basis events?

13. Were the severe accident management procedures used by operators? Were the procedures useful and if so how? Were there aspects of the Fukushima accident that went beyond what the procedures covered?

14. Given the lessons learned from the Fukushima accident, what changes do you anticipate making, if any, to procedures for handling severe accidents?

15. What changes do you anticipate making, if any, to personnel training for handling severe accident events?

16. What changes do you anticipate making, if any, for improving the safety culture?

Invited Participants:

Industry

- Naoki Anahara, Group Manager, Human Resources and Ethics Development Group, Nuclear Power and Plant Siting Administrative Department, TEPCO
- Akira Kawano, General Manager, Nuclear International Relations and Strategy Group, Nuclear Power and Plant Siting Administrative Department, TEPCO
- Shinya Mizokami, Deputy Manager, Nuclear Reactor Safety Engineering Group, Nuclear Asset Management Department, TEPCO
- Hiroshi Nakano, Group Manager, Operation Planning Group, Nuclear Power Plant Management Department, TEPCO
- Masahisa Ohtsuki, General Manager, Nuclear Power Plant Management Department, TEPCO
- Masaru Oowada, Chief Assistant, Human Resources and Ethics Development Group, Nuclear Power and Plant Siting Administrative Department, TEPCO
- Tatsuya Taminami, Group Manager, Nuclear Security and Industrial Safety Management Group, Nuclear Power Plant Management Department, TEPCO
- Masahiro Yamamoto, Group Manager, Quality and Safety Assessment Group, Nuclear Quality and Safety Management Department, TEPCO
- Yasunori Yamanaka, Group Manager, Nuclear Reactor Safety Engineering Group, Nuclear Asset Management Department, TEPCO

Government

- Kiyoharu Abe, Senior Technical Advisor, JNES
- Toshihiro Funahashi, Senior Staff, Emergency Response Training Group, Nuclear Emergency Response and Preparedness Department, Incorporated Administrative Agency, JNES

Academia/Professional Organizations and Societies

- Mitsumasa Hirano, Professor, Tokyo City University
- Kenkichi Hirose, Tokai Institute of Global Education and Research, Tokai University
- Nobuhide Kasagi, Principal Fellow, Professor Emeritus, The University of Tokyo

- Koji Okamoto, Department of Nuclear Engineering and Management, University of Tokyo

Session 4: Risk Assessments
Moderated by John Garrick, committee vice chair
Rapporteur: Barbara Hamrick, committee member

Questions:

1. What type of risk assessments were performed for the Daiichi and Daini units (i.e., probabilistic risk assessment [PRA] or deterministic risk assessment)?
2. What are the results of these PRAs?
3. Did the PRAs consider fire risk, flooding risk (including tsunamis), and seismic risk (each event separately or together)? What magnitude of these events was considered?
4. Do the PRAs consider long-term station blackouts?
5. Do the PRAs consider common-mode failures across multiple units?
6. Do the PRA results include consideration of events initiated from operating modes other than full-power operation?
7. Were the PRAs reviewed and revised throughout the life of the plants?
8. Do the plants use online risk monitors to indicate to the operators when the unit is in a high-risk state, for example, when equipment from one safety train is undergoing maintenance or testing?

Invited Participants:

Industry

- Toshihiko Fukuda, General Manager, Nuclear Quality and Safety Management Department, TEPCO
- Shinichi Kawamura, General Manager, Nuclear Seismic Engineering Center, General Manager, Nuclear Safety System Engineering, TEPCO
- Koichi Miyata, Group Manager, Nuclear Safety Group, Nuclear Quality and Safety Management Department, TEPCO
- Hitoshi Muta, Senior Researcher, Probabilistic Safety Assessment Group, Nuclear Energy System Safety
- Toshiaki Sakai, General Manager, Construction Engineering Center, Construction Department, TEPCO

- Shin Takizawa, Manager, Nuclear International Relations and Strategy Group, Nuclear Power and Plant Siting Administrative Department, TEPCO
- Tomoyuki Tani, Group Manager, Civil Engineering Group (Geological Survey), Nuclear Seismic Engineering Center, Nuclear Asset Management Department, TEPCO

Government

- Haruo Fujimoto, Director, Probabilistic Safety Assessment Group, Nuclear Energy System Safety Department, JNES
- Yoshinori Moriyama, Associate Vice President, JNES
- Masao Ogino, Senior Staff, Severe Accident Evaluation Group, Nuclear Energy System Safety Department, JNES
- Hiroshi Yamagata, Senior Coordinator for Severe Accident Measures, Secretariat of NRA

Academia/Professional Organizations and Societies

- Masayoshi Nakashima, Professor and Director, Disaster Prevention Research Institute, Kyoto University
- Takashi Sawada, Secretary General, Director, Atomic Energy Society of Japan
- Hiroe Tsubaki, Vice Director General and Director, Risk Analysis Research Center, Institute of Statistical Mathematics
- Akira Yamaguchi, Professor, Department of Energy and Environment Engineering, Osaka University

Session 5: Hydrogen Explosions
Moderator: Quinn Brewster, committee member
Rapporteur: Loring Wyllie, committee member

Questions:

1. What is the current knowledge regarding pathways for hydrogen entry into the reactor buildings? What is the basis for the identification of these pathways?
2. How much hydrogen was estimated to be released into the buildings?
3. What were the ignition sources for the explosions?
4. How did the explosions affect the building structures?
5. What accounts for the differences in structural damage to Units 1, 3, and 4? Are there damage maps from each of the explosions?

6. Was removal of panels in Units 2, 5, and 6 effective in preventing explosions?
7. What was the basis for reinforcing the structure beneath the Unit 4 spent fuel pool? Is this basis documented?
8. How did the secondary damage (i.e., blasts and debris) from the explosions impact recovery operations? How much of the debris around the reactor units was from the tsunami versus the explosions?
9. Did the explosions result in damage to the spent fuel pools or the fuel—for example, was any of the spent fuel damaged by portions of the crane that fell into the Unit 3 spent fuel pool?
10. What is the estimated impact of the reactor building upper-level destruction on offsite releases of radioactive material?
11. What impact, if any, will damage to the spent fuel have on recovery operations?

Invited Participants:

Industry

- Toshihiko Fukuda, General Manager, Nuclear Quality and Safety Management Department, TEPCO
- Rikiro Kikuchi, Group Manager, Architectural Engineering Group (Seismic Integrity), Nuclear Seismic Engineering Center, Nuclear Asset Management Department, TEPCO
- Koichi Miyata, Group Manager, Nuclear Safety Group, Nuclear Quality and Safety Management Department, TEPCO
- Shin Takizawa, Manager, Nuclear International Relations and Strategy Group, Nuclear Power and Plant Siting Administrative Department, TEPCO

Government

- Masao Ogino, Senior Staff, Severe Accident Evaluation Group, Nuclear Energy System Safety Department, JNES

Academia/Professional Organizations and Societies

- Michio Ishikawa, Former President and CEO, Japan Nuclear Technology Institute
- Masanori Naitoh, Director, Nuclear Power Engineering Center, The Institute of Applied Energy
- Takashi Sawada, Secretary General, Director, Atomic Energy Society of Japan

Site Visits

- November 29, 2012: Visit to Onagawa Nuclear Power Station, Miyagi Prefecture, Japan
- November 29, 2012: Visit to Fukushima Daini Nuclear Power Station, Fukushima Prefecture, Japan
- November 30, 2012: Visit to Fukushima Daiichi Nuclear Power Station, Fukushima Prefecture, Japan

WASHINGTON, D.C., FEBRUARY 7, 2013

Plenary Presentations

- NRC Regulatory Activities Following the Fukushima Dai-ichi Accident, Rob Taylor, Deputy Director of the Japan Lessons Learned Project Directorate, U.S. Nuclear Regulatory Commission (USNRC)

WASHINGTON, D.C., JUNE 24-25, 2013

Plenary Presentations

- Overview of Nuclear Regulatory Commission Responses to 9/11 Terrorist Attacks, Christiana Lui, USNRC Division Director, Office of Nuclear Security and Incident Response, Division of Security Policy
- Origin and Requirements for B.5.b Mitigating Strategies, Eric Bowman, USNRC Senior Project Manager, Office of Nuclear Reactor Regulation, Division of Policy and Rulemaking
- Spent Fuel Pool Study, Hossein Esmaili, USNRC Senior Reactor Systems Engineer, Office of Nuclear Regulatory Research, Division of Systems Analysis; Jose Pires, Senior Technical Advisor for Civil/Structural Engineering, RES/Division of Engineering (supporting speaker); Donald Helton, Senior Reliability and Risk Engineer, RES/Division of Risk Assessment (supporting speaker); Keith Compton, Senior Reactor Scientist, Office of Nuclear Regulatory Research (supporting speaker)

Breakout Session

- Application of Probabilistic Risk Assessment to Multiple Reactor Units, Karl Fleming, President, KNF Consulting Services LLC

Nonplenary Presentations

- Design-Basis Threat for Nuclear Plants and Spent Fuel Pools, Ralph Way, USNRC Senior Level Advisor, Office of Nuclear Security and Incident Response, Division of Security Operations
- Physical Security of Nuclear Plants and Spent Fuel Pools, Ralph Way, USNRC Senior Level Advisor, Office of Nuclear Security and Incident Response, Division of Security Operations
- Personnel Security for Nuclear Plants and Spent Fuel Pools, Mark Resner, USNRC Senior Security Specialist, Office of Nuclear Security and Incident Response, Division of Security Policy
- Physical Security of Independent Spent Fuel Storage Installations (ISFSIs), Doug Garner, USNRC Security Specialist, Office of Nuclear Security and Incident Response, Division of Security Policy
- Zirconium Fire Experiments on BWR and PWR Fuel, Ghani Zigh, USNRC Senior Level Advisor, Office of Nuclear Regulatory Research, Division of Systems Analysis
- Current Security Rulemaking (Vulnerability Assessments, ISFSI/Phase 1 Study), Phil Brochman, USNRC Senior Program Manager, Office of Nuclear Security and Incident Response, Division of Security Policy; Daniel Forsyth, USNRC Nuclear Engineer, Office of Nuclear Material Safety and Safeguards, Division of Spent Fuel Storage and Transportation

WASHINGTON, D.C., AUGUST 14, 2013

Plenary Presentations

- Discussion with Senior Reactor Operators About the Fukushima Nuclear Accident and Management of Severe Accidents, James Scarola, Chairman, Fukushima Response Steering Committee, Nuclear Energy Institute (Chief Nuclear Officer); Phillip Amway, Fukushima Fleet Technical Lead, Constellation Energy Nuclear Group (former Senior Reactor Operator); Derwood Tootle, SAM Project Manager, Hatch Nuclear Plant, Southern Nuclear (Senior Reactor Operator); Glen Morrow, Regulatory Assurance Manager, Dresden Nuclear Power Station, Exelon Generation (Senior Reactor Operator)

WASHINGTON, D.C., OCTOBER 1-3, 2013

Nonplenary Presentations

- U.S. Government's Response to the Fukushima Nuclear Accident, Dr. John P. Holdren, Assistant to the President for Science and Technology, Director of the White House Office of Science and Technology Policy

FORKED RIVER, NEW JERSEY, OCTOBER 28, 2013

Site Visit

- Visit to Oyster Creek Generating Station (Exelon Corporation)

BAXLEY, GEORGIA, NOVEMBER 15, 2013

Site Visit

- Visit to Edwin I. Hatch Nuclear Plant (Southern Company)

WASHINGTON, D.C., DECEMBER 5-6, 2013

Plenary Presentations

- Design-Basis Threats (DBTs) for Commercial Nuclear Reactors and Spent Nuclear Fuel, Dr. Patricia Holahan, Director, Division of Security Operations, Office of Nuclear Security and Incident Response (NSIR), U.S. Nuclear Regulatory Commission (USNRC); Dr. Ralph Way, Senior Level Advisor for Security, NSIR, USNRC; Mr. Jack Frost, Security Specialist, Reactor Security Licensing Branch, Division of Security Policy, NSIR, USNRC
- Need for Expedited Transfer of Spent Fuel from Pools to Dry Casks, William Reckley, Branch Chief, Policy and Support Branch, Japan Lessons Learned Project Directorate, Office of Nuclear Reactor Regulation (NRR), USNRC; Kevin Witt, Project Manager, Policy and Support Branch, Japan Lessons Learned Project Directorate, NRR, USNRC; Steven Jones, Senior Reactor Systems Engineer, Balance of Plant Branch, Division of Safety Systems, NRR, USNRC; Fred Schofer, Senior Cost Analyst, Rulemaking Branch, Division of Policy and Rulemaking, NRR, USNRC

Information-Gathering Conference Calls

- April 10, 2013: Nuclear Reactor and Plant Performance, Salomon Levy, S. Levy and Associates, Inc.; Craig Sawyer, consultant (formerly General Electric Company, retired)
- April 30, 2013: How Probabilistic Risk Assessment Is Used in Nuclear Plant Safety, John W. Stetkar, ACRS member; David H. Johnson, ABS Consulting; James R. Chapman, Scientech, Curtiss Wright Flow Control; Don Dube, Erin Engineering (formerly USNRC)
- May 17, 2013: Emergency Preparedness, Charles (Milt) Murray, Inspection and Regulatory Improvement Branch, Division of Preparedness and Response, Office of Nuclear Security and Incident Response, USNRC; Chad Gorman, Director, Chemical, Biological, Radiological, Nuclear, and Explosives (CBRNE) Office, Response Directorate, U.S. Department of Homeland Security, FEMA; Harry Sherwood, Professional Services Branch Chief, Technological Hazards Division National Preparedness Directorate, FEMA
- May 20, 2013: Human Performance, Robert Beall, Project Manager, Rulemaking Branch, Division of Policy and Rulemaking, Office of Nuclear Reactor Regulation; Christian Cowdrey, Reactor Engineer, Operator Licensing and Training Branch, Division of Inspection and Regional Support, Office of Nuclear Reactor Regulation; Kevin Williams, Branch Chief, New Reactor Licensing Branch, Division of Preparedness and Response, Office of Nuclear Security and Incident Response; David Hembree, Vice President, Emergency Response, Institute of Nuclear Power Operations (INPO); Steven Meng, Manager, Emergency Preparedness, INPO
- May 23, 2013: Use of Severe Accident Management Guidelines, Doug True and Jeff Gabor, ERIN Engineering and Research, Inc.
- June 21, 2013: Training Approaches to Prepare Personnel to Skillfully Handle Complex, Unanticipated, High-Risk, and High-Stress Situations, Randall J. Mumaw, Associate Technical Fellow, Human Factors, Aviation System Safety, Boeing; James A. Wall, Executive Director, Texas Center for Applied Technology
- November 13, 2013: Sara DeCair, Mike Boyd, and David Pawel, scientists in the Radiation Protection Division, Environmental Protection Agency; Jon Edwards, Director, Radiation Protection Division, Environmental Protection Agency

Appendix C

Detailed Accident Time Line

TABLE C.1 Time Line of Key Events in Units 1-3 at the Fukushima Daiichi Nuclear Plant

Event/Condition	Unit 1	Unit 2	Unit 3
Prior to earthquake		Operating at rated power level	
Earthquake (3/11/11 @ 14:46)[b]		T = 0[a] Reactor scram MSIVs close Loss of offsite AC power Emergency diesel generators (EDGs) start	
Tsunami warnings (Fukushima Prefecture) and estimated wave heights	←	14:49 (+3 min): 3 m 15:15 (+29 min): 6 m 15:30 (+44 min): >10 m	→
Tsunami arrival times (1st/2nd waves)		+41 min/+50 to +51 min (15:27/15:36-15:37)	
Loss of onsite AC power (EDGs) and DC power (batteries)[c]	AC lost at +51 min (15:37) DC lost at +60 min (15:46)	AC lost at +55 min (15:41) DC lost at +60 min (15:46)	AC lost at ~ +51 min (15:37) DC available until ~+36 h
Isolation condenser (IC) performance[d]	Failed on loss of AC and DC power	NA	NA

continued

TABLE C.1 Continued

Event/Condition	Unit 1	Unit 2	Unit 3
Reactor core isolation cooling (RCIC) performance[e]	NA	Real-time status uncertain; evidence of ~70 h running time	~20 h of running time; failed w/o restart at +20 h
High-pressure coolant injection (HPCI) performance	Unavailable due to loss of DC power	Unavailable due to loss of DC power	~16 h of running time beginning at +20 h
Reactor pressure vessel depressurization[f]	Depressurized due to assumed RPV failure at +12 h[g]	Depressurized at +75.2 h and +78.3 h[h]	Depressurization occurred at ~+42 h[i]
Time of max containment pressure (Max containment pressure/design pressure)[j]	+11.7 h (0.84 MPa/0.43 MPa)	~+80 h (~0.75 MPa/0.38 MPa)	~+42 h (0.64 MPa/0.38 MPa)
Estimated time of core damage[k]	+4 h to +7 h	+75 h to +85 h	+36 h to +40 h
First indication of offsite release of radioactive materials[l]	←	+8.2 to +14.1 h	→
Containment venting preparation/ success[m]	+9.7 h/~+24 h	+26.7 h/not successful	+29.5 h/+42 h
Hydrogen explosion[n]	+24.8 h	None[o]	+68.2 h
Injection of freshwater/ seawater[p]	+15.0/+28.8 h	None/+77.2 h	+42.6/+46.4 h
Restoration of offsite AC power	March 20	March 20	March 22

NOTES: ADS = automatic depressurization system; EDGs = emergency diesel generators; HPCI = high-pressure coolant injection system; IC = isolation condenser; MSIV = main steam isolation valve; RCIC = reactor core isolation cooling system; RPV = reactor pressure vessel; SRV = safety relief valve.

[a]Times are from the time of the earthquake in minutes (min) or hours (h).

[b]All control rods were inserted and several actions took place in all three units, including a loss of feedwater and condensate and closure of the main steam isolation valves, all prompted by the loss of offsite AC power. The emergency diesel generators started following the loss of offsite power and supplied power to the safety systems.

TABLE C.1 Continued

[c]The tsunami generally flooded emergency diesel generators, power panels, and backup batteries, resulting in the loss of AC and DC power except for some isolated systems and standalone battery-operated instrumentation. The immediate result was the loss of normal control room lighting, indicators, and controls. All units except Unit 6 lost AC power within 5 minutes after the tsunami flooded the plant (Investigation Committee, 2011, p. 108). Units 1, 2, and 4 also lost DC power shortly after the tsunami because of flooding of the switchgear and batteries. While the air-cooled Unit 2 emergency diesel generator was running at the time, the electrical switchgear located belowgrade was flooded and subsequently failed. Although there were intermittent signs of power on some indicators in Units 1 and 2, reliable DC power was only available by connecting arrays of scavenged vehicle batteries to selected systems and instrumentation in the control rooms. Unit 3 DC power remained available for emergency lighting and indicators for some time. The Unit 3 DC bus escaped flooding, and sufficient battery capacity was available to operate the RCIC, SRVs, and HPCI for up to 36 h. Ultimately, vehicle batteries had to also be employed in Unit 3 to operate critical systems after the installed backup battery was depleted.

[d]The IC system lost its ability to effectively cool the reactor in Unit 1 at approximately the time that AC and DC power was lost, due to system fail-safe control logic. When DC power to the logic circuit is lost, an interlocking operation is activated and all four isolation valves are designed to close automatically (TEPCO, 2012b, p. 195), effectively shutting off the IC. Without AC power, the valves inside containment cannot be reopened; thus, it was not possible to recover the IC system without an AC power source. A schematic of the IC system is provided in Chapter 2 (Figure 2.7) and more complete description of the automated fail-safe control logic is provided in Chapter 4 (Section 4.3.1.1).

[e]Unit 2 RCIC was manually started for the last time just prior to the loss of all electric power at ~+54 min. The loss of power at ~+54 min compromised the ability to monitor or control RCIC injection to the RPV in Unit 2. In Unit 3, where DC power was not lost, RCIC operated as intended until it failed at +20 h into the event and could not be restarted. The HPCI started automatically (on a low-reactor-water-level signal) an hour later and began to restore water level in the RPV. HPCI was manually tripped at +35.9 h into the event and attempts to restart it failed.

[f]As described in Sections 4.3.1-4.3.3, operators had limited options for depressurization given the blackout and the ensuing chaotic conditions caused by the destruction from the earthquake and flooding waters.

[g]Analysis results suggest that reactor water level reached the TAF at about 18:10 on March 11 and core damage started at about 18:50. (TEPCO, 2012b, pp. 190-191).

[h]The MAAP5 simulations performed by EPRI (2013) indicate that the RCIC system in Unit 2 operated in a degraded mode that maintained the core cooling for nearly 70 h. During the 70-h period the RPV pressure varied between ~7.5 MPa and ~5.3 MPa (design pressure is 8.24 MPa). The rise continued to the SRV setpoint and at ~+75 h into the event the RPV was depressurized to allow seawater injection. However, pressure increases between approximately +76 h and +84 h into the event compromised the continuity of seawater injection.

[i]The operators were able to control the amount of water added to the RPV in Unit 3 until +20 h into the event. The RCIC system operated under conditions for which it was designed (DC power available). However, at about +20 h, the RCIC failed and could not be restarted. On failure of RCIC, the HPCI system automatically started on low reactor water level, rapidly restoring the level. The significant steam extracted via the operation of the HPCI system led to a rapid decrease in RPV pressure. For a little under 10 h, the HPCI system operated at low

continued

TABLE C.1 Continued

RPV pressure with the turbine rotating at a very low speed. However, it is expected that RPV pressure would decrease with HPCI operation if HPCI is running at or near rated shaft speed, not at low speed. According to the EPRI evaluation, HPCI is not designed to operate at the low pressures at which it was operating (~1 MPa). Under low-pressure conditions it does not provide adequate coolant to remove all of the decay heat. When HPCI was manually tripped at approximately +36 h into the event and could not be restarted, the RPV pressure rose sufficiently to prevent any significant injection using the diesel-driven fire protection system. From +36 h to approximately +42 h into the accident, no injection was provided to the RPV. Following depressurization of the RPV by operators at ~+42 h, injection of freshwater was established using a fire truck pump staged away from the unit. At around +46 h, there was a brief interruption in injection when the water source was switched from freshwater—which had been exhausted—to seawater.

[j]Unit 1 containment reached its peak pressure of 0.84 MPa at +11.7 h into the event. The pressure was almost twice the design pressure (0.43 MPa). The overpressure and high temperature may have damaged the pressure boundary, allowing leakage of radionuclides and flammable gases into the reactor building. Unit 2 containment pressure gradually increased during the 3-day period of RCIC operation, approaching design pressure at about the time RCIC injections stopped (at ~+71 h). The sudden rise in containment pressure at +80 h in Unit 2 may have been due to the rapid production of hydrogen from zirconium oxidation and core concrete interaction, although the data and simulations are both inconsistent (ANS, 2012; Gauntt et al., 2012b) with a loss of RPV integrity at that time. Unit 2 containment pressure reached about twice design pressure at ~+80 h into the event. Uncontrolled leakage through the pressure boundary is believed to have occurred to relieve pressure as this was about the same time that observed dose rates begin to rise at the site boundary. The uncontrolled leakage of volatile fission products from Unit 1 and 2 containments would bypass the suppression pool, limiting the scrubbing of volatile fission product aerosols. The value shown in the table is from the ANS (2012) and INPO (2011) time lines.

[k]Early core degradation and fuel melting in the Unit 1 reactor (TEPCO, 2012b, p. 191) was caused by the loss of the IC heat removal path. As discussed in this report, the loss of all power caused by the tsunami led to automatic closing of valves which could not be reopened without AC and DC power. The entire core of Unit 1 may have penetrated the reactor pressure vessel. Analyses were performed by TEPCO (2012c), JNES, Sandia (Gauntt et al., 2012b), ORNL (Robb et al., 2013), and EPRI (2013) on the progression of the accident. EPRI analyses using the MAAP computer model indicates that it is possible that core debris melted through the bottom of the reactor pressure vessel as early as +10 h into the event. Their analyses also indicate that it is possible that the RPV depressurized to containment pressure before core debris melted through the bottom of the vessel. EPRI indicated that this depressurization could have occurred through the seizure of an SRV operating under severe accident conditions. The MELCOR analyses by Sandia and ORNL also indicated large uncertainties with respect to the exact level of damage of Units 2 and 3. It is believed that the Unit 2 core might have been contained, but the possibility of some core debris inside the containment cannot be precluded. Sensitivity studies by EPRI indicate the possibility of a significant fraction of the Unit 3 core debris being relocated into the containment.

[l]The source of this release has not been confirmed; however, its timing coincides with the slow (and unexplained) reduction in containment pressure in Unit 1 (see Note j). At about +8.2 h into the event, dose rates up to 120 mrem/h (1.2 mSv/h) were detected outside the personnel airlock door of Unit 1 reactor building. Dose rates inside the control room also increased at +8.2 h into the event.

TABLE C.1 Continued

[m]At about +11.7 h into the event, the drywell pressure in Unit 1 reached about twice the design pressure, causing a leak through the pressure boundary that never completely resealed. Data on the primary containment vessel for Unit 2 indicate that venting through the suppression chamber or drywell never occurred. Despite efforts to vent containment and depressurize the RPV into the suppression pool, the containment pressure never reached the burst pressure of the rupture disk in the hardened vent piping for the suppression chamber (the set point was 0.43 MPa gauge) prior to leakage through the pressure boundary. Unit 3 maximum containment pressures were considerably less than those of Units 1 and 2, although still greater than the design pressure.

[n]There were a total of three hydrogen explosions at Units 1, 3, and 4. At +87.2 h into the event, operators heard a loud noise around the torus and pressure suppression chamber of Unit 2. It was first thought to be a hydrogen explosion in Unit 2, but comparisons of seismic data from building accelerometer signals (TEPCO, 2012b) indicate that the noise heard was most likely the explosion of the Unit 4 reactor building. The noise was accompanied by the lowering of the torus pressure to 0 kPa (abs) but this is now believed to be an instrument failure as the measured pressures in the drywell and suppression chamber were drastically different after this time. The hydrogen explosion in Unit 4 (not included in the above table) which occurred at ~+87.2 h is believed to be a result of shared systems between Unit 3 and Unit 4 with the source of hydrogen in Unit 4 being backflow through common piping with Unit 3 at the stack whose containment was being vented.

[o]The open blowout panel in the reactor building of Unit 2 caused by the hydrogen explosion in Unit 1 may have prevented buildup of hydrogen to explosive levels. Unfortunately, the open blowout panel provided direct access to the environment of radionuclides escaping from the primary containment system as a result of leakage paths.

[p]Seawater injection was aligned earlier in each case and even started earlier especially in Unit 2, but flooding, the unavailability of fire trucks, lack of fuel for the fire trucks, RPV depressurization delays and the consequences of hydrogen explosions all contributed to much later injection times than planned and even then the injections were often interrupted. For example, in Unit 3, seawater injection started at +46.4 h, was suspended at +58.4 h, was restored at +60.5 h, was suspended again at +68.2 h, and was restored again at +73.7 h.

Appendix D

Operation and Support Organizations

Personnel involved in the accident response at the Fukushima Daiichi nuclear plant were stationed at a number of locations, including in

- The main control rooms (MCRs) at the plant,
- An onsite emergency response center at the plant (onsite ERC),
- An offsite ERC established at TEPCO headquarters in Tokyo (headquarters ERC), or
- An offsite ERC located about 5 km from the plant (Offsite Center [OFC])

The functions of these organizations are described briefly in this appendix.

D.1 MAIN CONTROL ROOMS

Personnel in the MCRs are responsible for operating the reactors during both normal and off-normal conditions. The Fukushima Daiichi nuclear plant has three MCRs, one each for Units 1 and 2, 3 and 4, and 5 and 6. There are two physically separated sets of independent controls for each reactor within each MCR.

Each MCR has one operating crew that is responsible for the two reactors being operated. During a normal shift the crew would consist of the following staff (INPO, 2011):

- Shift supervisor,
- Assistant shift supervisor,
- Two senior operators,
- Assistant senior operator,
- Two main shift operators, and
- Four auxiliary operators.

Shift supervisors in the MCRs (Figure D.1) are responsible for making reactor control and operation decisions in the event of an accident in accordance with the plant's emergency procedures. However, under certain circumstances, including for actions requiring the cooperation of other control rooms or that are expected to have large impact on reactor

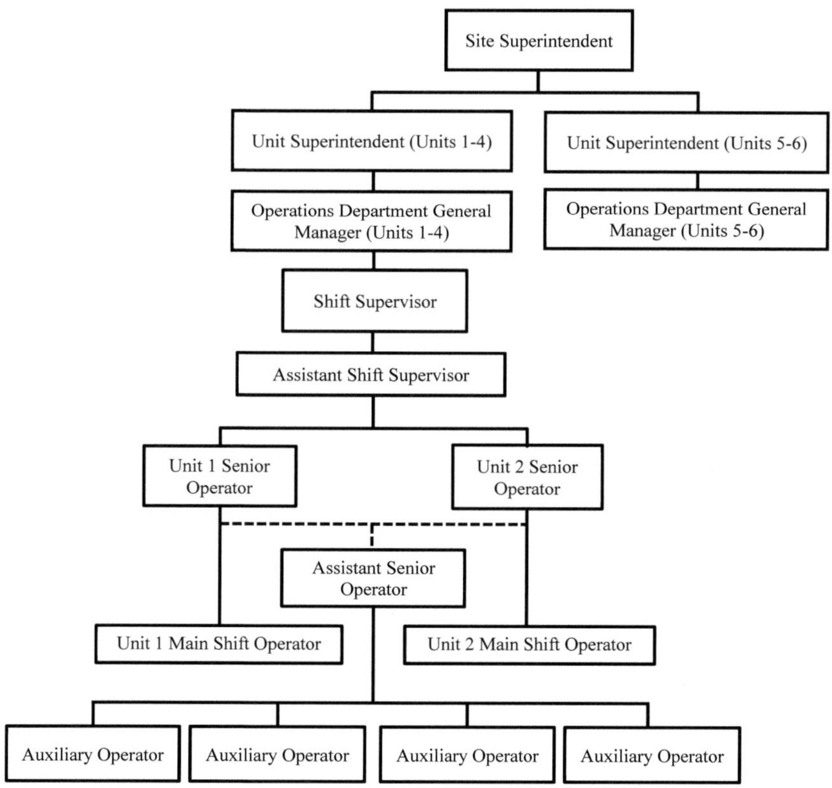

FIGURE D.1 Breakdown of responsibilities for operational staff at the Fukushima Daiichi nuclear plant. Details below the Operations Department General Manager level are shown only for Units 1 and 2. SOURCE: INPO (2011). Courtesy of the Institute of Nuclear Power Operations.

behavior, the shift supervisors are required to ask the onsite ERC for advice and direction.

At the time of the March 11, 2011, accident, the MCR crews were staffed as follows (INPO, 2011):

- The MCR for Units 1 and 2 had 11 operators and 1 trainee.
- The MCR for Units 3 and 4 had 8 operators and 1 trainee. The staffing was reduced in Unit 4 because it was in a maintenance outage.

Immediately after the earthquake, the crews then working at the MCRs were responsible for operating the reactors. Some members of other crews who were off duty at the time of the earthquake went to their control rooms to assist. Other members of those crews stayed in the ERC until it was time to relieve those on duty.

D.2 ONSITE EMERGENCY RESPONSE CENTER

The onsite ERC was housed in a seismically isolated building designed to withstand earthquakes and equipped with backup power and filtered ventilation. The building was one of the few administrative buildings at the plant that survived the earthquake and tsunami. The onsite ERC played a crucial role in coordinating and managing onsite response activities.

The onsite ERC was responsible for providing advice and direction to the MCR shift supervisor and crew. Key decision makers were seated around a large table in the middle of the ERC; these included

- Site superintendent, who serves as the emergency director and is in charge of the onsite ERC after a severe accident occurs;
- Unit superintendents;
- Deputy directors;
- Reactor chief engineers;
- Section chiefs of 12 function teams: communication, intelligence, public relations, health physics, engineering, recovery, operation, infrastructure, medical treatment, general affairs, guard guidance, and the procurement teams. An in-house firefighting team was organized under the recovery team (Investigation Committee, 2011, p. 93).

Members of each function team were stationed in booths behind their respective section chiefs to enable oral communications. When a function team obtained information that needed to be shared with all staff in the ERC they reported it to their section chief, who then announced it via microphone so that everyone in the room could hear it.

When a decision was made by the site superintendent or others at the main table or information was provided from the headquarters ERC (described in the next section) through a teleconference system, the leader of the relevant team communicated it to his team members. Members of the headquarters ERC are able to monitor discussions made at the main table of the onsite ERC, ask questions, and give advice via a teleconference system.

D.3 HEADQUARTERS EMERGENCY RESPONSE CENTER

An emergency response center was also established at the TEPCO headquarters in Tokyo. According to the Fukushima Daiichi NPS Nuclear Emergency Prevention Action Plan, the role of the headquarters ERC is to support the onsite ERC. The TEPCO president is chief of the headquarters ERC.

As part of its support function, the headquarters ERC is supposed to transmit information from the onsite ERC to the Ministry of Economy, Trade and Industry (METI) and to the Nuclear Disaster Response Headquarters at the Official Residence (i.e., prime minister's residence). The OFC (see next section) would also send information to METI and the Nuclear Disaster Response Headquarters at the Official Residence. This arrangement would, in theory, reduce the number of direct inquiries to the onsite ERC.

D.4 OFFSITE CENTER

The OFC is intended to coordinate TEPCO and central/local government activities. The OFC is located about 5 km from the plant. It is equipped with telephone lines, a video-conferencing system used primarily to connect to the Prime Minister's office, and a satellite circuit with six satellite telephones (one fixed, three portable, and two vehicle-mounted). The OFC never functioned as intended during the Fukushima nuclear accident for the reasons described in the main body of Chapter 4.

Appendix E

Recommendations from Other Organizations

As noted in Chapter 1, this National Academy of Sciences study is one of many investigations or assessments initiated in the wake of the Fukushima Daiichi accident. The reports from these activities have informed this committee's thinking about potential lessons learned for the United States. Key recommendations from these investigations/assessments have been captured and generalized in Table E.1 and compared with the recommendations from the present study.

The nuclear industry and its regulators have taken additional actions to implement lessons learned from the Fukushima Daiichi accident since the reports referenced in this table were published. Those actions are not reflected in this table.

TABLE E.1 Common recommendations from investigations and assessments of the Fukushima accident. Recommendations resulting from this report are shown with a star.

	ANS	CNSC	DIET	ENSI	IAEA	IC	INPO	JANTI	ONR	TEPCO	UCS	USNRC	NAS
RISK ASSESSMENT													
Review design bases for naturally occurring hazardous events	✓	✓		✓	✓	✓	✓	✓	✓	✓		✓	
Use risk to assess low-probability/high-consequence events	✓				✓	✓	✓	✓	✓	✓		✓	★
Consider multi-unit interactions in risk assessments	✓				✓		✓		✓				★
Review and assess risk of hydrogen-buildup during extreme events				✓	✓			✓	✓	✓		✓	★
Perform and incorporate results from extreme weather research					✓	✓							
Update or create severe accident management guidelines (SAMG)	✓	✓		✓	✓	✓	✓			✓			★
FACILITIES, EQUIPMENT, AND OPERATIONS													
Identify essential safety equipment and ensure it can support prolonged recovery operations				✓	✓				✓	✓	✓		
Secure backup power for extended periods		✓		✓				✓	✓	✓		✓	★
Build well-equipped emergency response centers with robust internal and external communication links		✓		✓	✓	✓	✓		✓	✓		✓	
Review and assess containment vent operation during extreme events				✓			✓			✓		✓	★
Review and assess reactor core cooling capability during extreme events				✓			✓	✓	✓	✓	✓		★
Improve environmental monitoring during event				✓		✓	✓		✓	✓		✓	★
Identify essential control systems that must be maintained during event		✓			✓								★
• monitoring of core and spent fuel pools				✓			✓			✓			★
• communications between systems and emergency response centers					✓								★
• ensure additional emergency resources will be made available during event				✓			✓		✓	✓	✓		

continued

TABLE E.1 Continued

	ANS	CNSC	DIET	ENSI	IAEA	IC	INPO	JANTI	ONR	TEPCO	UCS	USNRC	NAS
SPENT FUEL POOLS													
Increase robustness of spent fuel pools				✓				✓	✓	✓	✓	✓	
Reduce the number of spent fuel pools									✓		✓		
EMERGENCY PREPAREDNESS AND TRAINING													
Prepare for prolonged system blackouts				✓			✓	✓	✓	✓	✓	✓	★
Train for emergency events (general)		✓		✓	✓	✓	✓	✓		✓		✓	★
• develop procedures and train for venting for explosive gas buildup during events	✓			✓			✓		✓				★
• develop procedures and train for backup cooling options				✓			✓	✓	✓	✓			★
• develop procedures and train to have radiation protection/monitoring available for workers				✓	✓		✓			✓			
Maintain large pool of trained staff for sustained emergencies				✓	✓		✓			✓			★
SAFETY CULTURE AND OVERSIGHT													
Improve and maintain safety culture at nuclear power plants			✓	✓		✓	✓		✓	✓			★
Improve and maintain transparent regulatory oversight and independence		✓	✓	✓	✓	✓			✓			✓	★
Support and implement IAEA (or similar international) safety standards				✓	✓	✓			✓				
Allow dose limit flexibility during emergencies		✓					✓			✓			

OTHER													
Plan and train for communication with the public during event	✓			✓		✓	✓		✓	✓		✓	★
Maintain robust meteorological data throughout event for dispersion models			✓	✓	✓				✓				
Define clear evacuation guidelines and procedures			✓	✓		✓				✓			★
Establish clear lines of authority	✓		✓	✓			✓			✓			★
Improve international harmonization of information during emergencies (data sharing, coordination)				✓		✓			✓	✓			

NOTES: ANS = American Nuclear Society (ANS, 2012); CNSC = Canadian Nuclear Safety Commission (CNSC, 2011); DIET = Nuclear Accident Independent Investigation Commission, The National Diet of Japan (NAIIC, 2012); ENSI = Swiss Federal Nuclear Safety Inspectorate (ENSI, 2011); IAEA = International Atomic Energy Agency (IAEA, 2011); IC = Investigation Committee on the Accident at Fukushima Nuclear Power Stations of Tokyo Electric Power Company (Investigation Committee, 2012); INPO = Institute of Nuclear Power Operations (INPO, 2012); JANTI = Japan Nuclear Technology Institute (JANTI, 2011); ONR = Office for Nuclear Regulation (ONR, 2011); TEPCO = Tokyo Electric Power Company, Inc. (TEPCO, 2012b); UCS = Union of Concerned Scientists (UCS, 2011a); USNRC = U.S. Nuclear Regulatory Commission Near-Term Task Force (USNRC NTTF, 2011); NAS = National Academy of Sciences Committee on Lessons Learned from the Fukushima Nuclear Accident for Improving the Safety of U.S. Nuclear Plants (this report).

Appendix F

Regulator and Industry Actions in the United States

This appendix describes some key actions being taken by the U.S. nuclear industry and the U.S. Nuclear Regulatory Commission (USNRC) as a direct result of the Fukushima Daiichi accident.

F.1 REGULATOR ACTIONS

The USNRC took several near-term actions in the weeks following the Fukushima Daiichi accident:

- On March 18, 2011, the USNRC issued Information Notice 2011-05,[1] "Tohoku-Taiheiyou-Oki Earthquake Effects on Japanese Nuclear Power Plants." This notice was intended to "inform [U.S. nuclear plant licensees] of effects of the Tohoku-Taiheiyou-Oki Earthquake [Great East Japan Earthquake] on nuclear power plants in Japan. The [US]NRC expects that recipients will review the information for applicability to their facilities and consider actions, as appropriate, to avoid similar problems."
- On May 11, 2011, the USNRC issued Bulletin 2011-01,[2] "Mitigating Strategies," requiring that licensees "provide a comprehensive verification of their compliance with the regulatory requirements of Title 10 of the Code of Federal Regulations (10 CFR) Section 50.54(hh)(2)" and "to determine if 1) additional assessment of program implementation is needed,

[1] Available at http://pbadupws.nrc.gov/docs/ML1107/ML110760432.pdf.
[2] Available at http://pbadupws.nrc.gov/docs/ML1112/ML111250360.pdf.

2) the current inspection program should be enhanced, or 3) further regulatory action is warranted."

- On May 23, 2011, the USNRC issued Temporary Instruction 2515/183,[3] "Followup to the Fukushima Daiichi Nuclear Station Fuel Damage Event." This instruction was intended to "independently assess the adequacy of actions taken by licensees in response to the Fukushima Daiichi nuclear station fuel damage event. The inspection results from this TI will be used to evaluate the industry's readiness for a similar event and to aid in determining whether additional regulatory actions by the U.S. Nuclear Regulatory Commission are warranted."
- On March 23, 2011, the chairman of the USNRC directed the agency's executive director of operations to establish a task force composed of senior-level staff to conduct a review of the agency's processes and regulations and make recommendations to improve them.[4] A six-member task force, the Near-Term Task Force Review of Insights from the Fukushima Dai-ichi Accident (NTTF), was established to undertake this review.
- On April 29, 2011, the USNRC issued Temporary Instruction 2515/184,[5] "Availability and Readiness Inspection of Severe Accident Management Guidelines (SAMGs)." This instruction was intended to "[d]etermine that the severe accident management guidelines (SAMGs) are available and how they are being maintained" and also "[d]etermine the nature and extent of licensee implementation of SAMG training and exercises."

The NTTF delivered its report on July 17, 2011. The NTTF concluded that

> a sequence of events like the Fukushima accident is unlikely to occur in the United States and some appropriate mitigation measures have been implemented, reducing the likelihood of core damage and radiological releases. Therefore, continued operation and continued licensing activities do not pose an imminent risk to public health and safety. (USNRC NTTF, 2011, p. vii)

The NTTF also found that

> the Commission's longstanding defense-in-depth philosophy, supported and modified as necessary by state-of-the-art probabilistic risk assessment techniques, should continue to serve as the primary organizing principle of its regulatory framework. The Task Force concludes that the applica-

[3] Available at http://pbadupws.nrc.gov/docs/ML1107/ML11077A007.pdf.

[4] Available at http://www.nrc.gov/reading-rm/doc-collections/commission/comm-secy/2011/2011-0002comgbj-srm.pdf.

[5] Available at http://pbadupws.nrc.gov/docs/ML1111/ML11115A053.pdf.

tion of the defense-in-depth philosophy can be strengthened by including explicit requirements for beyond-design-basis events. (USNRC NTTF, 2011, p. viii)

The NTTF recommended a series of 12 broad actions "to clarify and strengthen the regulatory framework for protection against natural disasters, mitigation, and emergency preparedness, and to improve the effectiveness of the NRC's programs" (USNRC NTTF, 2011, p. viii). These were the following:

Clarifying the Regulatory Framework

1. The Task Force recommends establishing a logical, systematic, and coherent regulatory framework for adequate protection that appropriately balances defense-in-depth and risk considerations.

Ensuring Protection

2. The Task Force recommends that the NRC require licensees to reevaluate and upgrade as necessary the design-basis seismic and flooding protection of structures, systems, and components for each operating reactor.

3. The Task Force recommends, as part of the longer term review, that the NRC evaluate potential enhancements to the capability to prevent or mitigate seismically induced fires and floods.

Enhancing Mitigation

4. The Task Force recommends that the NRC strengthen station blackout mitigation capability at all operating and new reactors for design-basis and beyond-design-basis external events.

5. The Task Force recommends requiring reliable hardened vent designs in boiling water reactor facilities with Mark I and Mark II containments.

6. The Task Force recommends, as part of the longer term review, that the NRC identify insights about hydrogen control and mitigation inside containment or in other buildings as additional information is revealed through further study of the Fukushima Dai-ichi accident.

7. The Task Force recommends enhancing spent fuel pool makeup capability and instrumentation for the spent fuel pool.

8. The Task Force recommends strengthening and integrating onsite emergency response capabilities such as emergency operating procedures, severe accident management guidelines, and extensive damage mitigation guidelines.

Strengthening Emergency Preparedness

9. The Task Force recommends that the NRC require that facility emergency plans address prolonged station blackout and multiunit events.

10. The Task Force recommends, as part of the longer term review,

> that the NRC pursue additional emergency preparedness topics related to multiunit events and prolonged station blackout.
>
> 11. The Task Force recommends, as part of the longer term review, that the NRC should pursue emergency preparedness topics related to decisionmaking, radiation monitoring, and public education.
>
> Improving the Efficiency of NRC Programs
>
> 12. The Task Force recommends that the NRC strengthen regulatory oversight of licensee safety performance (i.e., the Reactor Oversight Process) by focusing more attention on defense-in-depth requirements consistent with the recommended defense-in-depth framework. (USNRC NTTF, 2011, p. ix)

The USNRC is using these Near-Term Task Force recommendations as the basis for developing regulatory actions. By the first anniversary of the Fukushima Daiichi accident, several regulatory actions had already been taken. These included requests for information from nuclear plant licensees, orders for immediate actions, and regulatory rulemaking. These actions are summarized in Table F.1 and described below.

TABLE F.1 Summary of USNRC Actions following the Fukushima Nuclear Accident Through May 2014

Regulatory Action	Summary Description	Method
Seismic and flood walkdowns[a]	Inspect existing protection features against seismic and flood design-basis events; correct degraded conditions	Information request
Seismic reevaluations[b]	Reanalyze potential seismic effects using present-day information to determine if safety upgrades are needed	Information request
Flooding reevaluations[c]	Reanalyze potential flooding effects using present-day information to determine if safety upgrades are needed	Information request
Emergency preparedness[d]	Assess staffing needs and communication capabilities to effectively respond to an event affecting multiple units	Information request
Spent fuel pool instrumentation[e]	Provide a reliable wide-range indication of water level in spent fuel storage pools	Regulatory order

continued

TABLE F.1 Continued

Regulatory Action	Summary Description	Method
Containment venting system[f]	Provide a reliable hardened containment vent system for BWRs with Mark I or Mark II containment systems	Regulatory order
Mitigation strategies[g]	Enhance the capability to maintain plant safety during a prolonged loss of electrical power	Regulatory order
SBO mitigation strategies[f]	Enhance the capability to maintain plant safety during a prolonged loss of electrical power	Regulatory rulemaking
Onsite emergency response procedures[h]	Strengthen and integrate different types of emergency procedures and capabilities at nuclear plants	Regulatory rulemaking
Filtration and confinement strategies[e]	Evaluate potential strategies that may further confine or filter radioactive materials if core damage occurs	Regulatory rulemaking
Regulatory framework[i]	Consider a revised risk-informed regulatory framework that affects reactor oversight process	Still in process

[a] http://pbadupws.nrc.gov/docs/ML1212/ML12129A108.pdf (June 27, 2012); http://pbadupws.nrc.gov/docs/ML1215/ML12156A052.pdf (July 6, 2012)
[b] http://pbadupws.nrc.gov/docs/ML1228/ML12286A029.pdf (November 12, 2012); http://pbadupws.nrc.gov/docs/ML1310/ML13106A331.pdf (May 7, 2013).
[c] http://pbadupws.nrc.gov/docs/ML1231/ML12311A214.pdf (November 30, 2013); http://pbadupws.nrc.gov/docs/ML1231/ML12314A412.pdf (January 4, 2013).
[d] http://pbadupws.nrc.gov/docs/ML1205/ML12053A340.pdf (March 12, 2012); http://pbadupws.nrc.gov/docs/ML1212/ML12125A412.pdf (May 2012).
[e] http://pbadupws.nrc.gov/docs/ML1205/ML12056A044.pdf (March 12, 2012); http://pbadupws.nrc.gov/docs/ML1222/ML12221A339.pdf (August 29, 2012).
[f] http://pbadupws.nrc.gov/docs/ML1314/ML13143A321.pdf (June 6, 2013)
[g] http://pbadupws.nrc.gov/docs/ML1205/ML12054A735.pdf (March 12, 2012); http://pbadupws.nrc.gov/docs/ML1222/ML12229A174.pdf (August 29, 2012); http://pbadupws.nrc.gov/docs/ML1224/ML12242A378.pdf (August 2012).
[h] http://www.regulations.gov/#!docketDetail;D=NRC-2012-0031.
[i] http://www.regulations.gov/#!docketDetail;D=NRC-2012-0173.

F.1.1 Seismic and Flooding Walkdowns

The USNRC concluded that U.S. nuclear plants needed to reconfirm their existing ability to resist seismic and flooding events. On March 12, 2012, the USNRC requested that plant licensees perform detailed inspections ("walkdowns") of their currently installed seismic and flooding protection features. Licensees were asked to ensure that the plant features met current requirements, and also identify, correct, and report any degraded conditions. The plants completed their walkdowns by November 2012. The USNRC has carried out follow-up inspections, and the agency's technical experts are reviewing licensee walkdown reports.

F.1.2 Seismic and Flooding Reevaluations

Licensees were directed to reevaluate earthquake and flooding hazards that could impact their plant sites. The evaluation is to determine whether plant structures, systems, and/or components need to be updated to protect against these hazards. The USNRC will review each step in the analysis process and take action to require plant changes as necessary.

F.1.3 Emergency Preparedness

The USNRC directed licensees to assess how many emergency staff they will need to respond to a large accident that could affect multiple reactors at their sites and to make changes to their emergency plans as necessary. The USNRC also directed licensees to ensure that plants can maintain communications equipment that their staff will need to effectively respond to such an accident (e.g., radios for response teams, cellular telephones, and satellite telephones).

F.1.4 Spent Fuel Pool Instrumentation Order

The USNRC issued an order on March 12, 2012, requiring that licensees install additional water-level instrumentation in the spent fuel pools at their plants. The instrumentation must remotely report at least three distinct water levels: (1) normal level, (2) low level but adequate to shield workers from radiation, and (3) a level near the top of the spent fuel rods where more water should be added without delay.

F.1.5 Containment Venting Systems Order

The USNRC issued an order on March 12, 2012, requiring all licensees of boiling water reactors with Mark I and Mark II containments to install reliable, hardened vents that can be used to vent containments. After issuing the order, additional USNRC evaluations examined the benefits of venting after reactor core damage occurs. In June 2013, the USNRC modified the order to ensure that those vents will remain functional if the reactor core is damaged.

F.1.6 Hardened Vents and Filtration Rulemaking

The USNRC is evaluating the need for filtered vents in Mark I and Mark II containments through the agency's rulemaking process.

F.1.7 Mitigation Strategies Order

The USNRC issued an order on March 12, 2012, requiring licensees of U.S. nuclear plants to implement strategies for coping without permanent electrical power sources for an indefinite period of time. These strategies are expected to utilize currently installed equipment (e.g., pumps that are powered by steam rather than by electrical power), portable equipment stored onsite, and equipment that can be shipped to the site.

F.1.8 Station Blackout Mitigation Strategies Rulemaking

The USNRC is conducting a rulemaking to permanently write into the agency's rules the requirements imposed by the Mitigation Strategies Order.

F.1.9 Onsite Emergency Response Capabilities Rulemaking

The USNRC is conducting a rulemaking to strengthen and integrate emergency response capabilities at U.S. nuclear plants. Currently, U.S. plants may have several categories of response procedures that they draw upon, depending on the nature of the incident. This rulemaking will establish standards to ensure that the plants can smoothly transition between these procedures while keeping plants' overall strategies coherent and comprehensive. The new rule will also require plants to improve strategies for large-scale events to promote effective decision making at all levels. The rule will include training, qualification, and evaluation requirements for the key personnel expected to implement the procedures and strategies.

F.1.10 Risk-Informed Regulatory Framework

The task force recommended that the NRC establish a logical, systematic, and coherent regulatory framework that appropriately balances multiple layers of protection and risk considerations to deal with beyond-design-basis events. This work is still in the process of being reviewed in connection with the Task Force Report on a Proposed Risk Management Regulatory Framework (USNRC, 2012a).

F.2 INDUSTRY ACTIONS

The U.S. nuclear industry, led by the Nuclear Energy Institute, Institute of Nuclear Power Operations, and Electric Power Research Institute, initiated a voluntary effort to integrate and coordinate the industry's response to the Fukushima Daiichi accident (NEI, 2012). The industry developed the "Diverse and Flexible Coping Strategies (FLEX) Implementation Guide" (NEI, 2012) to satisfy the USNRC's Mitigation Strategies Order.

FLEX is intended to augment plant coping capabilities (Figure F.1) for external beyond-design-basis (BDB) events. The strategy itself consists of the following four elements:

1. To have portable backup equipment capable of providing water and power to the reactor. Such equipment includes, for example, electrical generators, batteries, and battery chargers; compressors; pumps, hoses, and couplings; equipment for clearing debris; and equipment for temporary protection against flooding;
2. To stage this equipment in locations both onsite and offsite where it will be safe and deployable during a BDB external event;
3. To develop procedures and guidance for implementing FLEX;
4. To develop programmatic controls that will ensure personnel are well trained and equipment is maintained;

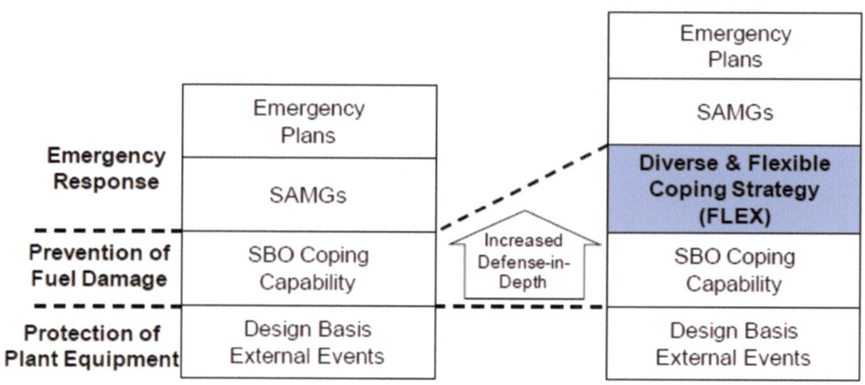

FIGURE F.1 FLEX is intended to augment coping capabilities for beyond-design-basis events involving the simultaneous loss of emergency AC power and ultimate heat sink at all units at a nuclear plant, thereby increasing defense in depth. SOURCE: NEI (2012).

Because each reactor has unique features, FLEX implementation is unit specific. Plants establish a baseline of current capabilities for coping with extreme events, assuming the following initial and boundary conditions:

- An external event impacts all units at a site.
- Initially the reactors are operating at power and then are safely shut down.
- No additional initiating events or failures are assumed to occur immediately prior to or during the event.
- All staff, at the minimum staffing levels, are available to help.

The plant then examines the likelihood and impacts of the following extreme external events:

- Large earthquakes,
- External flooding,
- Intense storms (e.g., hurricanes, high winds, tornados), and
- Extreme snow, ice, cold, and heat.

The plant then develops implementation procedures for the protection and deployment of equipment, procedural interfaces, and utilization of offsite resources for each of these events. Finally, plants use this analysis to identify needed enhancements to baseline capabilities.

NEI submitted a final draft of the FLEX plan to the USNRC in August 2012. The USNRC issued a statement and interim staff guidance (USNRC, 2012c) concluding that the FLEX plan successfully implements the Mitigation Strategies Order and is an acceptable approach to meet the December 31, 2016, compliance deadline in that order.

Appendix G

Hydrogen Control in Severe Accidents

This appendix describes regulatory actions to control hydrogen in nuclear plants since the 1979 Three Mile Island accident and why they were insufficient to prevent hydrogen explosions in the Fukushima Daiichi plant.

G.1 REGULATORY ACTIONS FOLLOWING THE THREE MILE ISLAND ACCIDENT

Immediately following the 1979 Three Mile Island accident, the USNRC established a lessons-learned task force to identify and evaluate safety concerns arising from the accident and recommend appropriate changes to licensing requirements and licensing processes for nuclear power plants. The task force made a number of recommendations (USNRC, 1980e), including two recommendations for controlling hydrogen produced by severe core accidents:

1. Provide inerting for all Mark I and Mark II BWR containments.
2. Provide the capability to add a hydrogen recombiner system (for hydrogen control) within a few days after an accident.

The inerting requirement was implemented in December 1981 as the first interim hydrogen rule for Mark I and Mark II reactors.[1] Plants that

[1] This resulted in an amendment to 10 CFR § 50.44 requiring inerted atmospheres in BWR Mark I and Mark II containments.

did not already have inerting systems were required to install them, and new plants were required to be equipped with hydrogen inerting systems.

These systems were used to displace air inside the containment with nitrogen to reduce oxygen concentration below 4 percent when the reactor was operating. This change was adopted worldwide, including at the Fukushima Daiichi plant. It has been widely accepted in the nuclear power and combustion communities that inerting resolved the hydrogen issue for plants with Mark I and Mark II containments (USNRC, 1987).

Hydrogen control and equipment survivability became important considerations in other containment designs (PWR plants with ice condenser containments and BWR plants with Mark III containments) that were coming online in the 1980s. In 1985, a rule required that plants having these containments must control combustible gas generated by up to 75 percent metal–water reaction to less than 10 percent hydrogen. New reactor designs were required to consider up to 100 percent metal–water reaction.

Three unresolved (generic) safety issues arose from the Three Mile Island Action Plan and subsequent research on hydrogen combustion inside containments:

- GSI-A48: Hydrogen Control Measures and Effects of Hydrogen Burns on Safety Equipment. Initiated by Three Mile Island Task Force findings and resolved in 1989 with changes to 10 CFR § 50.44 and results of research and testing programs. The exception was the large dry containment systems, which were treated by GSI-121.
- GSI-121: Hydrogen Control for Large, Dry PWR Containments. Initiated by USNRC staff as part of rulemaking for GSI-A48 and resolved in 1992. No new requirements were made for large dry containments and deliberate ignition systems were judged to not be cost-effective.
- GSI-189: Susceptibility of Ice Condenser and Mark III Containments to Early Failure from Hydrogen Combustion During a Severe Accident. Proposed in 2000 in response to industry requests to reconsider 10 CFR § 50.44 and long-standing concerns regarding station blackout leading to inoperable deliberate ignition systems. Resolved in 2007 through the addition of backup power systems.

Preventing containment failure by managing both pressure and thermal loads is critically important. The installation of severe accident–capable vents, availability of backup air and power sources, and revised accident management strategies are all steps that are currently being taken to address this critical issue.

G.2 IMPLICATIONS OF FUKUSHIMA DAIICHI ACCIDENT FOR HYDROGEN CONTROL

The accident at the Fukushima Daiichi Plant demonstrates that inerting primary containment is not sufficient to protect plants against hydrogen explosions. If the containment fails during a severe accident, the hydrogen generated by the metal–water reaction in the damaged reactor core can be released into the reactor building, mix with air, and burn. For this reason, the most effective control strategy is to manage the pressure and thermal loads on containment to prevent its failure. This requires the capability to safely vent hydrogen in a timely fashion with a minimum release of fission products into the environment.

The maximum amount of hydrogen generated in a severe core accident is almost three times the volume of nitrogen present initially in the primary containment. This quantity of hydrogen overwhelms the inerting effect of nitrogen. When the hot hydrogen–nitrogen–steam mixture leaks into the reactor building, the steam will begin to condense, and a flammable mixture will be formed.

The explosions at the Fukushima Daiichi plant significantly degraded the ability of personnel at the plant to mount an effective accident response. Substantial structural damage occurred to the Unit 1, 3, and 4 reactor buildings, and particularly Units 3 and 4, creating concerns about the integrity of their spent fuel pools as well. The explosions also created pathways into the environment for radioactive material leaks from containment. An intact BWR building acts as a filter to trap fission products released from the damaged core during a severe accident. Filtering is effective only if the reactor building remains intact and fission products can be removed by passing the exhaust gas through the filters in the standby gas treatment system.

In the 1980s, researchers at Oak Ridge National Laboratory examined severe accidents in boiling water reactor plants and the mitigating role of reactor buildings (i.e., secondary containment) on fission product releases. Greene (1990) specifically examined the potential for secondary containment failure due to combustion of hydrogen. He noted that reactor buildings have complex structures and relatively low failure overpressures (the pressure resulting from even a low-speed combustion event will substantially exceed the estimated failure pressure of the building outer walls); consequently, combustion of large amounts of hydrogen in a reactor building "would probably challenge the integrity of the secondary containment" (Greene, 1990). Greene identified two key mitigation strategies that focused on maintaining primary containment integrity: primary containment sprays and primary containment venting.

The explosions at the Fukushima Daiichi plant were indeed extremely destructive. The complex structure of the lower part of the reactor buildings is well suited to cause flame acceleration and potentially transition to detonation (see Sidebar 4.1 in Chapter 4). Ironically, having a strong structure with multiple compartments can greatly enhance the damage over a weaker structure—this result, although not intuitive, is now well established (NEA, 2000) and is an important consideration in combustion hazard analysis.

Based on what has been known about hydrogen behavior since 1980, the explosions and damage to reactor buildings at the Fukushima Daiichi plant should not have been surprising. They illustrate in dramatic fashion the importance of hydrogen control in severe accidents. Of course, the first line of defense in controlling hydrogen is to prevent the metal–water reaction in the core from occurring. The second line of defense is to manage the pressure and thermal loads on the containment to prevent failure. These are the primary goals of all accident management strategies. If these actions can be accomplished, then as a secondary result, hydrogen generation, releases, and explosion hazards will be minimized.

The Fukushima Daiichi accident prompted the Nuclear Energy Agency to produce a report on hydrogen generation, transport, and mitigation under severe accident conditions (NEA, 2014). The report summarizes the status of national requirements for hydrogen management and mitigation and computer codes for hydrogen risk assessment. The Natural Resources Defense Council considered a wide range of topics related to hydrogen explosions in severe accidents and issued a report giving their perspective on the issues (Leyse, 2014). These reports were issued just as the present report was being finalized.

Appendix H

Nuclear Plant Emergency Procedures and Guidelines

This appendix describes emergency procedures and guidance used at U.S. nuclear plants and how they are being revised in response to the Fukushima Daiichi accident.

H.1 EMERGENCY OPERATING PROCEDURES

Emergency Operating Procedures (EOPs) are "plant procedures that direct operators' actions necessary to mitigate the consequences of transients and accidents that have caused plant parameters to exceed reactor protection system set points or engineered safety feature set points, or other established limits" (USNRC, 1982, p. 3). For example, station blackout (i.e., loss of all AC power) situations or loss of ultimate heat sink can be handled within EOPs as long as reactor pressure and water level can be monitored and remain within acceptable ranges. An example of the successful use of EOPs in Japan is the response at the Fukushima Daini plant (see Sidebar 4.2 in Chapter 4). EOPs have always been a part of operational practice in the United States and are based around transient events or accidents that the plant was designed to handle, in some cases with operator actions, that is, the design-basis events—although a larger range of events was originally considered in the EOPs for boiling water reactors than for pressurized water reactors.

EOPs have long been part of the USNRC's safety requirements. These requirements are provided in 10 CFR Part 50 and in the technical specifications for each plant. Numerous technical reports (e.g., USNRC, 1980a,c, 1982, 1983) also help guide the development of EOPs. Training and both written and simulator exams for licensing reactor operators and senior reactor operators include EOPs. The shift supervisor, who is stationed in the control room, and the plant manager have command-and-control responsibilities for implementing EOPs. (Both individuals possess senior reactor operator licenses.)

H.2 SEVERE ACCIDENT MANAGEMENT GUIDELINES

Severe Accident Management Guidelines (SAMG) are intended to address "beyond-design-basis" situations in which the core has become or is becoming damaged. The goals of the SAMG are to stabilize a degraded core, maintain containment, and minimize the release of the core's fission products. SAMG are much less specific than the EOPs because they cover a wide range of possibilities of the reactor damage state after significant fuel damage occurs. The phenomenology of severe accidents in light-water reactors is too complex and highly dependent upon the timing of mitigation actions to be fully predictable in advance. An extensive discussion of the SAMG can be found in Sehgal (2012, ch. 6).

Events involving the loss of core cooling are considered to be beyond the nuclear plant's design basis and are covered by SAMG. The requirements in 10 CFR § 50.63[1] (Loss of all Alternating Current Power) address the conditions that can lead to the loss of core cooling. Licensees are required to provide an additional source of electrical power or otherwise demonstrate that the plant could cope with the loss of all AC power through other means for removing decay heat from the reactor for a specified period of time.

In events involving the loss of all AC power, operators would follow the procedures required under 10 CFR § 50.63(c)(ii)-(iii):

> (ii) A description of the procedures that will be implemented for station blackout events for the duration determined in paragraph (c)(1)(i) of this section and for recovery therefrom; and
>
> (iii) A list of modifications to equipment and associated procedures, if any, necessary to meet the requirements of paragraph (a) of this section, for the specified station blackout duration determined in paragraph (c)(1)(i) of this section, and a proposed schedule for implementing the stated modifications.

The procedures developed to address (ii), above, would address the maintenance of cooling functions using an alternative AC power source

---

[1]Available at http://www.nrc.gov/reading-rm/doc-collections/cfr/part050/part050-0063.html.

or coping strategies. The procedures would also address the restoration of onsite and offsite AC power sources.

A key difference between EOPs and SAMG is that the former are subject to regulatory oversight (see USNRC, 1982) whereas SAMG are a voluntary industry program. Another important difference is that SAMG anticipate that the engineering staff in the technical support center will be available to guide reactor operators in applying the guidance and evaluating trade-offs that inevitably occur in severe accident management, whereas EOPs enable control room staff to engage in immediate symptom-based responses. Transition points between EOPs and SAMG are defined, but some element of judgment is required to determine whether the transition criteria have been met. Consequently, operator training and education play an important role in making timely decisions.

SAMG make use of both standard and nonstandard plant systems. They include approaches to evaluate plant conditions, select the appropriate guidance, and evaluate the effectiveness of the selected guidance during a severe event. It also includes training plans for staff expected to be involved in any of the following three activities: (1) evaluation of plant damage, (2) making decisions on which strategies to implement, or (3) implementing the selected strategies.

NEI 91-04 (NEI, 1994) recommends that plants self-evaluate their strategies through use of periodic minidrills that ensure that personnel who would be involved in the emergency response are familiar with the implementation of SAMG. However, since SAMG are considered an industry initiative, the USNRC has no specific regulatory control. Instead, the USNRC has accepted the industry's commitment to assess its capabilities and implement appropriate improvements *within the constraints of existing personnel and hardware* (Taylor, 1996). In other words, the range of severe accident scenarios that could be managed with the training and steps outlined in the SAMG is limited to those situations that do not require additional resources in staffing or equipment.

Within the last decade, new requirements going beyond this limited approach have been created to respond to potential terrorist attacks. The events at the Fukushima Daiichi plant have further emphasized the need for a more comprehensive approach to severe accident management. Indeed, industry is in the process of developing and implementing new SAMG and associated physical resources.

H.3 EXTENSIVE DAMAGE MITIGATION GUIDELINES

Following the terrorist attacks of Sept 11, 2001, there was significant concern in the United States about attacks on nuclear power plants using hijacked airplanes or other means (e.g., NRC, 2004b). The USNRC and national laboratories analyzed terrorist attack scenarios on nuclear plants and their spent fuel pools and concluded that additional security and mitigation measures were needed (USNRC, 2010). The USNRC issued an Interim Compensatory Measure (ICM) Order in 2002 modifying the operating licenses of all plants. Section B.5.b of that order directed plant licensees to take certain actions:

> Section B.5.b of the ICM Order requires licensees to adopt mitigation strategies using readily available resources to maintain or restore core cooling, containment, and spent fuel pool cooling capabilities to cope with the loss of large areas of the facility due to large fires and explosions from any cause, including beyond-design-basis aircraft impacts.

The utilities, working through the Nuclear Energy Institute, developed detailed guidance for B.5.b response procedures, termed *Extensive Damage Mitigation Guidelines* (EDMGs), and additional equipment to be located at each site. The guidance assumed conditions far beyond design-basis accidents, including loss of all AC and DC power, denial of access to structures including the control room, and loss of plant control and monitoring capability.

EDMGs play a different role than EOPs. EDMGs are intended to provide operators with a "toolbox" of capabilities that can be used to respond to unpredictable damage from large fires and explosions. EDMGs also serve as a bridge between the plant operational command and control and the command and control that is provided by the plant's emergency response organization.

Little was publicly known about these B.5.b activities because they were initially protected as Safeguards Information. Additional details about the program became public knowledge after the March 2009 rulemaking that codified the B.5.b requirements contained in the order into regulations (10 CFR § 50.54(hh)(2)[2]) and the post-Fukushima acknowledgment (USNRC Bulletin 2011-01, "Mitigating Strategies"[3]) of the potential importance of B.5.b capabilities for responding to beyond-design-basis events.

Because the B.5.b order was determined to be Safeguards Information, the nuclear utilities in Japan were unaware of some of its content, although the Nuclear Safety Commission of Japan apparently was notified of its requirements. Even after the B.5.b. requirements became public knowledge, however, Japanese authorities did not recognize the change of policy and

[2] 10 CFR § 50.54. Conditions of Licenses. Available at http://www.nrc.gov/reading-rm/doc-collections/cfr/part050/part050-0054.html.

[3] Issued May 11, 2011. Available at http://pbadupws.nrc.gov/docs/ML1112/ML111250360.pdf.

therefore did not initiate any consultations on the requirements with Japanese nuclear utilities.

Many of the B.5.b capabilities and accident mitigation measures were needed or used at the Fukushima Daiichi and Daini plants following the March 11, 2011, earthquake and tsunami. The prepositioned equipment resources for B.5.b include portable generators, fire trucks or other portable water pumps, batteries, cables, tools, fuel, and firefighting equipment, all of which were part of these plant's responses. The mitigation strategies that the EDMGs are intended to cover are listed in Table H.1.

At least four of these boiling water reactor strategies were utilized at the Fukushima Daiichi plant, supporting the claim by the USNRC (2013a, p. 21) that

> the mitigating strategies implemented at U.S. nuclear plants following the terrorist attacks of September 11, 2001, to cope with large fires and explosions may have helped in responding to an extended loss of electrical power and core cooling capability that occurred at Fukushima if the equipment was stored in an area of the plant that was not inundated by the tsunami.

TABLE H.1 EDMGs Mitigation Strategies

BWR Mitigation Strategies	PWR Mitigation Strategies
Manual operation of RCIC or isolation condenser	Makeup to RWST
DC power supplies to allow depressurization of RPV & injection with portable pump	Manually depressurize SGs to reduce inventory loss
Utilize feedwater and condensate	Manual operation of turbine- (or diesel-) driven AFW pump
Makeup to condenser hotwell	Manually depressurize SGs and use portable pump
Makeup to CST	Makeup to CST
Maximize CRD	Containment flooding with portable pump
Procedure to isolate RWCU	Portable sprays
Manually open containment vent lines	
Inject water into drywell	
Portable sprays	

NOTE: AFW = auxiliary feed water, BWR = boiling water reactor, CRD = control-rod drive, CST = condensate storage tank, PWR = pressurized water reactor, RCIC = reactor core isolation cooling system, RPV = reactor pressure vessel, RWCU = reactor water cleanup, RWST = reactor water storage tank, SG = steam generator.
SOURCE: NEI (2012).

H.4 POST-FUKUSHIMA CHANGES

By definition, severe accidents are considered to result in plant conditions that are beyond design basis and outside of the traditional regulatory scope. Nevertheless, the USNRC does have the ability to inspect individual plants to verify that licensees have implemented SAMG. The USNRC used this authority following the Fukushima Daiichi accident to collect information on the implementation, training, and maintenance of SAMG. The USNRC Near-Term Task Force noted that, while some plants have maintained this important safety program, others have treated the volunteer initiative in a

> significantly less rigorous and formal manner, so much so that the SAMG inspection would have resulted in multiple violations had it been associated with a required program. (USNRC NTTF, 2011, p. 64)

The Task Force recommended that the USNRC initiate a rulemaking that would place SAMG under its oversight authority.

The industry has also taken a series of actions following the Fukushima Daiichi accident (see Appendix F). The 2012 report "The Way Forward" (NEI et al., 2012) outlines a set of goals and actions that the industry has committed to undertake to improve nuclear safety and apply lessons learned from the Fukushima Daiichi accident. These efforts are voluntary, remaining subject to inspection but outside of regulatory requirements. The industry is currently actively engaged with the USNRC in discussing how the industry response will fit in with the proposed changes in the regulatory framework mentioned above.

H.4.1 Diverse and Flexible Coping Strategies (FLEX)

An important component of the industry's response is the FLEX program, a set of prepositioned capabilities designed to extend the coping period in the event of an extended AC power loss and other adverse situations such as occurred at the Fukushima Daiichi plant. These capabilities are intended to be used in conjunction with revised SAMG. The USNRC reviewed FLEX and ordered that each U.S. nuclear plant develop a site-specific plan to mitigate severe accidents of the type experienced at Fukushima Daiichi using FLEX-type capabilities (USNRC, 2012d). The order requires a phased approach with the following elements (USNRC, 2012d, Attachment 2, p. 4):

> - The initial phase requires the use of installed equipment and resources to maintain or restore cooling, containment and spent fuel pool (SFP) cooling capabilities.

- The transition phase requires providing sufficient, portable, onsite equipment and consumables to maintain or restore these functions until they can be accomplished with resources brought from off site.
- The final phase requires obtaining sufficient offsite resources to sustain those functions indefinitely.

The FLEX implementation guide (NEI, 2012) contains these elements and was endorsed in USNRC Interim Staff Guidance (USNRC, 2012c) as being an acceptable means of complying with the Mitigation Strategies Order. The only caveat was that for the initial phase of the response, a determination of appropriate response time had to be made and used in the selection of storage location and readiness of equipment. The USNRC will review each plant's FLEX installation and guidance as they are being completed (which will be no later than the end of 2016) and will issue a safety evaluation report.

H.4.2 Revision of SAMG

The Electric Power Research Institute commissioned a revision to the Severe Accident Management Guidance Technical Basis Report (EPRI, 2012b). This report is the first update of the original 1991 version, adding additional Candidate High Level Actions in Volume 1 and providing supporting technical information in Volume 2. New material addresses using seawater injection for reactor core cooling, common-cause failures due to external events, cooling spent fuel pools, setting priorities in multiunit events, containment isolation failure, and hydrogen combustion within plant buildings. The intent, as with the original report, is to guide owners' groups in developing new SAMG.

Efforts are currently under way to develop revised versions of the SAMG for both generic and plant-specific guidance. The Emergency Procedures Group of the BWR Owners' Group has been meeting quarterly since the Fukushima Daiichi accident and completed Revision 3 of the generic guidelines in 2013. These are integrated guidance for emergency procedures and severe accidents, referred to as Emergency Procedure Guidelines/Severe Accident Guidelines. According to the Nuclear Energy Institute (Williamson et al., 2013), this revision utilizes both FLEX and EDMGs capabilities and guidance to provide core and spent fuel pool cooling and maintain containment functions. Individual plants are developing EOPs and SAMG based on this generic guidance but tailored to their specific situations. The generic guidance is in the process of being implemented for each plant, and industry workshops are being held in the United States, Europe, Mexico, Japan, and Taiwan to assist with this process. The USNRC has formally requested that the guidelines be submitted so that they can be reviewed by staff in 2014 to support ongoing rulemaking activities.

H.4.3 Response in Japan

TEPCO has proposed a program of countermeasures similar to FLEX. The strategies are to

> consider capabilities for accident control assuming situations where almost all station facilities used to control the accident lose their functions. This is in addition to the basic approach of assuming a certain scale of an external event, including tsunamis which caused the Fukushima accident, and taking complete countermeasures against it to prevent accidents from occurring. (TEPCO, 2012b, p. 471)

Examples of the type of equipment and guidance documents are provided by Kawano (2012). The descriptions of the equipment and capabilities are plant specific and designed to address the situations encountered at the Fukushima Daichi plant following the March 11, 2011, earthquake and tsunami.

Appendix I

Probabilistic Risk Assessment

This appendix describes probabilistic risk assessment (PRA) and its current uses at Japanese and U.S. nuclear plants

I.1 RISK ASSESSMENT

Numerous definitions exist on the meaning of risk and risk assessment. A working definition of risk is the "set of triplets" definition (Kaplan and Garrick, 1981). It has been used in many applications, but particularly by the nuclear power industry and the U.S. Nuclear Regulatory Commission. According to this definition, the question "What is the risk?" is really three separate questions:

1. What can go wrong?
2. How likely is that to happen?
3. What are the consequences if it does happen?

Risk can be defined mathematically using the following expression:

$$R = \{(S_i, L_i, X_i)\}_c$$

where

R denotes the risk attendant to the system or activity of interest;

S_i denotes the *i*th risk scenario (a description of the *i*th "what can go wrong" scenario);

L_i denotes the likelihood that the *i*th scenario will happen, with uncertainty; and

X_i denotes the consequences if the *i*th scenario does happen.

The outer brackets in {(Si, Li, Xi,)} imply "the set of" triplets and the subscript c indicates that the set of triplets is "complete" (i.e., all, or all of the important, triplets are included in the set). In other words, "risk" is a set of scenarios, likelihoods, and consequences. In practice, these can be assembled into a variety of forms to represent the risk of the system being evaluated.

I.2 PROBABILISTIC RISK ASSESSMENT

PRA is a process of probabilistic evidential and inferential analysis of the response of events, systems, or activities to different challenges based on the fundamental rules of logic and plausible reasoning. The risk measure is most often a frequency whose uncertainty is represented by a probability distribution. This is often referred to as the "probability of frequency" format. Frequency is based on observations, which could include something as abstract as a thought experiment, whereas probability calibrates the credibility of the frequency based on the supporting evidence. PRA is a thought process for answering the three basic risk questions stated previously.

PRAs for light-water reactors are classified according to their completeness, or *scope*:

- Level 1 assesses the risk of core damage generally in the form of core-damage frequency. Level 1 is sometimes referred to as the plant model.
- Level 2 assesses the magnitude and timing of releases of radioactive material from reactor containment and is sometimes referred to as the containment model or the plant plus containment model.
- Level 3 assesses the consequences of containment releases, for example, injuries, fatalities, and economic losses, and is sometimes referred to as the site model or the combination of the plant, containment, and site model.

Level 3 PRAs are frequently referred to as a "full-scope PRAs," but there is confusion at times as to whether it does or does not include the full treatment of external events, uncertainty analysis, and low-power and shutdown risk. In this report the terms "full scope" or "Level 3" are used interchangeably to mean the full range of internal and external events, low-power and shutdown risk, as well as a comprehensive treatment of the uncertainties involved taken to the endpoint of injuries, fatalities, and economic damage. If multiple reactor units are present at a site, then full-scope and Level 3 PRAs would include multiunit risks, not just the risks from individual units.

I.3 USE OF PRA IN JAPANESE NUCLEAR PLANTS

The information in this section is distilled from discussions with representatives from the Japanese government, industry, and academia at the committee's November 2012 meeting in Tokyo.

PRAs for Japanese nuclear plants were not required to be performed by rule prior to the March 2011 earthquake and tsunami; however, the Japanese regulatory agency (Nuclear and Industrial Safety Agency [NISA][1]) did require plant owners to perform PRAs to support license issuance and renewal. Moreover, in 1992 the Nuclear Safety Commission strongly recommended that nuclear plant operators identify effective measures for risk reduction based on PRAs of representative boiling water reactors and pressurized water reactors in Japan.

Also in 1992, the Ministry of Economy, Trade and Industry (METI) requested that nuclear plant operators perform periodic safety reviews (PSRs), in which the operators were to perform level 1.5 probabilistic safety analyses (PRAs) and introduce additional measures to ensure safety if the result of the PRA suggested that it was appropriate to do so. The Nuclear Power Engineering Corporation and the Japanese Nuclear Energy Safety Organization (JNES) had established, in addition to methodologies for Level 1 and 2 probabilistic safety analyses (PSAs), procedures to perform seismic PRAs before 2002 when the second round of PSRs were to be performed. But NISA decided in 2003 to leave the execution of PRAs to the plant operators' discretion, asserting that the execution of PRAs was outside the legal framework for licensing nuclear plants.

The Nuclear Safety Commission established a working group in July 2001 for reviewing seismic design guidelines. After deliberations lasting almost 5 years, a revised Seismic Design Examination Guideline was published (in 2006). The new guideline specified a design-basis seismic motion having a return period of about 10,000 years based on the probabilistic seismic hazard evaluation. Plant operators were requested to make efforts to reduce the risks from such hazards as low as practically achievable consistent with guidance given in the Report on Safety Goals published in December 2003.

Based on a proposal by JNES, the Standards Committee of the Atomic Energy Society of Japan (AESJ) compiled the requirements for seismic PSA and the specific methods to satisfy the requirements of "AESJ seismic PSA

[1] This agency was abolished and a new organization, the Nuclear Regulation Authority, took over its regulatory responsibilities in September 2012. See Chapter 2.

guidelines" before 2006. After publication of the NSC's new guidelines, all nuclear plant operators in Japan were requested by NISA to review the validity of the design-basis earthquake for their plants based on the new guideline and a seismic PRA. The process had not been completed before 2011.

With respect to tsunamis, in 1999 the Federation of Electric Power Companies asked the Japan Society of Civil Engineers (JSCE) to study a method to assess the characteristics of tsunamis for nuclear plant design in Japan. In response, the JSCE published "Tsunami Assessment Method for Nuclear Power Plants in Japan" in 2002.[2] The paper proposed a deterministic method for evaluating tsunami hazards. The JSCE subsequently began an effort to develop a probabilistic method for evaluating tsunami hazards. It published a draft report in 2009 and a final report in 2011. Both reports are in Japanese and could not be reviewed by the committee. The AESJ initiated a PRA for tsunami hazards at nuclear power plants in May 2011 and published a final report in December 2011. This report also is in Japanese.

The PRAs performed in support of license issuance were generally Level 1 with some consideration of Level 2 parameters (referred to as a Level 1.5 PRA). PRAs performed to support license renewals were Level 1 and were updated every 5 years. All were single-unit PRAs except for shared systems. According to Japanese nuclear industry representatives, sharing of systems is atypical at Japanese plants.

PRAs included internal events only, but they covered both full-power and shutdown operations. Unlike in the United States, online maintenance of safety systems is not performed in Japan; maintenance is only performed when a reactor is shut down. Consequently, PRAs were not performed to assess risk during online maintenance.

The PRAs performed by TEPCO at the Fukushima Daiichi and Fukushima Daini plants predicted a core-damage frequency of about 1×10^{-6} per reactor-year during full-power operations and generally less than that (approximately 1×10^{-7} per reactor-year) during most phases of shutdown. The scenarios considered include station blackouts; however, because the PRAs were performed on a unit-by-unit basis, the scenarios assumed that power would be available from a neighboring unit. As noted in Chapter 4, this was not the case for Units 1, 2, and 3 at the Fukushima Daiichi plant in the aftermath of the March 2011 earthquake and tsunami.[3]

Although PRAs performed by the Japanese nuclear industry did not include external events, AESJ had developed PRA guidelines for earthquakes in 2006. There were no PRA guidelines for tsunamis at the time of

---

[2] A 2006 English translation (JSCE, 2006) of this document is available.

[3] However, as discussed in Chapter 4, Unit 5 was cross-tied to an operating emergency diesel generator in Unit 6.

the March 2011 earthquake and tsunami. However, at the time of the committee's November 2012 meeting in Tokyo, PRA guidelines for tsunamis were in development, PRA guidelines for earthquakes had been updated, and PRA guidelines for fire events were under consideration.

The JNES, an independent administrative agency of the Japanese government, leads the work on PRA methods and practices. The nuclear industry's regulator does not maintain staff specializing in PRA. The private sector and academia perform research relating to the science of PRA, with the private sector's contributions pertaining mainly to reactor design.

TEPCO's PRA expertise resides in the technical specification groups at its plants, which are responsible for onsite risk management. These groups are generally responsible for producing the 5-year license renewal PRAs. There is no dedicated PRA staff at the plants or in company headquarters.

TEPCO officials noted that scenarios from a plant PRA may inform operator training, but there is not a one-to-one correspondence between PRA scenarios and training exercises. These officials also noted that several different plants share a single training center, so plant-specific scenarios are not practical.

The committee requested more detailed information from TEPCO about the scope, format, and results of its plant-specific PRAs and the treatment of uncertainties. However, none of the plant-specific PRA documents had been translated into English so the committee was not able to review them independently.

Representatives from government, industry, and academia expressed reticence about the usefulness of Level 2 and 3 PRAs; they noted that the consensus in Japan was that the methodologies used to treat uncertainties were still quite immature. According to these representatives, more deterministic approaches are preferred over PRA in Japan; many representatives emphasized that PRA is just one of many tools to assess and mitigate risk.

I.4 USE OF PRA IN U.S. NUCLEAR PLANTS

PRAs are not required by rule for existing U.S. nuclear plants; however, they exist for all plants and are used extensively in decision making about plant operations. Most of these PRAs are Level 1 with some Level 2 considerations included to have a basis for determining large early-release frequencies of fission products. The PRAs include external-event analysis, but their scopes vary and in most cases are somewhat limited, particularly with respect to the use of probabilities to define external-event frequencies. Only a few plants have PRAs that include external flood risks or low-power shutdown risks. A few plants also have Level 3 PRAs, but those PRAs are generally dated. Level 1 plant PRAs are mature and comprehensive.

There is currently no regulatory requirement for nuclear plant PRAs to be periodically updated unless a commitment to do so is included as a part of a plant's license conditions. Nevertheless, most plants update their PRAs approximately every 3 years. Also, if the PRA is used as a basis for a license renewal, or to support a risk-informed change to a plant's licensing basis, the industry regulator, the U.S. Nuclear Regulatory Commission, requires that it be current (USNRC, 2009b).

Under 10 CFR Part 52,[4] Level 1 and Level 2 PRAs are required for new nuclear plants. The PRAs must include the consideration of internal and external events[5] and address all plant operating modes (i.e., full power to shutdown). All new plants licensed under 10 CFR Part 52 are also required to update their PRAs on a regular basis. These updated PRAs are subject to review by the USNRC. Design-specific PRAs are also required by the USNRC for certifications of new nuclear plant designs.[6]

Except as noted above, the USNRC does not conduct detailed reviews of or issue safety evaluation reports on PRAs. The USNRC staff does, however, perform audits of PRAs to develop lines of inquiry during routine inspections. A plant PRA only needs to be adequate for the licensee to justify a general characterization of plant risk; the USNRC does not require licensees to expand the scope or improve the quality of their PRAs except as needed to support licensing actions. In addition, the USNRC only undertakes a detailed review of a PRA when a licensee submits it as part of a "risk-informed" change to the plant's licensing basis (USNRC, 2012a). Of course, the licensee may never choose to make a risk-informed change in the plant, in which case the opportunity for a detailed review is never triggered.

PRAs are used by licensees and the USNRC to evaluate the impact on risk of plant modifications and online or outage maintenance. PRAs are also used to support the licensees' inspection and surveillance activities and to risk-inform USNRC oversight, inspection, and enforcement activities. PRAs are often used as a basis for selecting equipment to be monitored under the "Maintenance Rule" (10 CFR § 50.65[7]) and to support determinations regarding the risk significance of plant transients and the safety implications of reportable events.

PRAs have become increasingly important in developing risk-informed information to support license amendments. They are also used to update a plant's technical specifications and the safety parameter displays in the

[4] Licenses, Certifications, and Approvals for Nuclear Power Plants. Available at http://www.nrc.gov/reading-rm/doc-collections/cfr/part052/.

[5] Work is currently under way to strengthen external-event analyses, particularly with respect to fires, earthquakes, and floods.

[6] It is important to note that the safety of operational nuclear plants in the United States will be dominated by the currently existing plants for many years to come.

[7] Requirements for Monitoring the Effectiveness of Maintenance at Nuclear Power Plants. Available at http://www.nrc.gov/reading-rm/doc-collections/cfr/part050/part050-0065.html.

control room. Perhaps one of the most important applications of PRAs is for training. The use of plant-specific PRAs in training varies from plant to plant; operators in many plants are now being trained on plant-specific simulators using actual accident sequences derived from that plant's PRA.

The USNRC has developed independent risk models for each nuclear plant under the Standardized Plant Analysis Risk (SPAR) program (USNRC, 2007). SPAR models are compared with licensees' PRAs; the results of these comparisons are used to make revisions to the SPAR models or to document unresolved technical issues. SPAR models are used in the USNRC's inspection and oversight programs and to support the Accident Precursor Program, Incident Investigation Program, and generic safety issue resolution process. SPAR models are also used to perform risk-informed reviews of license amendments.

The USNRC is currently examining the use of Level 3 PRAs for nuclear plant regulation (Borchardt, 2012b). USNRC staff are developing a Level 3 PRA for an existing nuclear plant (Vogtle) in Georgia. This PRA is planned to be completed over the next several years.

Appendix J

Human Reliability Analysis

This appendix describes the application of human reliability analysis (HRA) in probabilistic risk assessment (PRA).

HRA refers to the theory and practice of modeling human contributions to the overall reliability of complex systems (Kirwan, 1994). This includes modeling and quantifying human errors that increase the likelihood or severity of an accident, as well as modeling and quantifying recovery actions that can reduce the likelihood of an accident or mitigate its consequences.

Human error has been shown repeatedly to be a significant contributor to the risk associated with nuclear power plant operations. Researchers from Idaho National Laboratory, for example, found that human error was a significant contributor in over 75 percent of significant operating events that occurred during a 6-year period (1992-1997), highlighting the importance of accurately estimating the human contribution to overall risk (Gertman et al., 2002).

HRA is typically performed as part of PRAs to quantify the likelihood that people will fail to take actions that

- Are required for accident prevention or mitigation (*errors of omission*),
- Will cause or exacerbate the consequences of an accident (*errors of commission*), or
- Will terminate or mitigate the consequences of an accident (*recovery actions*).

The Fukushima Daiichi accident reaffirms the important role that people play in responding to severe nuclear accidents, and beyond-design-basis accidents more generally. The accident exposed some of the difficult situational challenges that arise during severe accidents as well as the psychological and team processes that influence recovery actions. It is essential that human performance be portrayed accurately in nuclear plant PRAs. Some of the specific needed improvements in this regard are described in the following sections.

J.1 NEED TO MORE REALISTICALLY MODEL COMPLICATING SITUATIONAL FACTORS

In responding to severe accidents at nuclear plants, operators are likely to face complex, unanticipated conditions (e.g., multiple interacting faults, failed or degraded sensors, goal conflicts, and situations not fully covered by procedures) that require them to engage in active diagnosis, problem solving, and decision making to determine what actions to take. This is commonly referred to as "knowledge-based performance."

There is a need for HRA methods that more accurately model the kinds of complicating situational factors that are likely to arise in severe accidents, and beyond-design-basis accidents more generally, and the psychological processes that underlie performance in these situations. Fortunately, there is growing agreement on this point in the PRA community. Indeed, several research and development thrusts have been initiated by the nuclear power industry and the USNRC to improve HRA methods along this front (e.g., Lois et al., 2009; Bye et al., 2011; Roth et al., 2012; USNRC and EPRI, 2012; Whaley et al., 2012; Chang et al., 2014).

J.2 NEED TO MODEL THE BROADER DISTRIBUTED RESPONSE TEAM

The set of human actors responding to a severe accident goes beyond the individuals in the control room and the field. It also includes the advisory and command-and-control organizational structure that influences and directs operator actions. For the Fukushima Daiichi accident, this included personnel at the plant's emergency response center, the headquarters emergency response center in Tokyo, as well as government personnel who monitored and sought to influence the actions at the plant (see Chapter 4). This highlights the importance of more accurately modeling the multiple decision makers involved in accident response as part of HRA (Helton et al., 2010).

J.3 NEED TO CONSIDER TEMPORAL, PHYSICAL, AND PSYCHOLOGICAL STRESSORS

It is also important to more realistically model the physical and psychological stressors that are likely to influence performance in severe accidents (Siu et al., 2013). The Fukushima Daiichi accident extended over multiple days and imposed severe mental and physical fatigue on control room operators, field staff, and personnel in the plant's emergency response center. Control room operators and field personnel were also exposed to physical stressors (e.g., loss of lighting and high radiation) as well as psychological stressors associated with risk to their lives and those of their co-workers and families.

Realistic assessments of the actual environmental factors that plant staff could encounter following a beyond-design-basis event are important for identifying gaps in preparation. For example, ensuring that plant staff will have adequate access to personal protective equipment and training for its use is important in assessing how effectively they can perform. Likewise, assessing potential radiation levels during a severe accident and how they may affect personnel access and ability to perform functions is important. The assessment of how plant staff may be injured during an external event, such as being injured by falling debris in an earthquake, needs to consider the potential for injury not just in vital areas such as the control room, but also in other parts of the plant, because the need to provide care to injured plant personnel may affect the ability of uninjured personnel to perform emergency response tasks.

J.4 NEED FOR GREATER EMPHASIS ON SEARCHING VS. SCREENING

This is also a need to place greater emphasis on searching vs. screening in conducting HRAs/PRAs to avoid prematurely screening out important sources of risk. Siu et al. (2013) point out that current PRA screening practices would likely lead to the screening out of beyond-design-basis scenarios analogous to the Fukushima Daiichi accident on the grounds that they are highly unlikely. They persuasively argue that there is a need to develop improved screening methods to reduce the possibility that important scenarios (or classes of scenarios) are prematurely screened out:

> [T]he ultimate success of screening depends on the pre-screening identification of all potentially important scenarios. Care is needed to ensure that this identification process is not unduly biased by prior expectations regarding what's likely to be important. (Siu et al., 2013, p. 8)

From a human reliability analysis perspective, there is a need to ensure that the types of situations that arise in real-world accidents and that challenge human performance are explicitly searched for and considered as part of HRAs/PRAs. Methods for systematic search of plausible complicating scenarios already exist that can provide a foundation to build upon (e.g., USNRC, 2000c).

J.5 NEED FOR GREATER APPRECIATION OF PEOPLE AS A SOURCE OF RESILIENCE AND RECOVERY

The Fukushima Daiichi accident highlights the key role that people play in accident recovery. As discussed in Chapter 4, the majority of the physical systems that had been counted on to mitigate the accident at the Fukushima Daiichi plant were unavailable because of the loss of onsite power. Recovery ultimately depended on the ingenuity of the people on the scene to develop and implement alternative mitigation plans in real time (see Chapter 4). Humans are too often treated as the "weak link" in systems; indeed, the emphasis in HRA/PRA is on uncovering ways people can fail (human errors) and estimating failure probabilities.

There is growing evidence that people are a source of system resilience because of their ability to adapt creatively in response to unforeseen circumstances (Hollnagel et al., 2006; Reason, 2008; Pariès, 2011). The Fukushima Daiichi accident reaffirmed that people are the last line of defense in a severe accident. It is therefore important that their role in recovery be better modeled in HRA and more effectively supported.

Appendix K

Tsunami Hazards in the Atlantic Ocean Basin

The U.S. Nuclear Regulatory Commission has sponsored the U.S. Geological Survey to carry out tsunami hazard assessments for the East Coast of the United States. The most recent assessment was published in 2008 (Atlantic and Gulf of Mexico Tsunami Hazard Assessment Group, 2008) and summarized in a 2009 special issue of Marine Geology titled *Tsunami Hazard Along the U.S. Atlantic Coast*. The assessment examined coincidental earthquakes and tsunamis (ten Brink et al., 2009), the distribution of landslides along the East Coast (Chaytor et al., 2009), and tsunami source probabilities (Geist and Parsons, 2009).

The lead paper in the special issue was written by ten Brink (2009), who concluded that

> [a]ssessment of tsunami hazard to the Atlantic coast of the United States poses a scientific challenge because of the paucity of both historical events and pre-historic tsunami evidence. The Atlantic coast . . . is highly vulnerable to tsunami damage because major population centers and industrial facilities are located near the shoreline at low-lying elevations.

A study of tsunami hazards of Canadian coastlines was recently published by Leonard et al. (2013). They concluded that

> The cumulative estimated tsunami hazard for potentially damaging run-up (≥1.5 m) of the outer Pacific coastline is ~40–80% in 50 years, respectively one and two orders of magnitude greater than the outer Atlantic (~1–15%) and the Arctic (<1%). For larger run-up with significant damage potential

(≥3 m), Pacific hazard is ~10–30% in 50 years, again much larger than both the Atlantic (~1–5%) and Arctic (<1%).

The U.S. Nuclear Regulatory Commission has contracted with the U.S. Geological Survey for a review of geological methods for estimating inundation hazards to U.S. nuclear power plants from riverine floods, tsunamis, and storm surges. The report from this project is planned to be completed in 2014 (Jim O'Connor, U.S. Geological Survey, oral communication).

Several large tsunamis have occurred in the Atlantic Ocean Basin during prehistoric and historic times. These are described in the following sections.

K.1 LISBON EARTHQUAKE AND TSUNAMI (1755)

A large earthquake struck Lisbon, Portugal, at approximately 9:40 a.m. local time on All Saint's Day (November 1) in 1755. The earthquake and associated tsunami destroyed much of the city, killing as many as 30,000-40,000 people of the city's approximately 200,000 inhabitants. The direct cost of the earthquake and tsunami was 32-48 percent of the Portuguese gross domestic product (Pereira, 2006). The tsunami was also recorded in Spain (Bay of Cadiz) and Morocco (Blanc, 2009).

The earthquake may have been caused by a thrust fault at the margins of the African and Eurasian plates to the west of the Strait of Gibraltar. Zitellini et al. (1999) identified such a fault on the seafloor about 200 km southwest of Cape St. Vincent (this cape forms the southwestern-most point of Portugal and is near the site of the naval Battle of Trafalgar in 1805). The northwest-southeast trending fault is approximately 50 km long.

Seafloor topography played a large role in determining the tsunami's propagation to the west (Barkan et al., 2009). Although the tsunami did reach Newfoundland, the presence of the nearby Gorringe Bank, Josephine Seamount, the Mid-Atlantic Ridge, and the Azores prevented the tsunami from reaching the East Coast of the United States with the exception of southern Florida. A potential tsunami source to the south near the Bay of Cadiz could present a larger hazard to the East Coast of the United States. This source may be capable of generating a large tsunami, but no tsunamis from this source have been observed historically.

K.2 GRAND BANKS TSUNAMI (1929)

The most recent large Atlantic Ocean Basin tsunami occurred on November 18, 1929, on the southern edge of the Grand Banks, 280 km south of Newfoundland (Fine et al., 2005). The tsunami was caused by a submarine slide, which was triggered by an unusually large (for this region) magnitude 7.2 earthquake. The earthquake was located offshore Newfoundland (at 44°30′N, 57°15′W) and was felt in both New York and Montreal.

The submarine slide had a thickness of several hundred meters and flowed for at least 4 hours at speeds of 60-100 km/h. It had a run-out distance of more than 520 km and transported about 200 km^3 of sediment. The slide's resultant turbidity current cut 12 telegraph cables on the continental slope and in the ocean basin (the timing of cable cutting was used to estimate slide propagation speeds; see Heezen and Ewing, 1952).

The tsunami killed 28 people in Newfoundland and was recorded on the Atlantic coasts of Canada and United States and in the Azores, Bermuda, and Portugal (Fine et al., 2005). In Newfoundland, tsunami run-up heights reached 13 m.

The excitation of a tsunami by a slump or landslide requires the movement of a substantial volume of sediment as a coherent mass. Only the energy generated during the first few minutes of rapid movement will be transferred to the water column. Locat et al. (2009) concluded that the landslide associated with the Grand Banks event had reached its peak velocity over the first few minutes and then slowed.

K.3 STOREGGA TSUNAMIS (PREHISTORIC)

Several large prehistoric submarine slides, referred to as the Storegga Slides, have been identified in the Norwegian Sea (Harbitz, 1991). The first of these slides occurred 30,000-50,000 years Before Present (B.P.), the second took place 6,000-8,000 B.P., and the most recent occurred about 6,000 B.P. Altogether, these three slides moved 5,580 km^3 of sediment and had run-out distances of as much as 850 km. Associated onshore tsunami deposits have been found in eastern Greenland, eastern Iceland, Norway, and northern Scotland. At many sites the tsunami deposits occur between layers of peat.

The ruptures that produced these slides occurred at the continental shelf break about 100 km offshore Norway (Figure K.1). Masson et al. (2006) examined the second slide, which they estimate to have occurred at 8,200 B.P. The area of the slide was 95,000 km^2; it extended 810 km with a thickness of 430 m. The volume of this slide was 2,400-3,200 km^3 and actually cut into the continental shelf at its top. The primary triggering mechanism was probably an earthquake accompanied by elevated pore pressures in weak layers of sediment from the rapid deposition of glacial sediments during deglaciation and possible outgassing of uncovered methane hydrates (Harbitz, 1991; Masson et al., 2006).

The Storegga slides created tsunamis of considerable size. Harbitz (1991) estimates that tsunami run-up heights on the Norwegian coast were

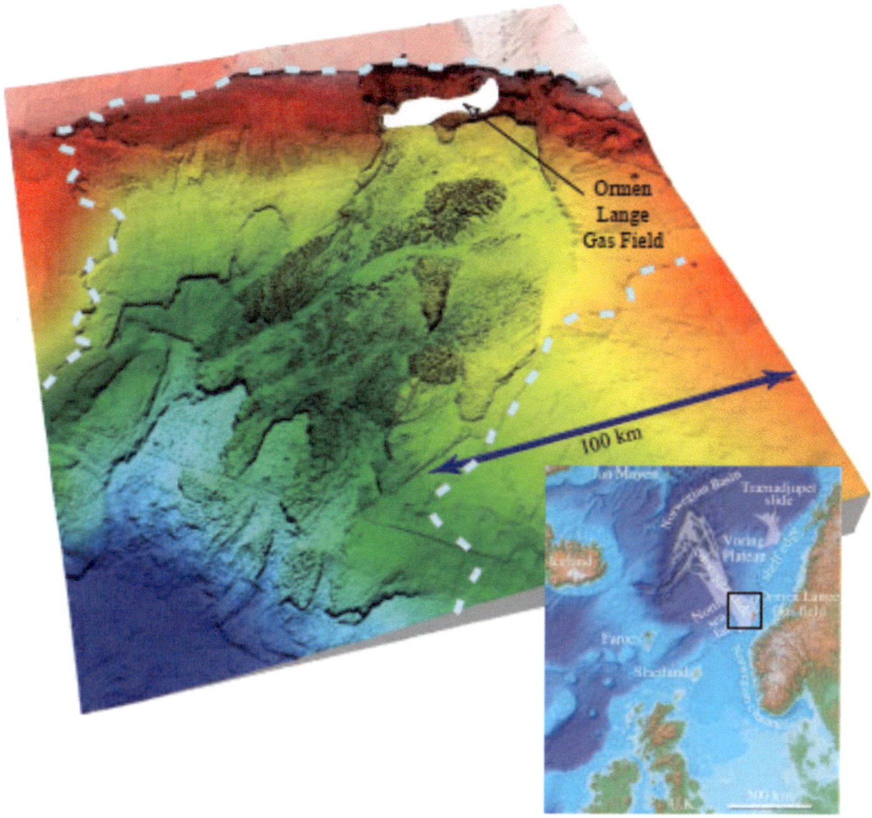

FIGURE K.1 Map of the ocean floor off the west coast of Norway showing the locations of the Storegga Slides. SOURCE: Masson et al. (2006).

10-15 m for the first slide and 5 m for the second. Run-up heights for the first tsunami were also estimated to be 5.6 m at eastern Greenland, 7.8 m for eastern Iceland, and 5.0 m for northern Scotland. The heights for the second slide were smaller; 3.1 m for eastern Greenland, 4.6 m for eastern Iceland and 3.7 m for northern Scotland. Tsunami deposits from the slides have been identified on the west coast of Norway, east coast of Scotland and northeast England, and the Faeroe and Shetland Islands (e.g., Smith et al., 2004; Bondevik et al., 2005).

Appendix L

Factoring the Costs of Severe Nuclear Accidents into Backfit Decisions

The Fukushima nuclear accident demonstrates that the economic costs of a severe nuclear accident can be considerable. The current cost estimates for the Fukushima accident include

- *Support for accident evacuees.* TEPCO (2014) estimated that, as of January 15, 2014, its compensation payments to the evacuees and businesses affected by radiological releases from the Fukushima Daiichi plant would be more than ¥5 trillion ($50 billion).[1]
- *Offsite decontamination.* Japan's National Institute of Advanced Industrial Science and Technology estimates that decontamination in Fukushima Prefecture will cost ¥2.5-5.1 trillion (~$25-51 billion) (*Mainichi*, 2013).
- *Onsite decommissioning.* TEPCO (2014) estimates its site cleanup costs at Fukushima Daiichi at ¥2 trillion (~$20 billion).
- *Replacing power from idled nuclear plants.* The undamaged units at the Fukushima Daiichi plant (Units 5 and 6) will never operate again (TEPCO, 2014). In addition, all of Japan's other nuclear power plants have been idled for about 3 years as a result of the accident. Japan's utilities have paid an estimated ¥7.3 trillion ($73 billion) for fuel in fiscal year 2012, double the amount in FY 2010, in large part because of the need to buy liquefied natural gas to replace the power from the shutdown nuclear

[1] Additional detail provided in *Asahi Shimbun* (2013).

power plants (Matsuo and Yamaguchi, 2013).[2] At that rate, over 3 years, the increased cost of generating electric energy would be about ¥10 trillion (~ $100 billion)

- *Other costs.* The Institut de Radioprotection et de Sûreté Nucléaire (IRSN) estimated the cost of a Fukushima-scale accident in France (Pascucci-Cahen and Momal, 2012[3]). The estimate included about €166 (~$215 billion) in costs for image or reputation losses, including loss of food exports, reductions in other exports, and loss of tourism. The estimated total loss of tourism in Japan is about ¥1 trillion ($10 billion) (World Travel and Tourism Council, 2011); this loss is attributable to the earthquake and tsunami as well as the nuclear accident.

The total cost of the Fukushima Daiichi accident could therefore exceed ¥20 trillion (~$200 billion).

It is instructive to compare these costs to the estimates developed by the USNRC staff for a hypothetical accident at the Peach Bottom nuclear plant in Pennsylvania. These cost estimates were used in the staff's backfit analysis for filtered vents (Borchardt, 2012a, Enclosure 5C, Table 7, Case 2):

- A collective population dose to workers and the public out to 50 miles (accounting for reductions due to evacuation) of 0.53 million rem, which, valued at $2,000/rem, translated into damage of about $1 billion;
- A loss of the use of offsite land and property due to radioactive contamination of $1.9 billion;
- A loss of onsite value of $3.2 billion (Borchardt, 2012a, Enclosure 5C, Table 8). This includes the loss of use of an average of 1.75 nuclear power reactors at a BWR plant site (Borchardt, 2012a, Enclosure 1, p. 15, Table 1).

The total estimated costs for the hypothetical accident at the Peach Bottom plant are therefore about $6 billion.

The USNRC staff estimated that most of the offsite damage ($2.5 billion) and $1.2 billion of the onsite damage could be prevented by the installation of filtered vents. After multiplying the savings of $3.7 billion by a probability of 2×10^{-5} accidents per reactor year (i.e., one accident every 50,000 years on average) and by 17.6 years (the assumed remaining 25 years of reactor life discounted by 3 percent per year), the savings per

[2] In March 2012, the Japan Atomic Industrial Forum estimated the extra cost at $40 billion/year; see http://www.world-nuclear.org/info/Country-Profiles/Countries-G-N/Japan/#.UeHlrFOzLEh.

[3] The technical analysis behind this IRSN analysis was published in 2013. An English translation ("Methodology Used in IRSN Nuclear Accident Cost Estimates in France," IRSN, PRP-CRI/SESUC/2014) was made available to the committee prior to its publication.

reactor would amount to only $1.3 million—much less than the estimated $15 million cost for installing filtered vents. Installation of filtered vents therefore failed the backfit cost-benefit test.

The cost estimates for the accident at the Fukushima Daiichi plant (~$200 billion) are about 33 times higher than the USNRC cost estimate for a hypothetical accident at the Peach Bottom plant (~$6 billion). The primary reasons for these differences are the following:

1. The relatively low USNRC estimate of costs associated with the calculated contamination of 354 km^2 (140 square miles) above 15 curies/km^2, which is approximately equal to the threshold that has been used for long-term evacuation in Japan (which affected about 625 km^2 of land).[4]
2. The USNRC assumed that the operation of other nuclear power plants would not be affected, unlike the situation in Japan where virtually all nuclear power plants have been shut down.

Differences between accident costs in Japan and the United States can be expected—as can differences in accident costs for different sites in the United States. Nevertheless, the large differences noted above serve to illustrate that cost estimates—and associated backfit rule decisions—are sensitive to the assumptions made in developing those estimates.

The point of this appendix is not to critique the USNRC's analysis—the committee did not perform an in-depth review of this analysis because it is outside the statement of task for the study. The committee offers this example to demonstrate that severe accidents such as occurred at the Fukushima Daiichi plant can have large costs and other consequences that are not considered in USNRC backfit analyses. These include national economic disruption, anxiety and depression within affected populations, and deterioration of social institutions arising from a loss of trust in governmental organizations.

The USNRC is launching a multiyear process for updating its regulatory guidance for backfit analyses. One focus of the update is to improve calculations of the economic consequences of a reactor accident, taking into account lessons learned from the accident at Fukushima Daiichi. The USNRC is also reevaluating how qualitative factors are used in the backfit analysis process.

---

[4] IRSN (2012a, Fig. 6-24). See also Table 6-11, which shows the areas contaminated above this level outside the 20-km radius around the plant as 320 km^2.

Appendix M

Access to Timely and Reliable Information to Support Decision Making During a Nuclear Power Plant Accident

The Fukushima accident revealed that permanently installed radiation monitoring instruments at and around nuclear power plants should be able to operate on batteries for long periods of time, at least a week, with plans in place to replace or recharge them thereafter. Also, because multiple parallel pathways for releases of radiation exist, instrumentation that gives continuous readout of the quantities of radionuclides being discharged from these pathways under accident conditions would improve accuracy of the information used to support decision making. A problem anticipating which pathways will be effective is that plant damage can create new pathways that would not be obvious from the "as built" status of the plant.

Additionally, there is a need for instruments to be quickly available at and near nuclear power plant sites that can measure (i) the quantities of radioactive iodine and cesium in the plume, whether or not the plume is elevated off the ground as a result of an initial rise due to its temperature; and (ii) what the initial cross-wind dimensions of the plume are whatever the direction in which it is being blown.[1] This information would be essential

[1] Concepts for evaluation include (1) imaging Compton gamma-ray spectrometers (Kataoka et al., 2013), (2) drones equipped with radiation detectors capable of distinguishing gamma and beta energies (Pöllänen et al., 2009), and (3) resonance-enhanced multiphoton ionization detected by microwave scattering (Shneider and Miles, 2005; Dogariu and Miles, 2011).

initialization for atmospheric models that project dose rates and cumulative doses to the population in different directions and at different distances.[2]

Thus there appears to be potential for reducing uncertainty in activity estimates and forecasts of plume behavior, although dose predictions will always carry significant uncertainties. Examples of quantities whose measurement might be used to assay the quantities of radioisotopes in the plumes are the intensities, directions, and energy spectra of gamma rays from the plume and measurements of the concentration profiles of cesium and iodine in the plume, which could be far above natural background levels for these isotopes.[3] The Defense Advanced Research Projects Agency (DARPA), an agency of the Department of Defense responsible for developing new technologies for use by the military, is already seeking novel approaches to low-cost, high-efficiency, packaged radiation detectors for identifying hidden threats, ranging from special nuclear material (SNM) to radiological sources (Federal Business Opportunities, 2013).

In improving capabilities of forecasting plume behavior, plants may consider extending the distance requirement for meteorological monitoring programs for providing atmospheric transport and diffusion estimates, which is currently within the 10-mile emergency planning zone (USNRC, 2007).

---

[2] The capabilities described here would go beyond existing capabilities. Currently, state and local agencies rely on field monitoring teams and appropriate instrumentation, the specifics of which vary from site to site. These capabilities are supplemented by the Federal Radiological Monitoring and Assessment Center and other federal resources, including the U.S. Department of Energy's capabilities to provide aerial mapping of the depositions from the plume. In addition, the U.S. Environmental Protection Agency's Radnet monitors and monitors established by the nuclear power plants for routine environmental monitoring can provide real-time air monitoring.

These instruments could be multipurpose so that they can be used for a wide variety of emergency response activities including monitoring plumes from chemical releases.

[3] For example, if a release of the same magnitude as occurred at Fukushima Daiichi took place over a period of 1 to 10 hours, concentrations of 0.01 to 100 ppb of radioactive cesium and iodine would be expected near the plant depending upon the plume cross-section. Consider that a hypothetical release of cesium-137 and iodine-131 then would be 0.25 megacurie (MCi) (2.8 kg) and 1.9 MCi (15 g), respectively. Although it would be an insignificant contributor to the dose because of its long half-life (17 million years), there also would be 17 times as much iodine-129 as iodine-131 by mass (0.26 kg). Thus, in total, there would be about 1,025 atoms of cesium and 1,024 atoms of iodine in the plume.

Appendix N

Conversions and Units

Distance-Related Conversions		
kilometers (km) and miles (mi)	1 km = 0.62 mi	1 mi = 1.6 km
km^2 (square kilometers) and mi^2 (square miles)	1 km^2 = 0.39 mi^2	1 mi^2 = 2.59 km^2
m (meters) and ft (feet)	1 m = 3.28 ft	1 ft = 0.30 m
m^3 (cubic meters) and ft^3 (cubic feet)	1 m^3 = 35.3 ft^{t3}	1 ft^3 = 0.03 m^3
km/h (kilometers per hour) and mph (miles per hour)	1 km/h = 0.62 mph	1 mph = 1.6 km/hr
Radiation-Related Conversions		
mSv (millisieverts), mrem (millirem), and mGy[a] (milligray)	1 mSv = 100 mrem = 1 mGy	1 mrem = 0.01 mSv = 0.01 mGy
Bq (becquerels) and Ci (curies)	1 Bq = 2.7×10^{-11} Ci	1 Ci = 3.7×10^{10} Bq
Other		
MJ (megajoules) and kWh (kilowatt hours)	1 MJ = 0.28 kWh	1 kWhr = 3.6 MJ
MPa (megapascals) and psi (pounds per square inch)	1 MPa = 145 psi	1 psi = 0.007 MPa
Celsius and Fahrenheit	°C = (5/9)*(°F – 32°)	°F = (9/5)*°C + 32°
metric tons and pounds (lbs)	1 metric ton (tonne) = 2,204 lbs	

Prefixes	
pico-	10^{-12}
micro-	10^{-6}
milli-	10^{-3}
kilo-	10^{3}
mega-	10^{6}
giga-	10^{9}
tera-	10^{12}
peta-	10^{15}

[a] Millisieverts and millirem are units of effective dose, whereas Gray is a unit of absorbed dose. They are numerically equivalent when exposure is from gamma rays and x-rays.

Unit of Measure	Abbreviation	Type of Measure
Becquerel	Bq	radiation activity
Celsius	C	temperature
centimeters	cm	distance
Fahrenheit	F	temperature
feet	ft	distance
gallon	gal	volume
gallons per minute	gpm	flow rate
Gray	Gy	absorbed radiation dose
Joule	J	energy
kilogram	kg	mass
kilometers	km	distance
kilopascals	kPa	pressure
kilovolts	kV	electrical potential
kilowatt	kW	electrical power
kilowatt-hour	kWh	energy
liters per minute	Lpm	flow rate
megapascals	MPa	pressure
megawatts electric	MWe	electrical power
meters	m	distance
millimeters	mm	distance
millirem	mrem	effective radiation dose
millisievert	mSv	effective radiation dose
newton	N	force
pound	lb	mass
pound-force	lbf	force
pounds (-force) per square inch	psi	pressure
volts	V	electrical potential

Appendix O

Acronyms

€	Euro
¥	Yen
ABWR	Advanced boiling water reactor
AC	Alternating current
ACRS	Advisory Committee on Reactor Safeguards
ADAMS	Agencywide Documents Access and Management System
ADS	Automatic depressurization system
AEA	Atomic Energy Act
AEC	Atomic Energy Commission
AESJ	Atomic Energy Society of Japan
AFW	Auxiliary feedwater
ALARA	As Low As Reasonably Achievable
AM	Accident management
ANS	American Nuclear Society
AOP	Abnormal operating procedures
ASME	American Society of Mechanical Engineers
ATWS	Anticipated transient without scram
B.5.b	Section B.5.b of Order EA-06-137
BDB	Beyond design basis
B.P.	Before present
BWR	Boiling water reactor
CANDU	Canada deuterium uranium
CDC	U.S. Centers for Disease Control and Prevention

CERCLA	Comprehensive Environmental Response, Compensation, and Liability Act
CFR	Code of Federal Regulations
CME	Coronal mass ejection
CNSC	Canadian Nuclear Safety Commission
CS	Core spray
CST	Condensate storage tank
CV	Containment vessel
DARPA	Defense Advanced Research Projects Agency
DART™	Deep-Ocean Assessment and Reporting of Tsunamis
DBA	Design-basis accident
DBT	Design-basis threat
DC	Direct current
DHS	U.S. Department of Homeland Security
DIET	National Diet of Japan
DOD	U.S. Department of Defense
DOE	U.S. Department of Energy
DRYAMB	Dry ambient pressure (containment)
D/W	Drywell
EAL	Emergency action level
EC	European Commission
ECCS	Emergency core cooling system
EDG	Emergency diesel generator
EDMGs	Extensive damage mitigation guidelines
ENSI	Eidgenössisches Nuklearsicherheitsinspektorat (Swiss Federal Nuclear Safety Inspectorate)
ENSREG	European Nuclear Safety Regulators Group
EOC	Emergency operations center
EOPs	Emergency operating procedures
EPRI	Electric Power Research Institute
EPZ	Emergency planning zone
ERC	Emergency response center
ERDA	Energy Research and Development Administration
EU	European Union
FEMA	U.S. Federal Emergency Management Agency
FLEX	Diverse and Flexible Coping Strategies
FRMAC	Federal Radiological Monitoring and Assessment Center
GE	General Electric Company
GRIPS	National Graduate Institute for Policy Studies (Japan)

GPS	Global Positioning System
GRS	Gesellschaft für Anlagen- und Reaktorsicherheit (GRS), mbH
GWe	Gigawatt electric
HERP	Headquarters for Earthquake Research Promotion (Japan)
HHS	U.S. Department of Health and Human Services
HO	Hydraulically operated
HPCI	High-pressure coolant injection
HPCS	High-pressure core spray
HQ	Headquarters
HRA	Human reliability analysis
HSPD	Homeland Security Presidential Directive
IAEA	International Atomic Energy Agency
IC	Isolation condenser
ICECND	Ice condenser (containment)
ICM	Interim compensatory measure
ICRP	International Commission on Radiological Protection
INES	International Nuclear and Radiological Event Scale
INPO	Institute of Nuclear Power Operations
IOM	Institute of Medicine (U.S. National Academies)
INSAG	International Nuclear Safety Group (International Atomic Energy Agency)
IPEEE	Individual plant examination of external events
IRSN	Institut de Radioprotection et de Sûreté Nucléaire (France)
JAEA	Japan Atomic Energy Agency
JAEC	Japan Atomic Energy Commission
JAES	Japan Atomic Energy Society
JANSI	Japan Nuclear Safety Institute
JANTI	Japan Nuclear Technology Institute
JAPC	Japan Atomic Power Company
JNES	Japanese Nuclear Energy Safety Organization
JSCE	Japan Society of Civil Engineers
JST	Japan Standard Time; Japan Science and Technology Agency
KI	Potassium iodide
K-NET	Kyoshin Network
KiK-Net	Kiban Kyoshin Network
L	Length

LOCA	Loss-of-coolant accident
LPCI	Low-pressure coolant injection
LPCS	Low-pressure core spray
LR	License renewal
LWR	Light-water reactor
MAAP	Modular accident analysis program
MCR	Main control room
MELCOR	Methods for Estimation of Leakages and Consequences of Releases
METI	Ministry of Economy, Trade and Industry (Japan)
MEXT	Ministry of Education, Culture, Sports, Science and Technology (Japan)
MHLW	Ministry of Health, Labor and Welfare (Japan)
MIT	Massachusetts Institute of Technology
MN	MegaNewton
MO	Motor-operated
MOE	Ministry of the Environment (Japan)
MOU	Memorandum of understanding
MRFA	Maximally reasonably foreseeable accident
MSIV	Main steam isolation valves
MSL	Main steam line
MW	Megawatt
MWe	Megawatt electric
MWt	megawatt thermal
N/A	Not applicable
NAE	National Academy of Engineering (U.S. National Academies)
NAIIC	Nuclear Accident Independent Investigation Commission (Japan)
NARAC	National Atmospheric Release Advisory Center
NAS	National Academy of Sciences (United States)
NCRP	National Council on Radiation Protection and Measurements
NEA	Nuclear Energy Agency (Organisation for Economic Co-operation and Development)
NEI	Nuclear Energy Institute
NEPA	National Environmental Policy Act
NERHQ	Nuclear Emergency Response Headquarters
NISA	Nuclear and Industrial Safety Agency (Japan)
NIST	National Institute of Standards and Technology
NOAA	U.S. National Oceanic and Atmospheric Administration

NPP	Nuclear power plant
NPS	Nuclear power station
NRA	Nuclear Regulation Authority (Japan)
NRC	National Research Council (U.S. National Academies)
NRF	National Response Framework
NSC	Nuclear Safety Commission (Japan)
NTSB	National Transportation Safety Board
NTTF	Near-Term Task Force Review of Insights from the Fukushima Dai-ichi Accident (U.S. Nuclear Regulatory Commission)
O.P.	Onahama Peil (Onahama Port Construction Standard Surface)
OECD	Organisation for Economic Co-operation and Development
OL	Operating license
ONR	Office for Nuclear Regulation (United Kingdom)
ORNL	Oak Ridge National Laboratory
OSHA	U.S. Occupational Safety and Health Administration
P	Pressure
PAGs	Protective action guidelines
PAZ	Precautionary action zone
PCV	Primary containment vessel
PRA	Probabilistic risk assessment
PSA	Probabilistic safety analysis; Probabilistic safety assessment
psi	Pounds per square inch
psia	Pounds per square inch absolute
psig	Pounds per square inch gage
PWR	Pressurized water reactor
R/B	Reactor building
RCIC	Reactor core isolation cooling
REM	Roentgen equivalent man
REP	Radiological emergency preparedness
RFP	Reactor feedwater pump
RHR	Residual heat removal
RIR	Risk-informed regulations
RPV	Reactor pressure vessel
RWCU	Reactor water cleanup unit
RWST	Reactor water storage tank
S/C	Suppression chamber

SAM	Severe accident management
SAMG	Severe accident management guidelines
SBO	Station blackout
SCJ	Science Council of Japan
SDF	Self-Defense Forces (Japan)
SFP	Spent fuel pool
SG	Steam generator
SGTS	Standby gas treatment system
SI	Système Internationale (International System of Units)
SNF	Spent nuclear fuel
SNL	Sandia National Laboratories
SNM	Special nuclear materials
SPAR	Standardized plant analysis risk
SPEEDI	System for Prediction of Environment Emergency Dose Information
SPDS	Safety parameter display system
SPSA	Seismic probabilistic safety assessment
SRV	Safety relief valve
SSAMG	Shutdown severe accident management guidelines
SSCs	Structures, systems, and components
T	Temperature
T/B	Turbine building
TAF	Top of active fuel
TEPCO	Tokyo Electric Power Company
TMI	Three Mile Island
TOPOFF	Top officials' exercises
UCS	Union of Concerned Scientists
UNSCEAR	United Nations Scientific Committee on the Effects of Atomic Radiation
USEPA	U.S. Environmental Protection Agency
USFDA	U.S. Food and Drug Administration
USGAO	U.S. Government Accountability Office
USNRC	U.S. Nuclear Regulatory Commission
V	Volt
VDC	Volt DC
WANO	World Association of Nuclear Operators
WENRA	Western European Nuclear Regulators' Association
WHO	World Health Organization